Advanced Technologies in Bio/Hydrometallurgy for Recovery and Recycling of Metals

Advanced Technologies in Bio/Hydrometallurgy for Recovery and Recycling of Metals

Editor

Stefano Ubaldini

MDPI

Basel • Beijing • Wuhan • Barcelona • Belgrade • Novi Sad • Cluj • Manchester

Editor
Stefano Ubaldini
Consiglio Nazionale delle Ricerche (CNR)—Istituto di Geologia Ambientale e Geoingegneria (IGAG),
Roma, Italy

Editorial Office
MDPI
St. Alban-Anlage 66
4052 Basel, Switzerland

This is a reprint of articles from the Special Issue published online in the open access journal *Metals* (ISSN 2075-4701) (available at: https://www.mdpi.com/journal/metals/special_issues/bio_hydrometallurgy_recovery_recycling).

For citation purposes, cite each article independently as indicated on the article page online and as indicated below:

Lastname, A.A.; Lastname, B.B. Article Title. *Journal Name* **Year**, *Volume Number*, Page Range.

ISBN 978-3-0365-8688-5 (Hbk)
ISBN 978-3-0365-8689-2 (PDF)
doi.org/10.3390/books978-3-0365-8689-2

Contents

About the Editor

Stefano Ubaldini

Dr. Stefano Ubaldini is senior research scientist of the Institute of Environmental Geology and Geoengineering of the Italian National Research Council (IGAG-CNR).

He is Member of the International Scientific Advisory Board of the Institute of Geotechnics of the Slovak Academy of Science, Member of the Board of Directors and of the Scientific Council of the High Tech Recycling International Interuniversity Research Center, and head of the Bio-Hydrometallurgy, Cyanidation and Electrochemical Laboratories of the IGAG-CNR.

He is Member, in the framework of the European Innovation Partnership on Raw Materials (EIP-RMs), of the Governance of the EIP-RMs as Sherpa member of the High Level Steering Group (HLSG)/Sherpa Group.

Dr. Ubaldini's main area of expertise is in hydrometallurgy and bio-hydrometallurgy. His activity has been carried out mainly in the framework of the primary and secondary raw materials for over 30 years.

He has participated to 15 EC projects and managed R&D projects with funds from industrial and governmental resources; he was scientist responsible in EC projects: H2020, Raw Materials & Recycling, Brite-Euram, VII FP, Life-Environment and INCO-Copernicus and INTAS.

Dr. Ubaldini is author of more than 250 publications on national and international journals and proceedings of international congresses.

His current research interests are primary and secondary raw materials, with particular reference versus the hydrometallurgy and biometallurgy of low-grade georesources, with the main aim to recover base and precious metals and to develop sustainable and innovative technologies for processing of industrial wastes (spent catalysts, industrial tailings, batteries, WEEE etc.) and mining residues, coming also from abandoned sites, and treatment of waste waters.

Preface

This Special Issue is composed of many topics, presenting recent developments in Advanced Technologies in Bio/Hydrometallurgy for the Recovery and Recycling of Metals and offering new solutions represented by environmentally sustainable practices.

I am ecstatic for the success it has had, with the final result and with the quality and originality of the contributions. As Guest Editor, I hope that the articles published in this Special Issue contribute to the advancement and future development of research in this field.

I would like to warmly thank all the authors for their contributions and all the reviewers for their efforts in ensuring a high-quality publication. Sincere thanks also goes to the Editors of Metals for their continuous help, and to the Metals Editorial Assistants for their valuable and inexhaustible engagement and support during the preparation of this volume. In particular, my sincere thanks goes to Mr. Toliver Guo, Assistant Editor, for his help and support during the publication process of this Special Issue.

Stefano Ubaldini
Editor

metals

Editorial

Advanced Technologies in Bio/Hydrometallurgy for Recovery and Recycling of Metals

Stefano Ubaldini

Consiglio Nazionale delle Ricerche (CNR)—Istituto di Geologia Ambientale e Geoingegneria (IGAG), Area della Ricerca di Roma RM 1, Montelibretti, Via Salaria Km 29300, 00015 Roma, Italy; stefano.ubaldini@cnr.it

1. Introduction and Scope

Studying innovative and sustainable technologies for the recovery and reuse of raw materials (RMs) from primary and secondary resources is fundamental for economic and industrial development in compliance with environmental protection and in the context of a circular economy.

Bio/Hydrometallurgy consists of leaching and recovery unit operations. Leaching is the solubilization of metals from a solid phase using chemicals or biological agents, whereas recovery is the extraction of metals from polymetallic leachate using physicochemical processes, electrowinning, or biological processes.

Bio-hydrometallurgical processes are new solutions offering environmentally sustainable practices in the mining and environmental sectors for the extraction of base, precious, and toxic metals and rare earth elements (REE) and for the valorization of secondary resources rich in critical raw materials (CRMs), which are fundamental in modern technological applications.

The economic concerns of these processes, which are closely linked to the choice and optimization of the experimental parameters, are of great importance.

The articles included in this Special Issue contribute to the improvement of innovative bio- and/or hydrometallurgical processes applied for the recovery of valuable and critical metals.

I would like to thank the authors who accepted this invitation, which has helped us to produce a high-impact, high-quality Special Issue on "Advanced Technologies in Bio/Hydrometallurgy for Recovery and Recycling of Metals".

Citation: Ubaldini, S. Advanced Technologies in Bio/Hydrometallurgy for Recovery and Recycling of Metals. *Metals* **2023**, *13*, 1105. https://doi.org/10.3390/met13061105

Received: 26 May 2023
Accepted: 29 May 2023
Published: 12 June 2023

2. Contributions

Researchers around the globe investigating advanced bio/hydrometallurgical technologies for the recovery and recycling of metals have been invited to submit research papers, thereby allowing readers to recognize the common points between the different technologies. Among the submitted manuscripts, twelve articles have been published in this Special Issue.

The papers are all of high scientific value, and the experimental activities carried out fall into various disciplinary sectors, thus confirming the importance of the study of the titular topic in different scientific and technological fields. An overview of the published papers is given below.

The results concerning the leaching of metals from computer PCBs via electrochemical hydrochlorination using alternating current (AC) with an industrial frequency (50 Hz) have been presented in [1]. Leaching was carried out with a disintegrator-crushed computer motherboard and mixed computer PCBs with a particle size (d) of <90 μm. The leaching efficiency of metals including Fe, Sn, Mn, Al, Cu, Zn, Pb, Ni, Ti, Sb, Cr, Co, and V was evaluated with respect to certain process parameters, such as AC density, experiment duration, hydrochloric acid concentration in the electrolyte solution, solid/liquid ratio, electrolyte temperature, and the loading option for the raw material.

The research results showed that AC superimposition significantly intensifies the leaching of metals.

An innovative cryo-mechano-hydrometallurgical process (named LIBAT) was demonstrated on a pilot scale for the treatment of EOL lithium primary batteries with Li(0)-MnO_2 chemical properties [2]. The process enabled the recycling of steel scraps from external cases after cryomechanical dismantling and the recovery of Mn and Li products after hydrometallurgical processing. The environmental impacts and the economic sustainability of the process were evaluated and compared with an innovative pyrometallurgical approach allowing for Li recovery, confirming the benefits of the proposed process due to a reduction in energy consumption.

The extraction and recovery of metals from spent hydrodesulfurization catalysts have been studied, for which lab- and pilot-scale results were achieved in the overall process [3]. These catalysts contain valuable metals such as cobalt (Co), molybdenum (Mo), nickel (Ni), and vanadium (V). The total recovery yields were nearly 61% for Mo and 68% for V, which were much higher compared with their initial concentrations in the spent Co-Mo catalysts.

The effect of varying process parameters during the bio-catalyzed leaching of metals from end-of-life printed circuit boards (PCBs) was investigated in [4]. Fragmented PCBs were subjected to indirect bioleaching in a stirred tank reactor while pulp density, pH, and initial ferric iron content were varied. An iron-oxidizing *Acidithiobacillus ferrooxidans* 61 microbial strain was used to generate the lixiviant through oxidizing Fe(II) to Fe(III). The achieved results offer possibilities for further studies at higher pulp densities that would advance the bioleaching approach such that it facilitates economical and environmentally friendly technology for the urban mining of non-ferrous metals.

A new scheme for leaching metals from computer-printed circuit boards (PCBs) pre-crushed in a disintegrator has been proposed in [5]. The processes of chlorine production and hydrochlorination are implemented in one reactor under the action of an alternating current (AC) at an industrial frequency (50 Hz). This technology for recycling electronic waste (e-waste) demonstrates high economic efficiency and remarkable environmental safety.

A mycorrhizal-assisted phyto-mining (MAP) system composed of *Helianthus annuus*, the arbuscular mycorrhizal fungus *Rhizophagus intraradices*, and Zn-rich volcanic ashes was applied in bioreactors for the recovery of CRMs (Sr and P) and SRMs (Cr, Zn, Cu, Mn, Rb, and Ni) from mining wastes of the Los Cóndores mine (Argentina) [6]. It proved to be a promising and cost-effective biotechnology applicable in agronomical practices, and the recovery of CRMs and SRMs from plant biomass via hydrometallurgy was demonstrated.

Cyanobacteria *Arthrospira platensis* was employed for Y(III) recovery from contaminated wastewater through biosorption and bioaccumulation processes [7]. Yttrium is an element of critical importance for industry and technology. Experiments performed using *Arthrospira platensis* showed a great potential for application in the recovery of rare earth elements from wastewater.

In one study, bioleaching was proposed as a method for the removal of nickel from bulk copper–nickel sulfide concentrate and the acquisition of a concentrate containing copper as a chalcopyrite [8]. This approach is based on the variable refractoriness of sulfide minerals in ferric sulfate solutions and oxidation by acidophilic microorganisms. It was demonstrated that the bio-beneficiation of bulk sulfide concentrates is a promising field of biohydrometallurgy, which may be highly attractive for the improvement of commercial applications.

The authors of [9] studied the leaching of zinc for its recovery from electric arc furnace dust with sulfuric acid solutions at an ambient temperature and atmospheric pressure along with the reuse of the leaching process residue as a raw material for ceramic materials. This research developed a new form of environmental hydrometallurgy in which metallic elements of interest are valorized and the production of waste is avoided, thereby reducing the deposition of hazardous waste in landfills and the extraction of raw materials for the manufacture of construction materials.

One study evaluated the feasibility of obtaining recoverable copper from different waste dumps in the mining district of Linares (Spain) using hydrometallurgical techniques [10]. A significant percentage of copper was extracted from the waste dumps, demonstrating that the waste, after its processing, can be used in bituminous mixtures, rendering it a product of a closed cycle of materials in which no waste is produced.

Mycorrhizal-assisted phytoextraction (MAP; *Helianthus annuus*–arbuscular mycorrhizal fungus *Rhizophagus intraradices*–Zn-volcanic ashes) was employed for the recovery of secondary and critical raw materials (SRMs and CRMs, respectively) from Joda West (Odisha, India) mine residues according to a novel multidisciplinary management strategy [11]. The final recovery of SRMs and CRMs obtained via hydrometallurgical techniques, with final purification via selective electrodeposition, proved the viability and cost-effectiveness of this approach. The results are promising for MAP application on a larger scale, constituting an effective circular-economy-based approach.

The pressure-based hydrothermal oxidation of arsenic (III) ions in a H_3AsO_3-Fe^{2+}-Cu^{2+}-H_2SO_4 system formed during the processing of arsenic-containing sulfide non-ferrous metals was studied in [12]. This research is relevant because the low-grade polymetallic materials, such as copper–zinc and copper–lead–zinc, and poor arsenic-containing copper concentrates obtained using hydrometallurgical methods are becoming increasingly important due to the depletion of rich and easily extractable mineral resources and the need to reduce harmful emissions from metallurgy, especially given the high content of arsenic in ores.

3. Conclusions and Outlook

A variety of topics comprise this Special Issue presenting recent developments in advanced bio/hydrometallurgical technologies for the recovery and recycling of metals.

As Guest Editor, I am very pleased with the success of the issue, the final result, and the quality and originality of the contributions, and I hope that the published articles in this Special Issue contribute to the advancement and future development of research in this field.

I would like to warmly thank all the authors for their contributions and all the reviewers for their efforts in ensuring a high-quality publication. Sincere thanks are also due to the Editors of *Metals* for their continuous support and to the *Metals* Editorial Assistants for their valuable and inexhaustible engagement and support during the preparation of this Special Issue. In particular, I offer my sincere thanks to Mr. Toliver Guo, Assistant Editor, for his help and support during the publication of this Special Issue.

Conflicts of Interest: The author declares no conflict of interest.

References

1. Serga, V.; Zarkov, A.; Shishkin, A.; Elsts, E.; Melnichuks, M.; Maiorov, M.; Blumbergs, E.; Pankratov, V. Study of Metal Leaching from Printed Circuit Boards by Improved Electrochemical Hydrochlorination Technique Using Alternating Current. *Metals* **2023**, *13*, 662. [CrossRef]
2. Pagnanelli, F.; Schiavi, P.G.; Altimari, P.; Beolchini, F.; Amato, A.; Coletta, J.; Forte, F.; Moscardini, E.; Toro, L. Economic and Environmental Sustainability of an Innovative Cryo-Mechano-Hydrometallurgical Process Validated at Pilot Scale for the Recycling of Li Batteries. *Metals* **2023**, *13*, 497. [CrossRef]
3. Xhaferaj, N.; Ippolito, N.M.; Maggiore, F.; Ferella, F. Extraction and Recovery of Metals from Spent HDS Catalysts: Lab- and Pilot-Scale Results of the Overall Process. *Metals* **2022**, *12*, 2162. [CrossRef]
4. Vardanyan, A.; Vardanyan, N.; Aâtach, M.; Malavasi, P.; Gaydardzhiev, S. Bio-Assisted Leaching of Non-Ferrous Metals from Waste Printed Circuit Boards—Importance of Process Parameters. *Metals* **2022**, *12*, 2092. [CrossRef]
5. Serga, V.; Zarkov, A.; Blumbergs, E.; Shishkin, A.; Baronins, J.; Elsts, E.; Pankratov, V. Leaching of Gold and Copper from Printed Circuit Boards under the Alternating Current Action in Hydrochloric Acid Electrolytes. *Metals* **2022**, *12*, 1953. [CrossRef]
6. Scotti, A.; Silvani, V.A.; Juarez, N.A.; Godeas, A.M.; Ubaldini, S. The Role of Mycorrhizal-Assisted Phytomining in the Recovery of Raw Materials from Mine Wastes. *Metals* **2022**, *12*, 1828. [CrossRef]
7. Yushin, N.; Zinicovscaia, I.; Cepoi, L.; Chiriac, T.; Rudi, L.; Grozdov, D. Biosorption and Bioaccumulation Capacity of Arthrospira platensis toward Yttrium Ion. *Metals* **2022**, *12*, 1465. [CrossRef]
8. Panyushkina, A.; Fomchenko, N.; Babenko, V.; Muravyov, M. Effect of Temperature on Biobeneficiation of Bulk Copper-Nickel Concentrate with Thermoacidophilic Microbial Communities. *Metals* **2021**, *11*, 1969. [CrossRef]

9. Terrones-Saeta, J.M.; Suárez-Macías, J.; Moreno-López, E.R.; Corpas-Iglesias, F.A. Leaching of Zinc for Subsequent Recovery by Hydrometallurgical Techniques from Electric Arc Furnace Dusts and Utilisation of the Leaching Process Residues for Ceramic Materials for Construction Purposes. *Metals* **2021**, *11*, 1603. [CrossRef]
10. Terrones-Saeta, J.M.; Suárez-Macías, J.; Castañón, A.M.; Corpas-Iglesias, F.A. Evaluation of Copper Leaching for Subsequent Recovery from the Waste Dumps of the Linares Mining District and Their Use for Construction Materials. *Metals* **2021**, *11*, 1328. [CrossRef]
11. Scotti, A.; Milia, S.; Silvani, V.; Cappai, G.; Guglietta, D.; Trapasso, F.; Tempesta, E.; Passeri, D.; Godeas, A.; Gómez, M.; et al. Sustainable Recovery of Secondary and Critical Raw Materials from Classified Mining Residues Using Mycorrhizal-Assisted Phytoextraction. *Metals* **2021**, *11*, 1163. [CrossRef]
12. Karimov, K.; Rogozhnikov, D.; Dizer, O.; Tretiak, M.; Mamyachenkov, S.; Naboichenko, S. Pressure Oxidation of Arsenic (III) Ions in the H_3AsO_3-Fe^{2+}-Cu^{2+}-H_2SO_4 System. *Metals* **2021**, *11*, 975. [CrossRef]

metals MDPI

Article

Pressure Oxidation of Arsenic (III) Ions in the H_3AsO_3-Fe^{2+}-Cu^{2+}-H_2SO_4 System

Kirill Karimov *, Denis Rogozhnikov, Oleg Dizer, Maksim Tretiak, Sergey Mamyachenkov and Stanislav Naboichenko

Department of Non-Ferrous Metals Metallurgy, Ural Federal University, 620002 Yekaterinburg, Russia; darogozhnikov@yandex.ru (D.R.); oleg.dizer@yandex.ru (O.D.); mtretyak144@gmail.com (M.T.); s.v.mamiachenkov@urfu.ru (S.M.); s.s.naboichenko@urfu.ru (S.N.)
* Correspondence: kirill_karimov07@mail.ru; Tel.: +7-912-695-3175

Abstract: The processing of low-grade polymetallic materials, such as copper–zinc, copper–lead–zinc, and poor arsenic-containing copper concentrates using hydrometallurgical methods is becoming increasingly important due to the depletion of rich and easily extracted mineral resources, as well as due to the need to reduce harmful emissions from metallurgy, especially given the high content of arsenic in ores. Ferric arsenates obtained through hydrothermal precipitation are the least soluble and most stable form of arsenic, which is essential for its disposal. This paper describes the investigation of the oxidation kinetics of As (III) ions to As (V) which is required for efficient purification of the resulting solutions and precipitation of low-solubility ferric arsenates. The effect of temperature (160–200 °C), the initial concentration of Fe (II) (3.6–89.5 mmol/dm^3), Cu (II) (6.3–62.9 of mmol/dm^3) and the oxygen pressure (0.2–0.5 MPa) on the oxidation efficiency of As (III) to As (V) was studied. As (III) oxidation in H_3AsO-Fe^{2+}-Cu^{2+}-H_2SO_4 and H_3AsO-Fe^{2+}-H_2SO_4 systems was controlled by a chemical reaction with the apparent activation energy (Ea (≈84.3–86.3 kJ/mol)). The increase in the concentration of Fe (II) ions and addition of an external catalyst (Cu (II) ions) both have a positive effect on the process. When Cu (II) ions are introduced into the solution, their catalytic effect is confirmed by a decrease in the partial orders, Fe (II) ions concentration from 0.43 to 0.20, and the oxygen pressure from 0.95 to 0.69. The revealed catalytic effect is associated with a positive effect of Cu (II) ions on the oxidation of Fe (II) to Fe (III) ions, which further participate in As (III) oxidation. The semi-empirical equations describing the reaction rate under the studied conditions are written.

Keywords: arsenic; iron; copper; oxidation; kinetics; catalytic effect; semi-empirical equation

Citation: Karimov, K.; Rogozhnikov, D.; Dizer, O.; Tretiak, M.; Mamyachenkov, S.; Naboichenko, S. Pressure Oxidation of Arsenic (III) Ions in the H_3AsO_3-Fe^{2+}-Cu^{2+}-H_2SO_4 System. *Metals* **2021**, *11*, 975. https://doi.org/10.3390/met11060975

Academic Editor: Stefano Ubaldini

Received: 13 May 2021
Accepted: 14 June 2021
Published: 17 June 2021

1. Introduction

Under the conditions of depleting mineral reserves, copper and other non-ferrous metal producers are urged to employ various primary and industrial low-grade polymetallic materials, such as copper–zinc, copper–lead–zinc, poor arsenic-containing copper concentrates, middlings, etc. The overall decrease in the quality of mineral raw materials, along with the need to use arsenic-containing ores in processing, results in large volumes of various semi-products comprising this highly toxic element [1,2].

The current technologies for processing copper concentrates are mainly pyrometallurgical, resulting in the formation of a large amount of waste gases and dusts. Recycling of metallurgical dusts consists in extracting valuable components in the form of individual products, as well as in obtaining low-toxic compounds from harmful impurities [3–10]. Arsenic is a harmful impurity contained in recovered non-ferrous metals, which determines the importance of its removal from technological processes [11].

A large quantity of arsenic-containing wastes is formed during the processing of refractory gold concentrates [12], where arsenic is present in the forms of arsenopyrite, jarosite, scorodite, tennantite, etc. During pressure oxidation leaching of these materials, ferric arsenates are produced ($FeAsO_4 \cdot 2H_2O$, $FeAsO_4 \cdot 0.68$–$0.77H_2O$, $Fe(AsO_4)_x(SO_4)_y(OH)_z$ wH_2O, where $0.36 \leq x \leq 0.69$, $0.19 \leq y \leq 0.5$, $0.55 \leq z \leq 0.8$ and $0.2 \leq w \leq 0.45$). Their crystal structure depends on both the process temperature and the concentration of iron and other ions in the solution [13].

Considering the high environmental hazard of arsenic [14–16], its conversion to low-soluble, low-toxic compounds is an important research task [17]. Ferric arsenates obtained through hydrothermal precipitation are the least soluble and most stable form of arsenic, which is essential for its disposal [13,18]. According to [19], at temperatures below 100 °C and at a pH close to neutral, an amorphous solid phase precipitates, comprising a gel-like ferric hydroxide with adsorbed arsenate ions. As temperature increases and pH decreases, the crystalline structure of the precipitate improves; adsorbed arsenate ions react with iron (III) to yield a crystalline iron arsenate structure. This interaction occurs at 150–200 °C and at a Fe:As ratio of 1.5 and higher. The hydrothermal interaction affords $FeAsO_4 \cdot 2H_2O$, $Fe_3(AsO_4)_2SO_4OH$, $FeSO_4OH$, and $Fe_2(HAsO_4)_3$ xH_2O. At temperatures of 150 °C and above, over 95% of arsenic is removed. Thus, the precipitate obtained in experiments followed at 190 °C [19] still contain crystalline scorodite.

During the hydrometallurgical processing of industrial arsenic-containing materials, a large proportion of arsenic dissolves into the solution as As (III). The oxidation of arsenic (III) ions to As (V) is required for efficient purification of the resulting solutions and precipitation of low-solubility ferric arsenates.

The hydrothermal precipitation of iron–arsenic cakes from model solutions containing iron (III) and arsenic (V) ions has been extensively discussed in the literature [20–27]; however, little information is available on the precipitation of arsenic (III) and its oxidation to As (V). Research into the oxidation kinetics of As (III) to As (V) ions under hydrothermal conditions can contribute to purification optimization of solutions obtained by processing various arsenic-containing materials.

2. Materials and Methods

2.1. Materials and Apparatus

All chemicals were of analytical grade. The As (III) and Fe (III) solutions were prepared by dissolving solid As_2O_3 and $Fe_2(SO_4)_3 \cdot 9H_2O$ in deionized water. $CuSO_4 \cdot 5H_2O$, H_2SO_4 and high-purity oxygen were also used from AO Vekton (S-Petersburg, Russia). Sulfuric acid was then added dropwise to adjust the pH to a certain value.

Experimental solutions were prepared under iron (II) and arsenic (III) concentrations of 4.8 and 3.7 of g/dm^3, respectively, to avoid the formation of insoluble ferric arsenates during their hydrothermal oxidation.

Laboratory experiments were carried out using 1 dm^3 autoclaves (Parr Instrument, Moline, IL, USA) equipped with electrical heating, mechanical stirring, and sampling systems.

The temperature of the solution during the experiments was maintained constant at a set value ±1.0 °C. The oxygen flow rate was controlled using two mass-flow controllers (Bronkhorst High-Tech BV Inc., Ruurlo, The Netherlands).

2.2. Analysis

The concentration of As (III) in solutions was determined by titration with potassium bromate ($KBrO_3$) and Fe (II) with potassium dichromate ($K_2Cr_2O_7$). Liquid samples containing both As (III) and Fe (II) ions were titrated after separation in ion exchange columns [28]. To determine the Fe (II) concentration, the sample solution was transferred into an Erlenmeyer flask. Using a graduated cylinder, 25 mL of 1 M H_2SO_4 was added to the flask. Then, 10 cm^3 of the phosphoric acid solution and 8 drops of sodium diphenylamine sulfonate indicator were added to the flask. The intense purple color produced by the first drop of excess $K_2Cr_2O_7$ signals the end point for titration.

Samples from each experiment were taken at predefined time intervals and analyzed using inductively coupled plasma mass spectrometry (ICP-MS—NexION 300S quadrupole mass spectrometer, PerkinElmer Inc., Waltham, MA, USA).

A Pourbaix diagram was constructed using the HSC Chemistry Software Version 6.0 (Outokumpu Research Oy, Finland).

Oxidation efficiency of arsenic (III) in solution was estimated by Formula (1):

$$\alpha_{As(III)} = \frac{\left(C_{As_{total}} - C_{As(III)}\right)}{C_{As_{total}}} \times 100 \quad (1)$$

2.3. Experimental Procedure

An autoclave was loaded with 0.6 dm^3 of a solution with a known composition and heated to the required temperature. Then, oxygen was introduced, and a mechanical agitator was turned on. Portions of the solution were collected at determined intervals in a sealed vessel, cooled to atmospheric temperature and analyzed. In all experiments, pH = 1.0 was maintained, since, according to [29], this pH value has no effect on the process.

After the completion of the experiments, the autoclave was quickly cooled down to 70 °C with cold water and then depressurized. The liquid specimens were sampled for further analysis.

3. Results and Discussion

3.1. Influence of Iron Ions on Arsenic (III) Oxidation under Hydrothermal Conditions

3.1.1. Influence of Fe (II) and Fe (III) Ions

A solution of 3.7 of g/dm^3 As (III) was oxidized under pressured oxygen for 40 min. Table 1 shows the results of As (III) oxidation by pressured oxygen without the addition of Fe (II) ions. The ranges for the temperature, oxygen pressure, and pH value in the experiments were 150–200 °C, 0.5–1.0 MPa and $[H_2SO_4]_0$ = 4–20 of g/dm^3, respectively. The results of preliminary studies showed that the hydrothermal oxidation of arsenic (III) ions by oxygen in acidic media was insignificant. These data show that, without iron ions in the solution, the As (III) oxidation was hindered, even under hydrothermal conditions at a temperature of 200 °C and oxygen pressure of 1 MPa.

Table 1. Oxidation efficiency of As (III) to As (V) in the absence of iron ions in the solution.

No.	t (°C)	O_2 (MPa)	Cu (g/dm^3)	H_2SO_4 (g/dm^3)	Oxidation Efficiency of As (III)
1	150	0.5	0	4	2.1
2	180	0.5	0	4	5.5
3	200	0.5	0	4	7.2
4	150	1	0	20	3.4
5	180	1	0	20	6.2
6	200	1	0	20	9.1
7	150	1	4	10	3.8
8	180	1	4	10	6.9
9	200	1	4	10	8.9

It is known that Cu (II) ions can catalyze oxidation reactions, for example, the Fe (II) to Fe (III) ions oxidation. The introduction of Cu (II) into the solutions had little effect on the oxidation of As (III) ions without Fe (II) and Fe (III) ions (Table 1).

The As (III) oxidation efficiency as a function of Fe (III) ion concentration at a temperature of 180 °C, duration of 40 min, pH = 1.0, As (III) = 3.7 g/dm^3 with no oxygen introduced into the system, is shown in Figure 1.

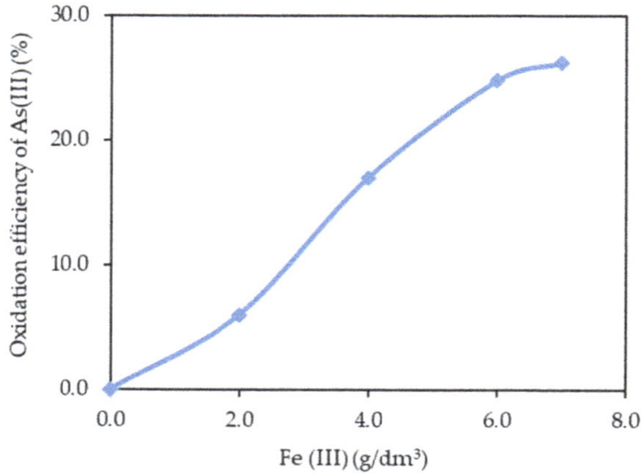

Figure 1. As (III) oxidation efficiency as a function of Fe (III) ion concentration with no oxygen introduced into the system.

Upon increasing the concentration of Fe (III) ions in the solution from 2 to 7 g/dm^3, the As (III) oxidation efficiency increased from 6.1 to 26.2%. Despite the lower standard oxidation potential of Fe (III)/Fe (II) (+0.77 V) than that of oxygen (+1.23 V, O_2/H_2O), the oxidation of arsenic by Fe (III) ions was more efficient.

Figure 2 demonstrates the effect of iron (II) ions on the degree of As (III) oxidation at 180 °C, duration of 40 min, pH = 1.0, oxygen pressure of 0.2 MPa.

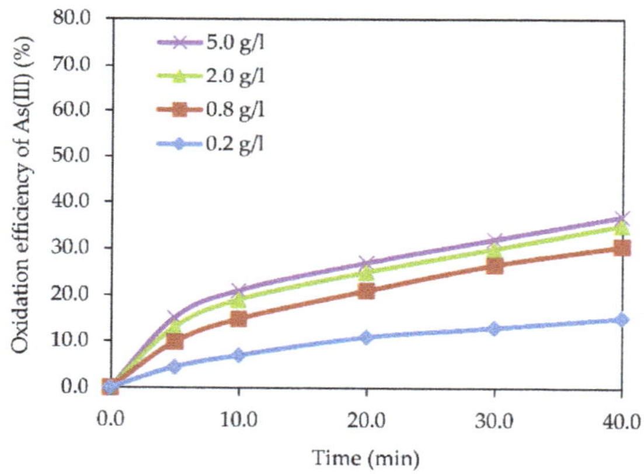

Figure 2. As (III) oxidation efficiency as a function of Fe (II) ion concentration.

Raising the initial concentration of iron (II) ions from 0.2 to 5.0 g/dm^3 had a positive effect on the As (III) ions oxidation efficiency, which increased from 15.3 to 36.8%.

Under hydrothermal conditions, Fe (II) ions are oxidized by oxygen to Fe (III), which further oxidizes As (III), according to the thermodynamic calculation presented in the following section.

3.1.2. Thermodynamics of As (III) Ions Oxidation in the Presence of Fe (III) Ions under Hydrothermal Conditions

The pH-dependences of the oxidation potentials of As (III) to As (V) and Fe (II) to Fe (III) couples are presented in a Pourbaix diagram (Figure 3).

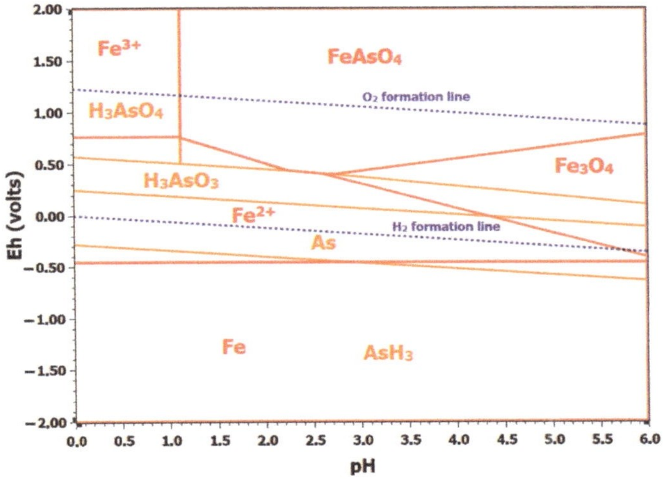

Figure 3. Pourbaix diagram of As-Fe-H_2O system.

According to the Pourbaix diagram (Figure 3), in acidic media (pH = 0–2), the complete oxidation of As (III) to As (V) occurred at Eh = 0.48–0.56 V in the As–H_2O system. At the same time, in the Fe–H_2O system, the oxidation of Fe (II) ions to Fe (III) occurred at Eh = 0.771 V. It should be noted that arsenic oxidation by Fe (III) ions was more efficient than that by oxygen (1.23 V, O_2/H_2O) [28].

The effect of iron (III) ions on arsenic oxidation may be associated with the formation of intermediate stable complexes of $FeH_2AsO_4^{2+}$ in acidic media.

The standard electrode oxidation potential of arsenic is Eh_0 = 0.56 V, the oxidation is described by the following reaction [30,31]:

$$H_3AsO_3 + H_2O = H_3AsO_4 + 2H^+ + 2e;\ Eh_0 = 0.56\ V \quad (2)$$

The stability constant of the $FeH_2AsO_4^{2+}$ complex is equal to 4.36×10^{-4}, while the complex dissociation is described by the following reaction:

$$FeH_2AsO_4^{2+} + H^+ = Fe^{3+} + H_3AsO_4;\ K_d = 4.36 \times 10^{-4} \quad (3)$$

At t = 473 K, the oxidation potential of arsenic is calculated as follows:

$$Eh = 0.56 + 0.0195\ lg(\frac{a_{H_3AsO_4} \times a^2{}_{H^+}}{a_{H_3AsO_3}}) = 0.56 - 0.0390pH + 0.0204\ lg(\frac{a_{H_3AsO_4}}{a_{H_3AsO_3}}) \quad (4)$$

The stability constant is equal to:

$$K_d = \frac{a_{H_3AsO_4} \times a_{Fe^{3+}}}{a_{Fe_2H_2AsO_4^{2+}} \times a_{H^+}} = 4.36 \times 10^{-4} \quad (5)$$

By expressing activity $a_{H_3AsO_4}$ as $a_{H_3AsO_4} = \frac{K_d \times a_{H^+} \times a_{Fe_2H_2AsO_4^{2+}}}{a_{Fe^{3+}}}$ and placing it into an equation to calculate the oxidation potential of As (III), we obtain:

$$Eh = 0.56 - 0.0406pH + 0.0204\ lg\ (\frac{K_d \times a_{H^+} \times a_{Fe_2H_2AsO_4^{2+}}}{a_{Fe^{3+}} \times a_{H_3AsO_3}}) \quad (6)$$

$$Eh = 0.49 - 0.0611pH + 0.0204\ lg\ (\frac{a_{Fe_2H_2AsO_4^{2+}}}{a_{Fe^{3+}} \times a_{H_3AsO_3}}) \quad (7)$$

Specifically, at pH = 1.5, the oxidation potential of arsenic is equal to:

$$Eh = 0.40 + 0.0204 \lg \left(\frac{a_{Fe_2H_2AsO_4^{2+}}}{a_{Fe^{3+}} \times a_{H_3AsO_3}}\right) \quad (8)$$

The oxidation potential of Fe (II) to Fe (III) does not depend on pH (Figure 3) and is described by the following equation:

$$Fe^{2+} = Fe^{3+} + e^{-} \quad (9)$$

At t = 453 K (Eh_0 = 0.771 V):

$$Eh = 0.771 + 0.0897 \lg \left(\frac{a_{Fe^{3+}}}{a_{Fe^{2+}}}\right) \quad (10)$$

The results of thermodynamic assessment explain the oxidation of As (III) by Fe (III) ions in acidic solutions, at t = 180 °C, without introducing oxygen into the system. When added to the solution, Fe (II) ions are oxidized by oxygen to Fe (III) ions, which oxidize As (III).

3.2. *Influence of Cu (II) Ions on Fe (II) and As (III) Oxidation in Hydrothermal Conditions*

As shown earlier, copper (II) ions have little effect on arsenic (III) oxidation without iron ions in the solution. However, according to the previous studies [19,32] Cu (II) catalyzes Fe (II) to Fe (III) transition and Fe (III) is an oxidizing agent for As (III). Figure 4 demonstrates the effect of copper (II) ions on the degree of Fe (II) and As (III) oxidation at a temperature of 180 °C, duration of 40 min, pH = 1.0, As (III) = 3.7 g/dm^3 and oxygen pressure of 0.2 MPa.

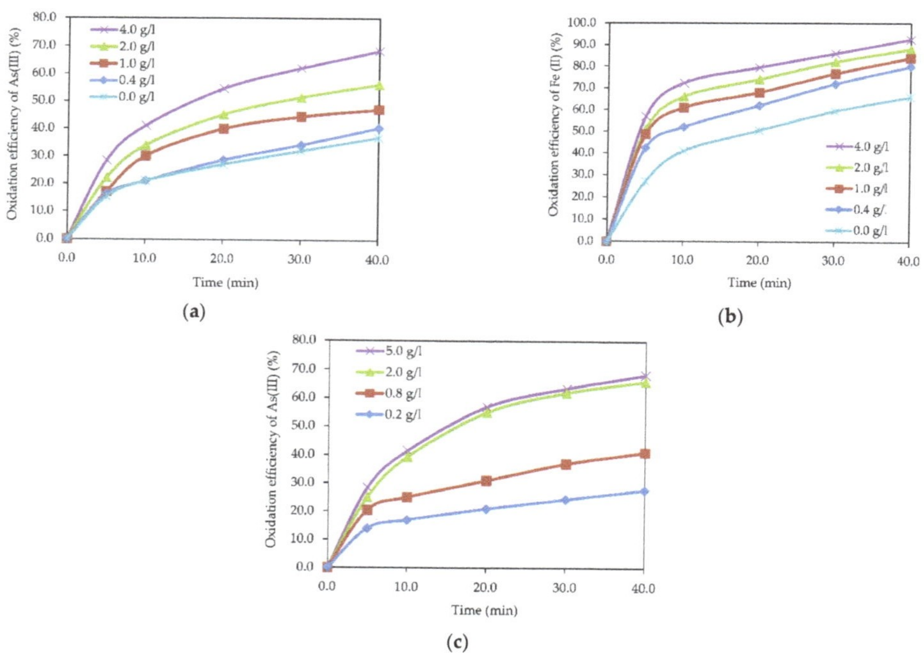

Figure 4. Oxidation efficiency of As (III) as a function of the copper concentration in the presence of iron (**a**), iron in the presence of copper (**b**), the oxidation efficiency of As (III) as a function of Fe (II) concentration in the presence of the copper (**c**).

According to the obtained data, addition of Cu (II) ions to the solution leads to a sharp increase in the As (III) oxidation efficiency in the presence of Fe (II) ions. If the concentration of Cu (II) in the solution increases from 0.4 to 4.0 g/dm^3, the oxidation efficiency of As (III) increases from 40.4 to 68.3%, and of Fe (II) from 66.2 to 92.6 % (Figure 4a,b).

An increase in the initial Fe (II) concentration from 0.2 to 5.0 g/dm^3 leads to an increase in the As (III) oxidation efficiency from 27.7 to 68.3%, relative to 15.3–35.1%, without adding copper to the solution. In the presence of Cu (II) ions, an increase in the iron concentration above 2 g/dm^3 has little effect.

An increase in the Cu (II) ion concentration from 0 to 4 g/dm^3 has a similar effect on the Fe (II) oxidation efficiency, which increases from 66 to 93%.

The positive effect of copper ions on As (III) oxidation is probably associated with an increase in the oxygen oxidizing power in the presence of variable valency ions (Co^{3+}, Fe^{3+}, Cu^{2+}), prone to complex formation or valency changes [19,28,32]. This explains the increase in the As (III) oxidation rate in the presence of Cu (II) ions. Cu (II) ions have a catalytic effect on Fe (II) to Fe (III) ion oxidation, necessary for the As (III) oxidation (reaction (11)). The catalytic effect of Cu (II) ions can be described by the following reactions:

$$Cu^{2+} + Fe^{2+} = Cu^{+} + Fe^{3+} \quad (11)$$

$$2Cu^{+} + 1/2O_2 + 2H^{+} = 2Cu^{2+} + H_2O \quad (12)$$

Cu^{2+} cations initiate the oxidation of Fe^{2+} to Fe^{3+} ions, which then oxidize As (III), as described by the following reactions:

$$H_3AsO_3 + 2Fe^{3+} + H_2O = 2Fe^{2+} + H_3AsO_4 + 2H^{+} \quad (13)$$

$$H_3AsO_4 + Fe^{3+} = FeH_2AsO_4{}^{2+} + H^{+} \quad (14)$$

3.3. *Influence of Temperature on As (III) Ion Oxidation*

Oxidation of As (III) in an oxygen atmosphere in the presence of Fe (II) ions with (4 g/dm^3) and without Cu (II) ions addition was performed across the temperature range of 160–200 °C, for 40 min, at pH = 1.0, As (III) = 3.7 g/dm^3, oxygen pressure 0.2 MPa and Fe (II) = 5.0 g/dm^3.

As shown in Figure 5, the As (III) oxidation efficiency at 160 °C slowly increased over time; however, it rapidly increased as the temperature increased to 200 °C, indicating a strong temperature effect on the As (III) to As (V) transition rate. Increasing the temperature from 160 to 200 °C leads to an increase in the As (III) oxidation efficiency from 18 to 45% and from 57 to 91%, without and with (4 g/dm^3) Cu (II) in the solution, respectively, in 40 min.

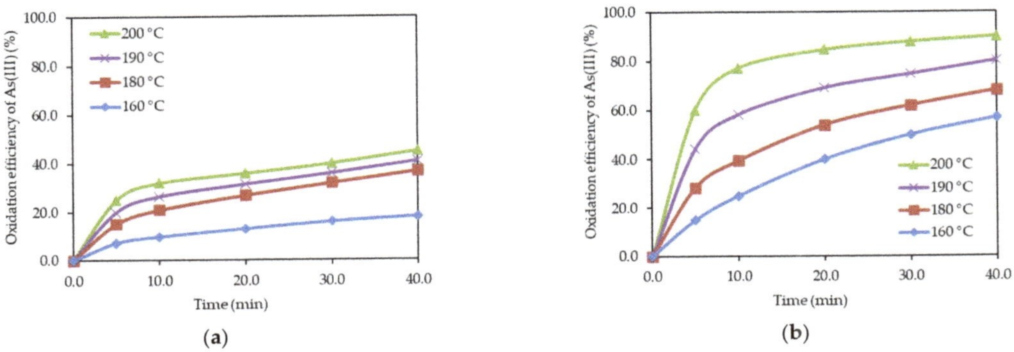

Figure 5. As (III) oxidation efficiency as a function of temperature with (**a**) and without (**b**) Cu (II) ions.

3.4. Influence of Oxygen Pressure on the Oxidation of Fe (II) and As (III) Ions

The influence of oxygen pressure on the As (III) and Fe (II) ion oxidation was assessed at a temperature of 180 °C, duration of 40 min, pH = 1.0, As (III) = 3.7 g/dm^3, oxygen pressure of 0.2–0.5 MPa and Fe (II) = 5.0 g/dm^3 (Figure 6).

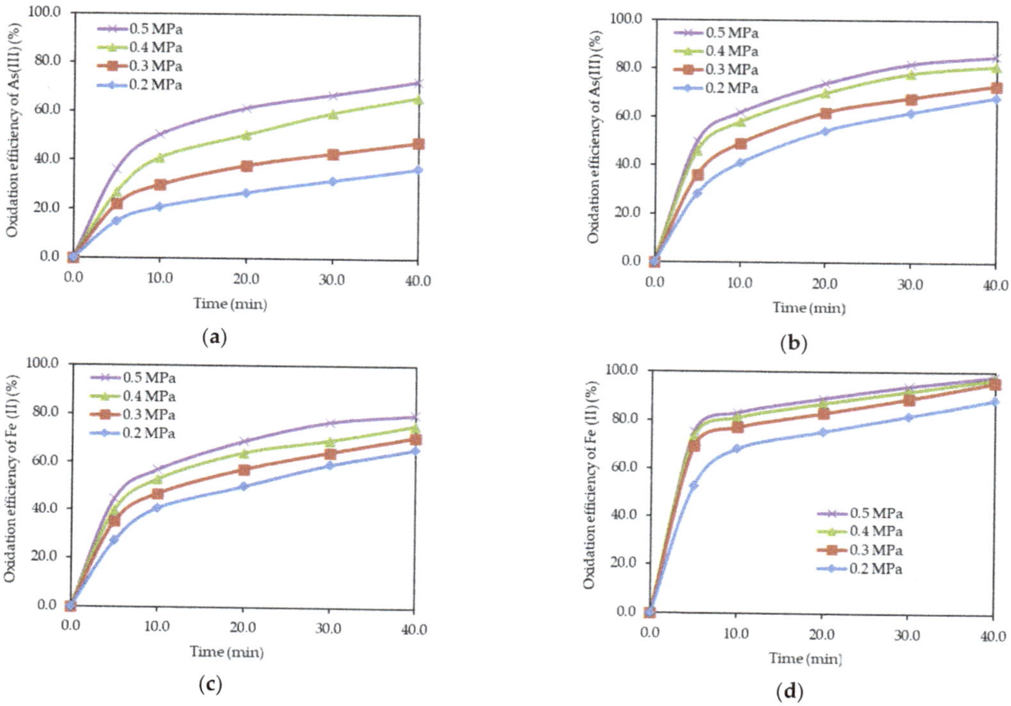

Figure 6. As (III) and Fe (II) oxidation efficiency as a function of temperature with (**a**,**c**) and without (**b**,**d**) Cu (II) ions.

Oxygen pressure has a significant effect on the As (III) and Fe (II) ions oxidation in solution. An increase in oxygen pressure from 0.2 to 0.5 MPa leads to an increase in the oxidation efficiency of As (III) and Fe (II), from 36.9 to 72.5% and 66.1 to 80.3%, respectively, within 40 min. Introduction of Cu (II) ions into the solution also had a positive effect on the oxidation efficiency. In addition, the influence of oxygen pressure on the process decreased; its increase from 0.2 to 0.5 MPa brought about an increase in the oxidation efficiency of As (III) from 68.3 to 85.2% and Fe (II) from 88.6 to 95.6%.

Increased oxygen pressure leads to a more intense oxidation of Fe (II) to Fe (III) ions, which further oxidizes As (III). Moreover, Cu (II) ions catalyze the Fe (II) to Fe (III) transition, hence being effective in reducing the oxygen pressure dependence of the process.

4. Kinetics Analysis

Since the As (III) oxidation efficiency is strongly influenced by temperature, Fe (II) concentration and oxygen pressure, this process can occur in both kinetic and diffusion regimes. This means that both the oxidizing agent diffusion into the reaction zone and the chemical reaction itself can be the rate-limiting stage of the process. An investigation of the process kinetics is necessary to determine the rate-limiting step.

A kinetics study of the oxidation of As (III) ions was performed at temperatures of 433–473 K (160–200 °C), duration of 40 min, pH = 1.0, As (III) = 13.3–100.6 mmol/dm^3, oxygen pressure of 0.2–0.5 MPa, Fe (II) = 3.6–89.5 mmol/dm^3, Cu (II) = 6.3–62.9 of mmol/dm^3.

The time-to-a-given-fraction method was used to calculate the kinetic parameters at different stages (As (III) oxidation efficiencies) [33]. The time required to achieve a certain degree of leaching and the apparent activation energy of E_a are related according to Equation (15):

$$Lnt_x = const - lnA + E_a/RT \quad (15)$$

The slope angle of the graph displaying the relationship between Lnt_x and 1/T allows the apparent activation energy to be calculated.

To determine the experimental activation energy of Cu-free As (III) oxidation, the inverse temperatures and natural logarithms of time were calculated (Figure 7a) at the following As (III) conversions in %: 5, 10, and 15. The same steps were implemented to calculate the experimental activation energy with copper ions (II) added to the solution (Figure 7b) at the following As (III) conversions in %: 15, 40, and 55. The experiment was repeated three times to obtain reliable values at each temperature; the averaged values are presented in Figure 7.

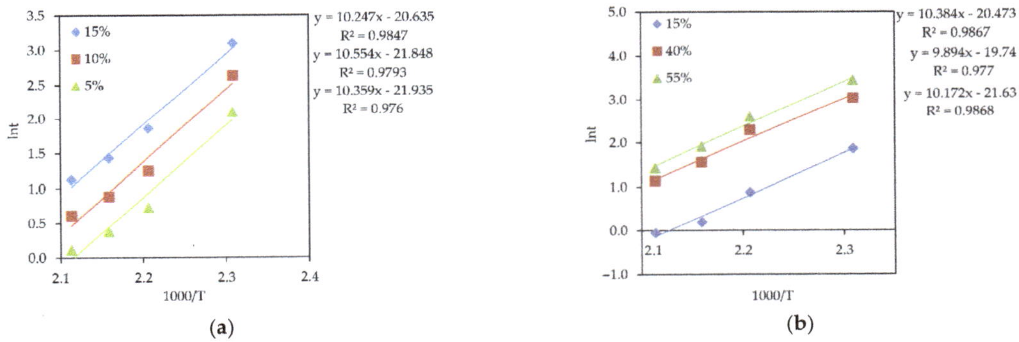

Figure 7. The correlation between Lnt_x and 1/T for various As (III) oxidation efficiencies, without (**a**) and with (**b**) Cu (II).

The average values of the experimental activation energy of As (III) oxidation are 84.4 kJ/mol for the H_3AsO_3-Fe^{2+}-Cu^{2+}-H_2SO_4 system and 86.4 kJ/mol for the H_3AsO_3-Fe^{2+}-H_2SO_4 system, which are characteristic of systems controlled by a chemical reaction. In the course of the process, the values of the experimental activation energy undergo minor variations, indicating no changes in the process mechanism.

The differential method of initial rates was used to calculate partial orders. The initial As (III) oxidation rate (at $\tau = 0$) was determined from the plot of As (III) concentration (mmol/dm^3) against time (min). The initial rate was equal to the negative slope of the As (III) concentration curve against time at $\tau = 0$ min. In the study, the concentration of only one component was varied, while the others were added in excess and kept constant. To calculate partial orders, we plotted the relationship between the natural logarithms of the As (III) initial oxidation rates (lnr_0) and the natural logarithms of the concentrations of the studied components (lnC_i). The slope angle of the resulting plot corresponds to the partial order.

A graphical determination of the partial order relative to the concentration of iron ions is presented in Figure 8.

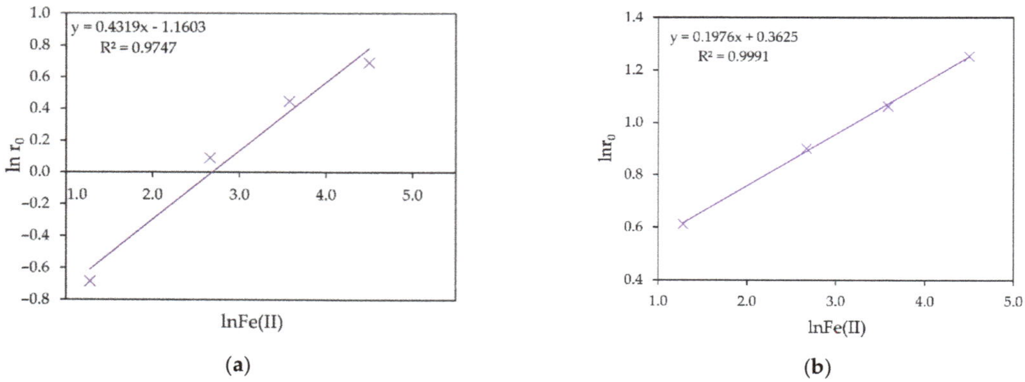

Figure 8. Dependence of $\ln r_0$ on lnFe (II) without (**a**) and with (**b**) Cu (II).

The experimental reaction order relative to the concentration of Fe (II) ions is 0.43 for the H_3AsO_3-Fe^{2+}-H_2SO_4 system and 0.20 for the H_3AsO_3-Fe^{2+}-Cu^{2+}-H_2SO_4 system (Figure 8a,b, respectively). The difference in the reaction orders relative to iron was caused, apparently, by the catalytic influence of the Cu (II) ions on the Fe (II) ion oxidation in the H_3AsO_3-Fe2$^+$-Cu^{2+}-H_2SO_4 system (Equations (11)–(14)).

A graphical determination of the partial order relative to the Cu (II) ions concentration is given in Figure 9.

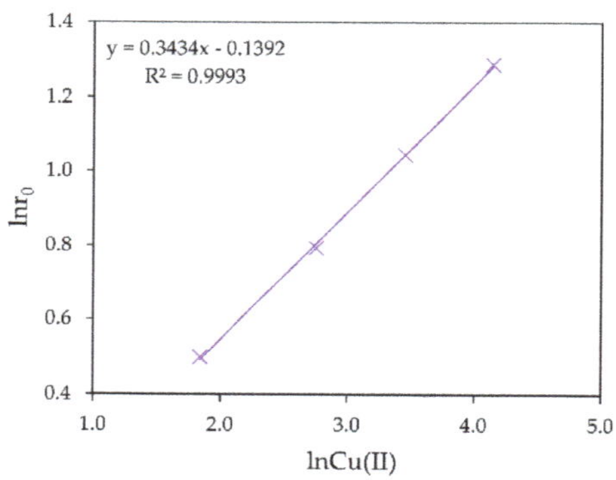

Figure 9. Lnr_0 versus lnCu (II).

The experimental reaction order relative to the concentration of Cu (II) ions was 0.34 for the H_3AsO_3-Fe^{2+}-Cu^{2+}-H_2SO_4 system (Figure 9). The positive effect of copper ions on the As (III) oxidation was associated with the catalytic effect of oxygen on the Fe (II) ions oxidation, as mentioned earlier.

The reaction order relative to oxygen pressure was determined graphically (Figure 10).

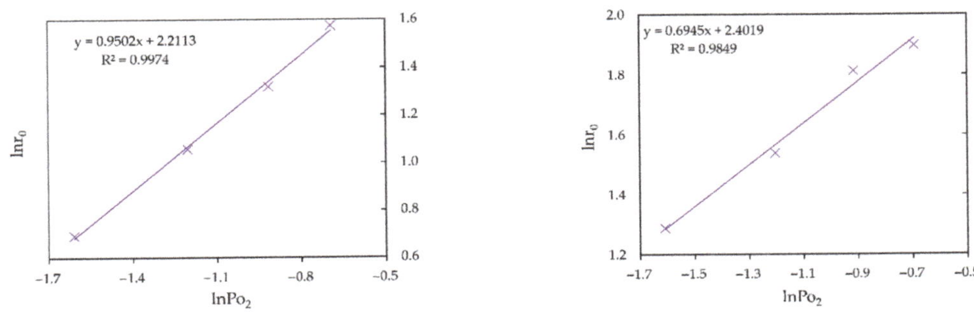

(a) (b)

Figure 10. Dependence of $\ln r_0$ on $\ln Po_2$ without (**a**) and with (**b**) Cu (II).

According to the data obtained, the reaction order relative to oxygen pressure was 0.95 for the H_3AsO_3-Fe^{2+}-H_2SO_4 and 0.69 for the H_3AsO_3-Fe^{2+}-Cu^{2+}-H_2SO_4 system (Figure 10a,b, respectively). Introduction of Cu (II) ions into the solution led to a decrease in the partial order and, hence, to a decreased dependence of the process on oxygen pressure.

A graphical determination of the partial order relative to the concentration of As (III) ions is shown in Figure 11a,b. The experimental reaction order relative to the initial concentration of As (II) ions was 0.46 for the H_3AsO_3-Fe^{2+}-Cu^{2+}-H_2SO_4 system and 0.48 for the H_3AsO_3-Fe^{2+}-H_2SO_4 system.

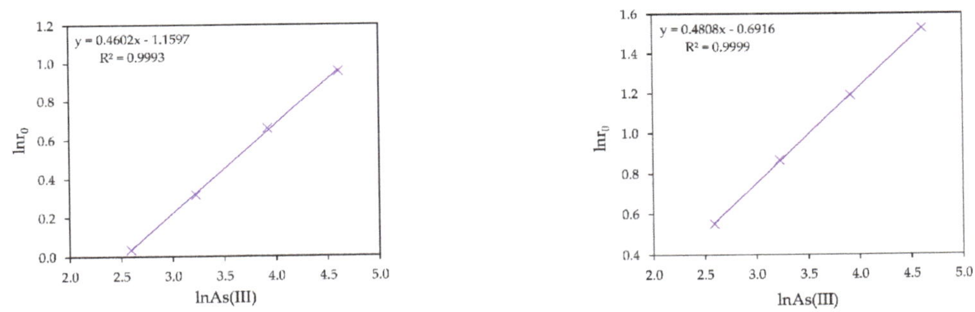

(a) (b)

Figure 11. Dependences of $\ln r_0$ on lnAs (III) without (**a**) and with (**b**) Cu (II).

To derive a general kinetic equation, we plotted all temperatures, concentrations, and oxygen pressures, which allowed the fixed slope angle of a = 1.08 to be determined. The "a" value obtained by the graphical method and the corresponding correlation coefficient R^2 are presented in Figure 12 for the H_3AsO_3-Fe^2-H_2SO_4 and H_3AsO_3-Fe^{2+}-Cu^{2+}-H_2SO_4 systems. The obtained value of the "a" coefficient corresponds to k_o. Some kinetic data at different parameters for oxidation efficiency of As (III) to As (V) are shown in Table 2.

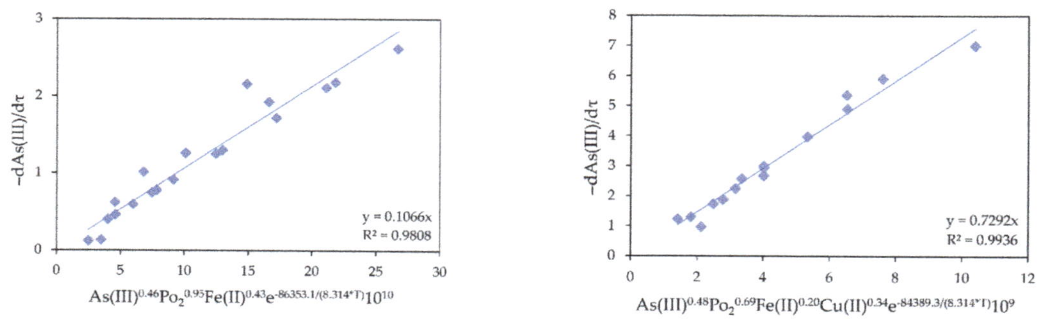

(a) (b)

Figure 12. Graphical determination of the k_o coefficient for H_3AsO_3-Fe^{2+}-H_2SO_4 (**a**) and H_3AsO_3-Fe^{2+}-Cu^{2+}-H_2SO_4 (**b**) systems.

Table 2. Kinetic data at different parameters for oxidation efficiency of As (III) to As (V).

No.	Po_2 (MPa)	Fe (II) (mmole/dm³)	*t* (K)	Cu (II) (mmole/dm³)	−dAs (III)/dτ
1	0.2	90.01	433	0.00	0.14
2	0.2	90.01	463	0.00	1.92
3	0.2	3.58	453	0.00	0.12
4	0.2	35.81	453	0.00	1.02
5	0.3	90.01	453	0.00	2.16
6	0.4	90.01	453	0.00	2.23
7	0.2	90.01	453	6.29	1.30
8	0.2	90.01	453	31.47	2.25
9	0.2	90.01	433	62.95	1.24
10	0.2	90.01	463	62.95	4.91
11	0.2	3.58	453	62.95	0.98
12	0.2	35.81	453	62.95	2.58
13	0.3	90.01	453	62.95	3.98
14	0.4	90.01	453	62.95	5.37

Based on the above results, the following general kinetic equations can be derived for the hydrothermal oxidation of As (III) ions in the presence of Fe (II) ions:

For the H_3AsO_3-Fe^{2+}-H_2SO_4 system:

$$-\frac{dAs(III)}{d\tau} = 0.107As(III)^{0.46}P_{O_2}{}^{0.95}Fe(II)^{0.43}e^{\frac{-86353.1}{8.314\times T}}10^{10} \quad (16)$$

For the H_3AsO_3-Fe^{2+}-Cu^{2+}-H_2SO_4 system:

$$-\frac{dAs\ (III)}{d\tau} = 0.729As(III)^{0.48}P_{O_2}{}^{0.69}Cu(II)^{0.34}Fe(II)^{0.20}e^{\frac{-84389.3}{8.314\times T}}10^{9} \quad (17)$$

One can assume that As (III) oxidation in the H_3AsO_3-Fe^{2+}-Cu^{2+}-H_2SO_4 and H_3AsO_3-Fe^{2+}-H_2SO_4 systems was controlled by a chemical reaction with the apparent activation energy ($E_a \approx$ 84.3–86.3 kJ/mol) for the following reasons: the increase in the concentration of Fe (II) ions and addition of an external catalyst (ions Cu (II)) had a positive effect on the process; elevated E_a values in the reaction in the order of 0.43–0.20. When Cu (II) ions were introduced into the solution, their catalytic effect was confirmed by a decrease in partial orders, Fe (II) ions concentration from 0.43 to 0.20 and the oxygen pressure from 0.95 to 0.69. The revealed catalytic effect was associated with a positive effect of Cu (II) on the oxidation of Fe (II) to Fe (III) ions (Equations (11)–(14)), which further participate in the As (III) oxidation. This also confirms that Fe (III) ions formed that oxidized As (III).

5. Conclusions

This study focuses on the pressure hydrothermal oxidation of arsenic (III) ions in the H_3AsO_3-Fe^{2+}-Cu^{2+}-H_2SO_4 system formed during processing of arsenic-containing sulfide non-ferrous metals. The main conclusions of the study are as follows:

- Upon hydrothermal oxidation, Fe (II) ions are oxidized by oxygen to Fe (III) ions, which act as As (III) oxidizing agents. The influence of Fe (III) ions on arsenic oxidation can be associated with the formation of stable intermediate $FeH_2AsO_4^{2+}$ complexes in acidic media, which reduce the standard oxidation potential of As (III) to 0.40 V.
- It has been shown that the positive effect of copper ions on As (III) oxidation is associated with an increase in the oxygen oxidizing power in the presence of variable valency ions, prone to complex formation or valency changes. This explains the increase in the As (III) oxidation rate in the presence of Cu (II) ions. In addition, Cu (II) ions promote the oxidation of Fe (II) to Fe (III) ions.
- General kinetic equations for the As (III) oxidation rate in the H_3AsO_3-Fe^{2+}-Cu^{2+}-H_2SO_4 and H_3AsO_3-Fe^{2+}-H_2SO_4 systems were determined. The experimental reaction orders relative to arsenic and iron concentrations and oxygen pressure were obtained.
- As (III) oxidation in the H_3AsO-Fe^{2+}-Cu^{2+}-H_2SO_4 and H_3AsO_3-Fe^{2+}-H_2SO_4 systems was controlled by a chemical reaction with the apparent activation energy (E_a (≈84.3–86.3 kJ/mol)) for the following reasons: the increase in the concentration of Fe (II) ions and addition of an external catalyst (Cu (II) ions) both have a positive effect on the process; at a reaction order of 0.43–0.20, E_a values are elevated. When Cu (II) ions are introduced into the solution, their catalytic effect is confirmed by a decrease in the specific orders of Fe (II) ions and the oxygen pressure. The revealed catalytic effect is associated with a positive effect of Cu (II) ions on the oxidation of Fe (II) to Fe (III) ions, which further participate in As (III) oxidation. This confirms that the formed Fe (III) ions act as oxidizing agents towards As (III).

Author Contributions: Conceptualization, D.R.; data curation, S.M. and S.N.; investigation, O.D. and M.T.; writing—review and editing, K.K. All authors have read and agreed to the published version of the manuscript.

Funding: This work was financially supported by the Russian Science Foundation Project No. 20-79-00321.

Institutional Review Board Statement: Not applicable.

Informed Consent Statement: Not applicable.

Data Availability Statement: Data are contained within the article.

Conflicts of Interest: The authors declare no conflict of interest.

References

1. Azcue, J.M.; Mudroch, A.; Rosa, F.; Hall, G. Effects of abandoned gold mine tailings on the arsenic concentrations in water and sediments of jack of Clubs Lake, B.C. *Environ. Technol.* **1994**, *15*, 669–678. [CrossRef]
2. Bissen, M.; Frimmel, F.H. Arsenic—A review. Part I: Occurrence, Toxicity, Speciation, Mobility. *Acta Hydroch. Hydrob.* **2003**, *31*, 9–18. [CrossRef]
3. Shibayama, A.; Takasaki, Y.; William, T.; Yamatodani, A.; Higuchi, Y.; Sunagawa, S.; Ono, E. Treatment of smelting residue for arsenic removal and recovery of copper using pyro–hydrometallurgical process. *J. Hazard. Mater.* **2010**, *181*, 1016–1023. [CrossRef]
4. Kovyazin, A.; Timofeev, K.; Krauhin, S. Copper smelting fine dust autoclave leaching. *Mater. Sci. Forum* **2019**, 615–620. [CrossRef]
5. Jarošíková, A.; Ettler, V.; Mihaljevič, M.; Penížek, V.; Matoušek, T.; Culka, A.; Drahota, P. Transformation of arsenic-rich copper smelter flue dust in contrasting soils: A 2-year field experiment. *Environ. Pollut.* **2018**, *237*, 83–92. [CrossRef] [PubMed]
6. Karimov, K.A.; Naboichenko, S.S. Sulfuric Acid Leaching of High-Arsenic Dust from Copper Smelting. *Metallurgist* **2016**, *60*, 456–459. [CrossRef]
7. Montenegro, V.; Sano, H.; Fujisawa, T. Recirculation of high arsenic content copper smelting dust to smelting and converting processes. *Miner. Eng.* **2013**, *49*, 184–189. [CrossRef]
8. Montenegro, V.; Sano, H.; Fujisawa, T. Recirculation of chilean copper smelting dust with high arsenic content to the smelting process. *Mater. Trans.* **2008**, *49*, 2112–2118. [CrossRef]

9. Xue, J.; Long, D.; Zhong, H.; Wang, S.; Liu, L. Comprehensive recovery of arsenic and antimony from arsenic-rich copper smelter dust. *J. Hazard. Mater.* **2021**, *413*, 125365. [CrossRef]
10. Karimov, K.A.; Naboichenko, S.S.; Kritskii, A.V.; Tret'yak, M.A.; Kovyazin, A.A. Oxidation Sulfuric Acid Autoclave Leaching of Copper Smelting Production Fine Dust. *Metallurgist* **2019**, *62*, 1244–1249. [CrossRef]
11. Safarzadeh, M.S.; Moats, M.S.; Miller, J.D. Recent trends in the processing of enargite concentrates. *Miner. Process. Extr. Metall. Rev.* **2014**, *35*, 283–367. [CrossRef]
12. Rogozhnikov, D.A.; Shoppert, A.A.; Dizer, O.A.; Karimov, K.A.; Rusalev, R.E. Leaching Kinetics of Sulfides from Refractory Gold Concentrates by Nitric Acid. *Metals* **2019**, *9*, 465. [CrossRef]
13. Fujita, T.; Fujieda, S.; Shinoda, K.; Suzuki, S. Environmental leaching characteristics of scorodite synthesized with Fe (II) ions. *Hydrometallurgy* **2012**, *111–112*, 87–102. [CrossRef]
14. Singh, R.; Singh, S.; Parihar, P.; Singh, V.P.; Prasad, S.M. Arsenic contamination, consequences and remediation techniques: A review. *Ecotoxicol. Environ. Saf.* **2015**, *112*, 247–270. [CrossRef]
15. Ungureanu, G.; Santos, S.; Boaventura, R.; Botelho, C. Arsenic and antimony in water and wastewater: Overview of removal techniques with special reference to latest advances in adsorption. *J. Environ. Manag.* **2015**, *151*, 326–342. [CrossRef]
16. Jadhav, S.V.; Bringas, E.; Yadav, G.D.; Rathod, V.K.; Ortiz, I.; Marathe, K.V. Arsenic and fluoride contaminated groundwaters: A review of current technologies for contaminants removal. *J. Environ. Manag.* **2015**, *162*, 306–325. [CrossRef]
17. Otgon, N.; Zhang, G.; Zhang, K.; Yang, C. Removal and fixation of arsenic by forming a complex precipitate containing scorodite and ferrihydrite. *Hydrometallurgy* **2019**, *186*, 58–65. [CrossRef]
18. Gomez, M.A.; Becze, L.; Celikin, M.; Demopoulos, G.P. The effect of copper on the precipitation of scorodite ($FeAsO_4{\cdot}2H_2O$) under hydrothermal conditions: Evidence for a hydrated copper containing ferric arsenate sulfate-short lived intermediate. *J. Colloid Interface Sci.* **2004**, *360*, 508–518. [CrossRef] [PubMed]
19. Monhemius, A.J.; Swash, P.M. Removing and stabilizing As from copper refining circuits by hydrothermal processing. *J. Miner. Met. Mater. Soc.* **1999**, *51*, 30–33. [CrossRef]
20. Krause, E.; Ettel, V.A. Solubilities and stabilities of ferric arsenate compounds. *Hydrometallurgy* **1989**, *122*, 311–337. [CrossRef]
21. Demopoulos, G.P.; Droppert, D.J.; Van Weert, G. Precipitation of crystalline scorodite ($FeAsO_4{\cdot}2H_2O$) from chloride solutions. *Hydrometallurgy* **1995**, *38*, 245–261. [CrossRef]
22. Dutrizac, J.E.; Jambor, J.L. The synthesis of crystalline scorodite, $FeAsO_4{\cdot}2H_2O$. *Hydrometallurgy* **1988**, *19*, 377–384. [CrossRef]
23. Zhang, W.; Lu, H.; Liu, F.; Wang, C.; Zhang, Z.; Zhang, J. Hydrothermal treatment of arsenic sulfide slag to immobilize arsenic into scorodite and recycle sulfur. *J. Hazard. Mater.* **2021**, *406*, 124735. [CrossRef]
24. Pozo, G.; Van Houtven, D.; Fransaer, J.; Dominguez-Benetton, X. Arsenic immobilization as crystalline scorodite by gas-diffusion electrocrystallization. *React. Chem. Eng.* **2009**, *5*, 1118–1128. [CrossRef]
25. Gomez, M.A.; Becze, L.; Cutler, J.N.; Demopoulos, G.P. Hydrothermal reaction chemistry and characterization of ferric arsenate phases precipitated from $Fe_2(SO_4)_3$-As_2O_5-H_2SO_4 solutions. *Hydrometallurgy* **2011**, *107*, 74–90. [CrossRef]
26. Filippou, D.; Demopoulos, G.P. Arsenic immobilization by controlled scorodite precipitation. *J. Miner. Met. Mater. Soc.* **1997**, *49*, 52–55. [CrossRef]
27. Gomez, M.A.; Becze, L.; Bluteau, M.C.; Le Berre, J.F.; Cutler, J.N.; Demopoulos, G.P. Autoclave precipitation and characterization of Fe(III)-AsO_4-SO_4 phases. In Proceedings of the Hydrometallurgy 2008: Proceedings of the 6th International Symposium, Phoenix, AZ, USA, 17–21 August 2008; pp. 1078–1086.
28. Guo, F.; Wang, Q.; Demopoulos, G.P. Kinetics of iron (III)-catalyzed oxidation of arsenic (III) in acidic solutions with SO_2/O_2 gas mixture using different iron sources. *Hydrometallurgy* **2019**, *189*, 105130. [CrossRef]
29. Song, K.Z.; Ke, P.C.; Liu, Z.Y.; Liu, Z.H. Co-oxidation of arsenic (III) and iron (II) ions by pressurized oxygen in acidic solutions. *Int. J. Miner. Metall. Mater.* **2020**, *27*, 181–189. [CrossRef]
30. Marini, L.; Accornero, M. Prediction of the thermodynamic properties of metal-arsenate and metal-arsenite aqueous complexes to high temperatures and pressures and some geological consequences. *Environ. Geol.* **2007**, *52*, 1343–1364. [CrossRef]
31. Lu, P.; Zhu, C. Arsenic Eh–pH diagrams at 25 °C and 1 bar. *Environ. Earth Sci.* **2011**, *62*, 1673–1683. [CrossRef]
32. Zhang, P.; Li, C.; Wei, C.; Chu, M.; Zhang, J.; Deng, Z.; Li, M.; Li, X. Effects of zinc and copper ions on ferric arsenate precipitation in hydrothermal scorodite. *J. Cent. South Univ. Sci. Technol.* **2019**, *50*, 2645–2655. [CrossRef]
33. Hidalgo, T.; Kuhar, L.; Beinlich, A.; Putnis, A. Kinetic study of chalcopyrite dissolution with iron (III) chloride in methanesulfonic acid. *Miner. Eng.* **2018**, *125*, 66–74. [CrossRef]

metals

MDPI

Article

Sustainable Recovery of Secondary and Critical Raw Materials from Classified Mining Residues Using Mycorrhizal-Assisted Phytoextraction

Adalgisa Scotti [1,2,*], Stefano Milia [2], Vanesa Silvani [3], Giovanna Cappai [2,4], Daniela Guglietta [2], Francesca Trapasso [2], Emanuela Tempesta [2], Daniele Passeri [2], Alicia Godeas [3], Martín Gómez [1] and Stefano Ubaldini [2]

1 Bio Environmental Laboratory, International Center for Earth Sciences, National Atomic Energy Commission, San Rafael Mendoza 5600, Argentina; mpgomez@cnea.gov.ar
2 Institute of Environmental Geology and Geoengineering, National Research Council of Italy, Research Area of Rome 1, 00015 Montelibretti, Italy; stefano.milia@igag.cnr.it (S.M.); gcappai@unica.it (G.C.); danielaguglietta@gmail.com (D.G.); francesca.trapasso@igag.cnr.it (F.T.); emanuela.tempesta@igag.cnr.it (E.T.); daniele.passeri@igag.cnr.it (D.P.); stefano.ubaldini@igag.cnr.it (S.U.)
3 Faculty of Exact and Natural Science, Institute of Biodiversity and Applied and Experimental Biology, National Scientific and Technical Research Council—University of Buenos Aires, Buenos Aires 1428, Argentina; vsilvani@bg.fcen.uba.ar (V.S.); godeas@bg.fcen.uba.ar (A.G.)
4 Department of Civil-Environmental Engineering and Architecture, University of Cagliari, 09123 Cagliari, Italy
* Correspondence: scotti@cnea.gov.ar

Abstract: In this work, mycorrhizal-assisted phytoextraction (MAP, *Helianthus annuus*–arbuscular mycorrhizal fungus *Rhizophagus intraradices*–Zn-volcanic ashes) was applied for the recovery of secondary and critical raw materials (SRMs and CRMs, respectively) from Joda West (Odisha, India) mine residues, within a novel multidisciplinary management strategy. Mine residues were preliminarily characterized by using advanced analytical techniques, and subsequently mapped, classified and selected using multispectral satellite Sentinel-2A images and cluster analysis. Selected mine residues were treated by MAP at laboratory scale, and the fate of several SRMs (e.g., Zn, Cr, As, Ni, Cu, Ca, Al, K, S, Rb, Fe, Mn) and CRMs (such as Ga, Ti, P, Ba and Sr) was investigated. Bioconcentration factors in shoots (BC_S) and roots (BC_R) and translocation factors (TF) were: 5.34(P) > BCS > 0.00(Al); 15.0(S) > BCR > 0.038(Ba); 9.28(Rb) > TF > 0.02(Ti). Results were used to predict MAP performance at larger scale, simulating a Vegetable Depuration Module (VDM) containing mine residues (1 m^3). Estimated bio-extracting potential (BP) was in the range 2417 g/m^3 (K) > BP> 0.14 g/m^3 (As), suggesting the eventual subsequent recovery of SRMs and CRMs by hydrometallurgical techniques, with final purification by selective electrodeposition, as a viable and cost-effective option. The results are promising for MAP application at larger scale, within a circular economy-based approach.

Keywords: hydrometallurgy; resource recovery; mycorrhizal-assisted phytoremediation; remote sensing; circular economy

Citation: Scotti, A.; Milia, S.; Silvani, V.; Cappai, G.; Guglietta, D.; Trapasso, F.; Tempesta, E.; Passeri, D.; Godeas, A.; Gómez, M.; et al. Sustainable Recovery of Secondary and Critical Raw Materials from Classified Mining Residues Using Mycorrhizal-Assisted Phytoextraction. *Metals* **2021**, *11*, 1163. https://doi.org/10.3390/met11081163

Academic Editor: Fernando Castro

Received: 21 June 2021
Accepted: 20 July 2021
Published: 22 July 2021

1. Introduction

A successful economy minimizes the production of waste and reuses it as a raw material. Resource constraints and environmental pressures will accelerate the transition from a linear "extraction-use-throw away" model of production and consumption to a circular one: moving towards a near-zero waste society has an environmental rationale and it represents a factor of competitiveness at the same time [1].

An integrated multidisciplinary approach for characterization and analysis of iron- and manganese-rich mine residues in Joda West (State of Odisha, India), and their valorization as a resource, was recently proposed by Guglietta et al. [1]. Such an approach would simultaneously promote both economic development and environmental protection. Moreover, the integration of different expertise from characterization to exploitation, passing through mining, reusing and monitoring, will consolidate the knowledge of the various processes. The application of innovative integrated technologies is urgently necessary to create synergistic efforts and results for sustainable exploitation of mining residues. The aim is to promote the transition to a more circular model that can strongly contribute, with products, processes and business models that are designed, to maximize the value and utility of resources while at the same time reducing adverse health and environmental impacts (Figure 1). In particular, this multidisciplinary strategy combines:

1. New sensitive technologies (such as remote sensing technologies) for mapping and classification of mining wastes containing minerals and metals;
2. New solutions for chemical, physical and morphological characterization;
3. An eco-innovative methodology for converting mining waste into resource: mycorrhizal-assisted phytoextraction (MAP) of mining wastes, and the consequent accumulation of secondary and critical raw materials (SRMs and CRMs, respectively) in biomass tissues;
4. Recovery of SRMs and CRMs from biomass through hydrometallurgy and/or electrochemical methods. The leaching extraction is selective and allows the recovery of all the contained elements because specific reagents are used for each of them, applying innovative processes (e.g., thiourea process, thiosulfate process, etc.). The metals are purified to a high degree of purity by electrochemical methods.

Figure 1. A scheme of the proposed multidisciplinary strategy for classification and recovery of secondary and critical raw materials (Adapted from ref. [2]).

The synergy between points 1 and 3 is of particular interest, since it couples low-cost, highly efficient remote sensing and bioremediation technologies. As to remote sensing classification (i.e., point 1), it highlighted mining deposits rich in Fe and Mn in a mine area and all different data were stored in a geodatabase [1], which is a powerful tool for mine waste management and for the selection of samples containing valuable minerals and metals to treat by phytoremediation technique. As to phytoremediation, it is widely recognized as an environmentally friendly technique that uses plants and beneficial microorganisms to remove pollution at greatly reduced costs and minimum adverse side effects [3]. Regarding the bioaccumulation of SRMs by using metal hyperaccumulator plants, it was widely reported that *Helianthus annuus* (Asteraceae family) can accumulate large amounts of Ni, Mn, Zn, Cu, Pb, As, and Cd [4–6] and CRMs such as Ga, Sr, Ti [1,7,8]. This capacity of accumulation in their biomass can be enhanced by association in their roots with the arbuscular mycorrhizal (AM) fungi belonging to the Phylum Glomeromycota [9]. In this mycorrhizal association, plants benefit from nutritional supply and growth improvement, and it also play a key role in stress alleviation, such as that imposed by excess heavy metals [10]. Some mechanisms responsible for the amelioration of toxicity in the host plants mediated by the AM fungi are known, such as sequestration and accumulation of different metals by extra- and intra-radical AM fungal structures (spores, mycelia, vesicles), and the modulation of genetic expression in the detoxification process of plants [10,11]. The AM symbiosis is frequently included in the bioremediation strategies of different metal polluted soils [4,12]. Various reports have shown that the AM association established between sunflowers (*Helianthus annuus* L.) and the AM fungus *Rhizophagus intraradices* is effective for the uptake and accumulation of various chemical elements [4,13]. During the bioremediation process, the elements bioaccumulated in plant biomass can also be subsequently recovered and reused. However, little is known about phytoaccumulation of certain CRMs and SRMs by using mycorrhizal sunflower plants when grown in contaminated soils. In previous works, we found significant accumulation of SRMs (i.e., Zn, Mn, Cu) and a CRM (i.e., Sr) from contaminated soil by using the MAP system comprised of "*H. annuus–R. intraradices*–Zn 350-volcanic ash" [1,14,15] at laboratory scale. However, some raw materials have not yet been investigated at a larger scale approach. The synergistic coexistence of soil, water, plants, and soil microorganisms has been also exploited in constructed wetlands, which imitate the hydrological system of natural wetlands and have become a simple, low-cost alternative for wastewater treatment [16–18]. The MAP system introduced into an innovative constructed wetland called Vegetable Depuration Module (VDM) [8] was proposed as a scaling tool to reach a technology readiness level (TRL) 6 (i.e., pilot test) by simulation of a relevant environment characterized by high concentrations of chemical elements [19]. Using this VDM, the first cycle of an engineered process for decontamination of a polluted environment composed of wastewater and substrate (soil) with an excess of chemical elements can be simulated and tested. In this study, the outcomes of spectral and mineralogical characterization of mine residues, and their mapping and classification by remote sensing technologies were used for the selection of samples rich in Fe and Mn. The selected mining residues were treated at lab-scale (TRL-2) with the MAP system (*H. annuus–R. intraradices*–Zn 350-volcanic ash) for the recovery of several SRMs (i.e., Fe, Mn, Zn, Cr, As, Ni, Cu, Rb, Al, K, S, Ca) and CRMs (i.e., Ba, Ga, P, Ti and Sr) from plant biomass. With a future perspective of on-site treatment of Joda West mine residues (TRL-6), results achieved at TRL-2 were used to estimate the bioextracting potential of the MAP system in the VDM, specifically focusing on less studied SRMs and CRMs.

2. Materials and Methods

2.1. Fe and Mn Mine Wastes

The Fe and Mn mine area is located in the village of Joda, in the Keonjhar district of Odisha (India) (Figure 2). In situ sampling campaign was carried out on November 29th in 2017 by scientific and technical staff of the Institute of Environmental Geology and Geoengineering (National Research Council, CNR, Italy) in collaboration with TATA Steel and the National Environmental Engineering Research Institute (NEERI-India). Thirty-six different samples were collected from waste, stock deposits and dumps. For each sample, additional information consisting of GPS coordinate, pictures and a brief description of the sampling area were gathered. Furthermore, for each sample about 200 g were selected for further laboratory analysis (chemical, physical and spectral analysis), and remote sensing analysis at National Research Council (CNR) in Italy.

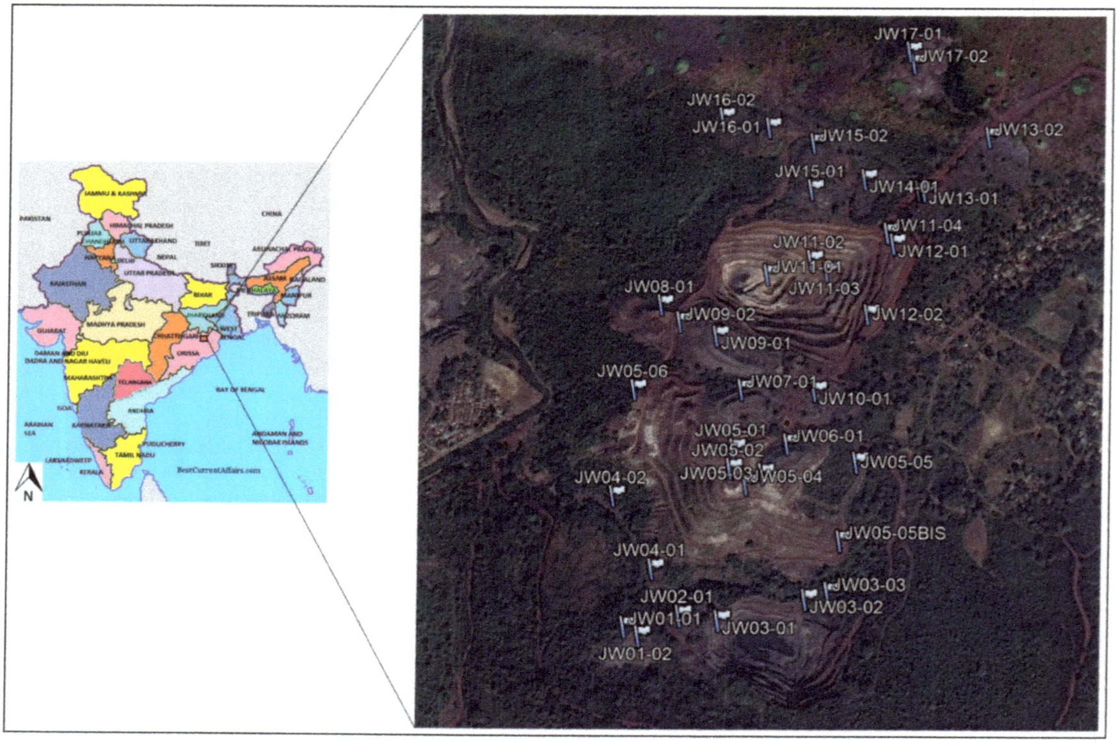

Figure 2. Location of the study area and in situ sampling campaign overlaid on Sentinel-2A image acquired on 29 November 2017.

2.2. *The Joda West Iron and Manganese Mine Geodatabase*

For each sample collected in the mining area, the corresponding spectral signature was extracted by the image and the different spectral classes were individuated by means of the interpretation of chemical and mineralogical analysis. Afterwards, the map of Fe and Mn residues was obtained by applying supervised pixel based classification and the algorithm used in this methodology was Spectral Angle Mapper (SAM) [1]. The Sentinel-2A map highlighted residue deposit areas with different mean percentages of Fe and Mn (Figure 3) as well as their surface alteration (i.e., presence of vegetation). In particular, Class 1 was characterized by presence of sparse vegetation that indicated residues accumulated over a long period; Class 2 was represented by a steady accumulation of extractive residues; Class 3 showed high percentages of Fe and Mn, but these residues were not of interest for the steel industry as they would need prior ore dressing; Class 4 contained residues with lower Fe and Mn content. For the phytoremediation process, Class 2 and Class 4 residue samples have been selected and analyzed.

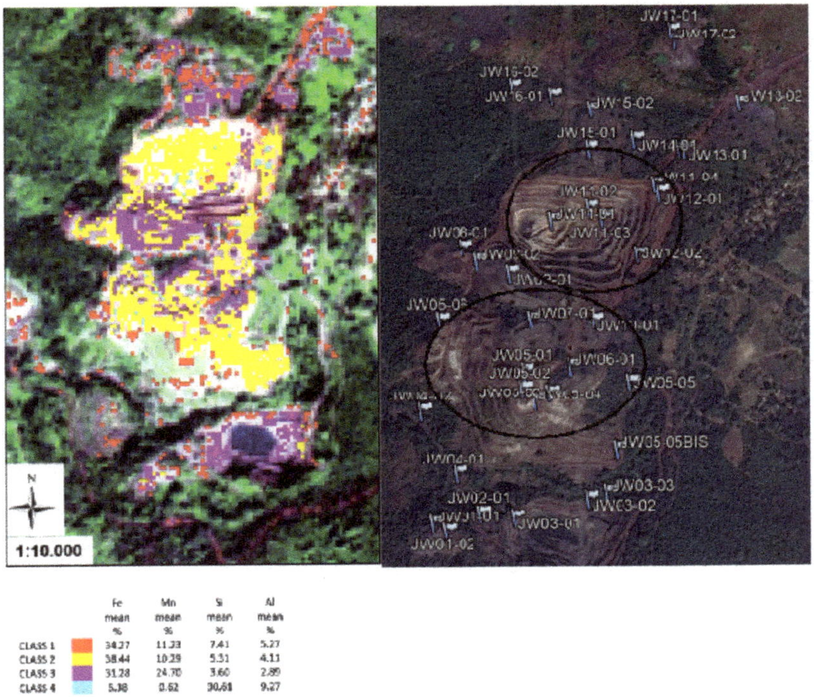

Figure 3. Spatial distribution of the mining residue classes (**left**) and samples overlaid (**right**) on Sentinel-2A image. Black circles indicate the areas of soil selected for MAP.

The in situ observations and the results of statistical, spectral, chemical and mineralogical analysis as well as of the satellite classification process were collected into a geodatabase to identify the mining residues stored in old dumps or fresh landfills in the Joda West mine. The following information was recorded for each sample: location (including coordinates); details of the study site (including land cover, land use, geomorphology); information and observations on the residue (including results of analysis, pictures, color, and area if known). The Joda West geodatabase was built using ESRI's ArcGIS 10.2 software (Keonjhar district, State of Odisha, India). The geodatabase is useful for progressively identifying the residue areas of spatial importance for optimized management, as well as for highlighting the mining residues rich in Fe and Mn that could be potentially used for bioremediation processes. In order to provide a useful tool for researchers as well as decision makers (stakeholders), the GIS layers were exported and reassembled as a project file in Google Earth. The final Joda West project file is in a Google Earth KMZ (keyhole Markup Language—zipped) format (Figure 4).

Figure 4. Example of the Joda West geodatabase: JW 16-02 sample.

2.3. *MAP Test at TRL-2 Scale*

2.3.1. Plant Material and AM Fungal Strain

The MAP system consisted of sunflowers (*Helianthus annuus L.*, hybrid cultivar DK4045, Syngenta seeds) colonized by the AM fungus *Rhizophagus intraradices* strain GA5 (provided by Bank of Glomeromycota In Vitro, Faculty of Exact and Natural Sciences, Buenos Aires University, www.bgiv.com.ar, accessed on 21 July 2021) grown in a substrate comprised of a homogeneous mixture of soil and volcanic ash (50:50, v/v) and supplemented with $ZnSO_4$ (as catalyst compound) [8]. The AM fungal strain was propagated using transformed carrot roots grown in minimal medium as described by Silvani et al. [20]. This fungal strain was previously reported to promote tolerance to different abiotic stresses in several plant species and to have potential as bioremediator [8,21].

2.3.2. TRL-2 Experimental Setup

Fe- and Mn-rich soil samples classified as Class 2 and 4 by Sentinel-2A remote sensing were used to test the MAP system at TRL-2 scale. For that purpose, homogeneous composite soil samples (CS) were prepared by mixing individual samples from selected areas of the Indian mine. The volcanic ash (VA) and granular pumice stone (PS) (diameter, 3–6 mm) were provided by Pumex S.P.A. (Lipari Island, Italy) and Europomice S.R.L. (Milan, Italy), respectively. The fertile commercial topsoil (FCT) was provided by Euroterriflora S.R.L. (Bucine, Italy). The physical-chemical properties and mineralogical characterization of CS, VA and FCT are described in Guglietta et al. [1]. To corroborate that non-indigenous AM fungal propagules were present in the mine soil samples and the commercial topsoil, a subsample (10 g) of each type of soil was wet sieved and decanted and then checked for AM fungal spores under a binocular microscope. Before planting, sunflower seeds were surface-sterilized with an ethanol solution (70%, v/v) and washed twice in sterile water.

The experimental design of the TRL-2 system consisted of 16 pots (height: 12 cm; diameter: 13.5 cm), as detailed below:

- $MA_{1\text{–}4}^{+}$. Four pots filled with CS and VA in a 1:1 (v/v) ratio, completing to 500 mL and 125 mL granular PS. Considering the CS only, 300–500 ppm $ZnSO_4$ were added. At least three sunflower seeds were sown and inoculated with a piece of 4-month old monoxenic culture (containing approximately 300 spores, mycelia and colonized roots).
- $MA_{1\text{–}4}^{-}$. Four pots filled with mixed CS and VA in a 1:1 (v/v) ratio, completing to 500 mL and 125 mL PS. Considering the CS only, 300–500 ppm $ZnSO_4$ were added. At least three sunflower seeds were sown in each pot.
- $B_{1\text{–}4}^{+}$. Four pots filled with 500 mL FCT and 125 mL PS. At least three sunflower seeds were sown and inoculated with a piece of 4-month old monoxenic culture (containing approximately 300 spores, mycelia and colonized roots).
- $B_{1\text{–}4}^{-}$. Four pots filled with 500 mL FCT and 125 mL PS. At least three sunflower seeds were planted in each pot.

On day 30, sunflower plantlets were thinned to one, leaving the most homogeneous with better leaves development in each pot. Plant cultures were maintained during 133 days in a room at controlled temperature (18–21 °C). As precautionary strategy, plants were irrigated only twice a week (50 mL per pot) in order to provide a sufficient amount of water for plant growth, and avoid the risk of any metals leaching.

2.3.3. Harvest and Sample Analyses

Representative soil samples (50 g) and plant biomass were taken from each pot on day 133. The soil samples were dried and micronized under 70 microns in size by vibrating rotary cup mill at 900 rpm motor speed and a standard 100 mL steel crew. The grinding containers incorporate an anvil ring and puck to pulverize the sample by an eccentric vibratory action. Physical and mineralogical characterization of soil samples were carried out by X-Ray Powder Diffraction (XRPD) as described in Guglietta et al. [1] and by Total Reflection X-Ray Fluorescence (TXRF). Sunflower plants were removed from each pot, and shoots and roots were separately harvested. Roots were carefully rinsed with distilled water, and then shoots and roots were dried in an oven (45 °C) to constant weight for registration of shoot and root dry weights, and subsequent quantification of chemical elements by the TXRF method (see: Section 2.3.4). To assess the mycorrhization level, subsamples of dried roots were stained with trypan blue solution following the modified technique of Phillips and Hayman [22]. The AM root colonization was checked according to Giovanneti and Mosse [23]. The frequency of mycorrhizal colonization was calculated as the percentage of root segments containing AM fungal structures. All measurements were taken under a Nikon Optiphot-2 light binocular microscope (Nikon Instruments Inc., Melville, LA, USA) at 100× magnification.

2.3.4. TXRF Analysis

The chemical determination of plant biomass and soil substrate from each pot was carried out by the TXRF method. For that, the entire plant sample was weighed and placed in a Teflon beaker with 3 mL sub-boiling HNO_3 and 1 mL H_2O_2, and then microwave digested. After digestion, the solution was transferred to a 10 mL volumetric flask (or 5 mL in case that the sample was smaller than 0.05 g) and made up to volume with milliQ water. A total of 10 μL of Ga 100 mg/L (Merck, Taufkirchen, Germany) was added as an internal standard to 1 mL of sample solution. Then, 5 μL of the solution was deposited on a quartz reflector and measured for 300 s in the TXRFS2 Picofox spectrometer (Bruker, Buenos Aires, Argentina) with Mo tube. For soils analysis, a 28 mm diameter pressed pellet was prepared with 3 g of dry sample. Elemental determination was carried out with a WDXRF S8 Tiger spectrometer (Bruker, Buenos Aires, Argentina), using the Quant Express application.

2.4. Bioextracting Potential (BP) in VDM

The estimation of BP of CRMs (Ba, Ga, P, Ti and Sr) and SRMs (Fe, Mn, Zn, Cr, As, Ni, Cu, Rb, Al, K, S, Ca) at TRL-6 by using the VDM containing contaminated soil was carried out considering the results achieved at TRL-2 experiment. The VDM is located at the Centro de Desarrollo Regional Los Reyunos of Universidad Tecnológica Nacional (San Rafael, Mendoza, Argentina) [8]. It consists of an external environment-isolated pool connected to collection chambers through a hydraulic system. The water runs through pipes connected to a reserve tank and a water pump that drives the vertical flow towards the pool, and the remaining fraction of percolated liquid is drained towards a collection chamber. The 15 cm-superficial layer, representing the growth substrate for the consortium *H.annuus-R. intraradices*, consisted of a homogeneous mixture of contaminated soil and volcanic ash (50:50; *v:v*) supplemented with 350 ppm of $ZnSO_4$ (Figure 5).

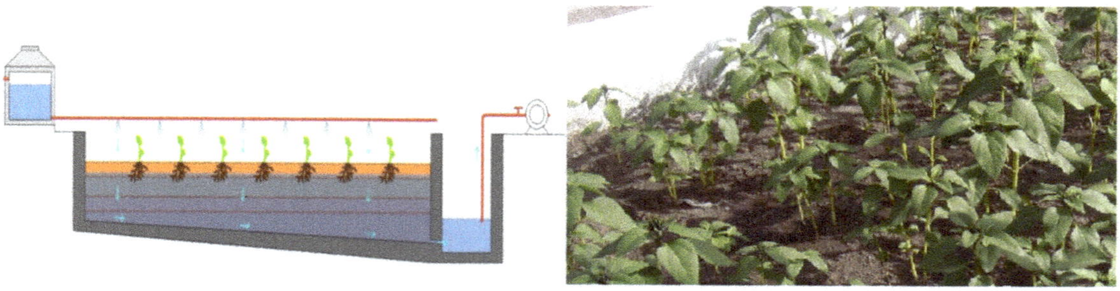

Figure 5. Diagram of the Vegetable Depuration Module (**left**) and the bioaugmentation system (**right**) (*Helianthus annuus—Rhizophagus intraradices)* with 2 m^3 volume of soil-ash (1:1) and 20 plants per m^2.

2.5. Calculations and Statistical Analyses

2.5.1. Bioconcentration Coefficients and Translocation Factor

The bioconcentration coefficient (BC) was calculated as the ratio between the concentration of the chemical element in plant tissue and soil (Equation (1)), for shoots and roots (BC_S and BC_R, respectively).

$$BC_{(S,R)} = C_{p(S,R)}/C_{soil} \quad (1)$$

where:

$C_{p(S,R)}$: concentration of element in the aerial (shoots, S) or radicular (roots, R) plant tissue (ppm)

C_{soil}: concentration of element in soil (ppm)

The translocation factor (TF) was calculated as the ratio between the concentration of the chemical element in shoots and roots (Equation (2)).

$$TF = C_{p(S)}/C_{p(R)} \quad (2)$$

where:

$C_{p(S)}$: concentration of element in the aerial plant tissue (shoots; ppm)

$C_{p(R)}$: concentration of element in the radicular plant tissue (roots; ppm)

BC and TF were calculated for the following SRMs (Fe, Mn, Zn, Cr, As, Ni, Cu, Rb, Al, K, S, Ca) and CRMs (Ba, Ga, P, Ti and Sr).

2.5.2. Bioextracting Potential (BP)

The estimation of BP of SRMs and CRMs by the VDM and the MAP system was calculated considering the concentration of each element in plant biomass obtained at

TRL-2 experiment, and the total biomass grown in the VDM with 1 m^3 of contaminated soil, as follows (Equation (3))

$$BP\ (mg) = [(W_{tot(S)} \times C_{p(S)}) + (W_{tot(R)} \times C_{p(R)})]/1000 \quad (3)$$

where:

$W_{tot(S,R)}$: Total weight (g) of aerial (shoots, S) and radical (roots, R) plant tissue.
$C_{p(S,R)}$: concentration (ppm) of element in the aerial (shoots, S) or radicular (roots, R) plant tissue.

3. Results

MAP Test at TRL-2 Scale

During the experiment at TRL-2 scale, the sunflower plants grew healthy in all treatments. The AM fungus *Rhizophagus intraradices* strain GA5 colonized sunflower roots in both MA^+ and B^+ treatments after 133 days with typical AM fungal structures (vesicles, arbuscules and hyphae; Figure 6A), whilst no root colonization was found in the non-inoculated plants in control treatments (B^- and MA^-), as expected (Figure 6B). The MA^+ treatment showed the lowest value of AM root colonization with 5.0 ± 2.2% (mean ± standard error), and sunflower plants of the B^+ treatment reached 42.0 ± 2.71%. The AM inoculation with *R. intraradices* strain GA5 did not affect the total dry plant biomass in the contaminated treatment, while total biomass of mycorrhized sunflower plants grown in blank soil samples was significantly higher than contaminated soil (Table A1).

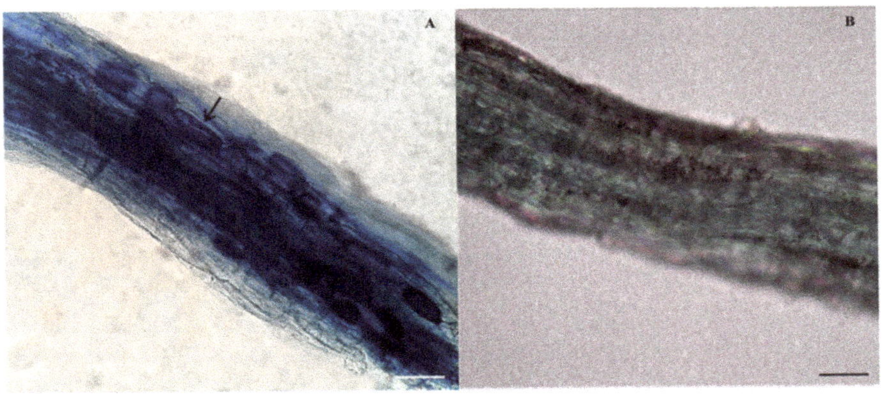

Figure 6. (**A**) Sunflower root fragment colonized by the AM fungus *Rhizophagus intraradices* (strain GA5) after 133 days (MA_1^+ treatment). Detail of the vesicles of AM fungus (black arrow); (**B**) uncolonized root of sunflower plant in control treatment (MA_1^- treatment); Bars: 200 μm.

The average concentrations of SRMs (Fe, Mn, Zn, Cr, As, Ni, Cu, Rb, Al, K, S and Ca) and CRMs (Ba, Ga, P, Ti and Sr) in soil samples taken from each pot of the contaminated and blank treatments, as well as in sunflower plant biomass after 133 days of growth are reported in Tables 1 and 2. The $BC_{(S,R)}$ and TF values of these elements for all treatments are summarized in Table 3.

Table 1. Mean concentration of CRMs in samples from pots containing contaminated soil (MA) and blank soil (B), as well as in biomass ($C_{p(S,R)}$), with and without mycorrhization (+, −). Standard error (SE) is reported in brackets.

Sample Type	Sample ID	N	Ga (ppm)	P (ppm)	Ti (ppm)	Sr (ppm)	Ba (ppm)
Contaminated	MA^-	4	12.88 (0.572)	731 (32.8)	2575 (94.65)	110 (3.47)	2550 (86.6)
	MA^+	4	20.13 (1.994)	711 (15.14)	2675 (25)	112(4.94)	2600 (40.8)
	$C_{p(S)}{}^+$	2	31.50 (9.596)	3803 (209.1)	9,5 (0.5)	410 (172)	7.85 (0.15)
	$C_{p(S)}{}^-$	2	13.70 (1.616)	2878 (158.6)	16 (1)	254 (9.60)	11.5 (0.5)
	$C_{p(R)}{}^+$	2	25.55 (3.889)	1228 (35.86)	446 (24.24)	79(4.04)	98.5 (1.51)
	$C_{p(R)}{}^-$	2	22.00 (1.010)	880 (27.27)	947.5 (53)	138 (7.07)	163.5 (6.5)
Blank	B^+	4	5.00 (0.34)	2125 (25.00)	1275 (25)	1100 (57.7)	431.25 (21.29)
	$C_{p(S)}{}^+$	2	3.02 (0.75)	3583 (200.1)	9 (3)	229 (1.01)	8.0 (1.0)
	$C_{p(R)}{}^+$	2	3.00 (0.10)	1412 (39.39)	509 (11.11)	106 (5.56)	28.5 (1.51)

Table 2. Mean concentration of SRMs in samples from pots containing contaminated soil (MA) and blank soil (B), as well as in biomass ($C_{p(S,R)}$), with and without mycorrhization (+, −). Standard error (SE) is reported in brackets.

Sample Type	Sample ID	N	Mn (%)	Fe (%)	As (ppm)	Zn (ppm)	Cr (ppm)	Ni (ppm)	Cu (ppm)	Rb (ppm)	Al (ppm)	K (%)	S (ppm)	Ca (%)
Contaminated	MA^-	4	4.93 (0.13)	11.2 (0.26)	12.5 (0.59)	1500 (108)	110 (7.67)	67.0 (2.24)	60.3 (1.65)	78.3 (1.49)	62,500 (1040.8)	2.30 (0.04)	1195 (113.1	1.12 (0.10)
	MA^+	4	5.55 (0.03)	12.2 (0.06)	23.2 (1.24)	1475 (25.0)	113.3 (2.93)	72.3 (1.49)	64.5 (2.53)	80.5 (1.76)	64,750 (250)	2.20 (0.04)	1125 (125)	1.05 (0.04)
	$C_{p(S)}{}^+$	2	0.052 (0.003)	0.03 (0.001)	4.05 (0.05)	1328 (37.9)	2.80 (0.20)	1.80 (0.10)	16.5 (0.51)	116 (4.04)	0.95 (0.05)	7.70 (0.40)	5945 (162.1)	2.25 (0.15)
	$C_{p(S)}{}^-$	2	0.05 (0.003)	0.03 (0.002)	1.50 (0.10)	1410 (39.4)	0.65 (0.05)	1.00 (0.10)	14.0 (1.01)	102 (2.53)	0.93 (0.07)	7.10 (0.20)	6189 (162.1)	2.06 (0.06)
	$C_{p(R)}{}^+$	2	0.455 (0.024)	1.19 (0.06)	6.50 (0.505)	1162 (32.3)	29.5 (1.52)	38.0 (2.02)	17.5 (0.51)	12.5 (0.51)	5550 (2272.4)	4.65 (0.15)	16,885 (672.3)	0.99 (0.02)
	$C_{p(R)}{}^-$	2	0.74 (0.04)	2.13 (0.13)	6.50 (0.505)	1717 (92.4)	10.00 (1.01)	14.0 (10.0)	36.0 (2.02)	109 (3.54)	21,375 (419.2)	5.20 (0.15)	13,772 (307.6)	2.46 (0.08)
Blank	B^+	4	0.06 (0.001)	1.43 (0.03)	8.78 (0.41)	210.8 (9.4)	86.50 (9.1)	42.3 (0.75)	108 (2.68)	212 (14.1)	13,625 (239.3)	1.33 (0.05)	3525 (110.9)	5.33 (0.06)
	$C_{p(S)}{}^+$	2	0.007 (0.0006)	0.01 (0.001)	1.65 (0.25)	54.50 (4.6)	3.45 (1.26)	2.80 (1.31)	15.0 (0.00)	76.0 (1.01)	8.9 (0.04)	5.40 (0.10)	2045 (49.5)	1.86 (0.11)
	$C_{p(R)}{}^+$	2	0.03 (0.002)	0.22 (0.01)	3.55 (0.02)	166.5 (4.6)	12.0 (1.01)	6.45 (0.25)	42.5 (2.53)	33.0 (2.02)	13,900 (303,04)	1.06 (0.03)	1725 (29.3)	4.32 (0.09)

Table 3. Average bioconcentration coefficient (BC) in shoots (S) and in roots (R) of plants grown on mycorrhized (+) and not mycorrhized (−) systems. Standard error (SE) is reported in brackets.

Sample ID	Parameter	Mn	Ga	Fe	P	As	Zn	Ti	Sr	Cr	Ni	Cu	Rb	Al	Ba	K	S	Ca
MA^+	$BC_S{}^+$	0.009 (0.001)	1.565 (0.632)	0.002 (0.0001)	5.349 (0.408)	0.175 (0.012)	0.900 (0.046)	0.004 (0.0002)	3.653 (3.356)	0.025 (0.002)	0.025 (0.002)	0.256 (0.018)	1.441 (0.082)	0.000 (0.000)	0.003 (0.0001)	3.500 (0.247)	5.284 (0.731)	2.138 (0.233)
	$BC_R{}^+$	0.082 (0.005)	1.270 (0.319)	0.098 (0.006)	1.726 (0.087)	0.281 (0.162)	0.788 (0.035)	0.167 (0.011)	0.704 (0.388)	0.261 (0.020)	0.526 (0.039)	0.271 (0.019)	0.155 (0.010)	0.086 (0.035)	0.038 (0.001)	2.114 (0.107)	15.008 (2.265)	0.936 (0.053)
	TF^+	0.114 (0.013)	1.233 (0.808)	0.023 (0.003)	3.098 (0.393)	0.623 (0.401)	1.142 (0.064)	0.021 (0.002)	5.190 (3.973)	0.095 (0.012)	0.047 (0.005)	0.943 (0.056)	9.280 (0.698)	0.0002 0.0001	0.080 (0.003)	1.656 (0.141)	0.352 (0.024)	2.284 (0.189)
MA^-	$BC_S{}^-$	0.011 (0.001)	1.064 (0.173)	0.003 (0.0002)	3.937 (0.394)	0.125 (0.014)	0.940 (0.100)	0.006 (0.001)	2.310 (0.161)	0.006 (0.001)	0.015 (0.002)	0.232 (0.023)	1.297 (0.057)	0.000 (0.000)	0.005 (0.0004)	3.087 (0.143)	5.181 (0.626)	1.835 (0.224)
	$BC_R{}^-$	0.150 (0.012)	1.709 (0.154)	0.190 (0.016)	1.204 (0.091)	0.522 (0.065)	1.144 (0.144)	0.368 (0.034)	1.257 (0.104)	0.091 (0.016)	0.209 (0.022)	0.598 (0.050)	1.387 (0.072)	0.342 (0.012)	0.064 (0.048)	2.283 (0.106)	11.529 (1.349)	2.187 (0.270)
	TF^-	0.073 (0.008)	0.623 (0.157)	0.014 (0.002)	3.271 (0.576)	0.231 (0.034)	0.821 (0.067)	0.017 (0.002)	1.837 (0.164)	0.065 (0.012)	0.071 (0.012)	0.389 (0.050)	0.936 (0.054)	0.000 (0.000)	0.070 (0.006)	1.352 (0.078)	0.449 (0.022)	0.839 (0.051)

There was a marked difference for $BC_{(S,R)}$ and TF in the elements under study when blank and contaminated soil were compared. The $BC_{(S,R)}$ and TF values when sunflower plants were colonized followed the order: P = 5.34 > S > Sr > K > Ca > Ga > Rb > Zn > Cu > As > Ni = Cr > Mn > Ti > Ba > Fe > Al = 0.00 for $BC_{(S)}$, while for $BC_{(R)}$ was: S = 15.00 > K > P> Ga > Ca > Zn > Sr > Ni > As > Cu > Cr > Ti > Rb > Fe > Al > Mn > Ba and for TF was: Rb = 9.28 > Sr > P > Ca > K > Ga > Zn > Cu > As > S > Mn > Cr > Ba > Ni > Ti = Fe > Al. Significant differences in BC and TF values for all the elements, except for Ga, were observed between soil contaminated with or without AM inoculation for at least one of the parameters studied (Table 4). No differences in the metal concentrations in blank soil were observed in presence or absence of the AM fungus (Table A2). The effect of AM colonization on TF values was different for the elements under study. Mn, Fe, Zn, Cr, As, Cu, Rb, Al, Ba, K, Ca, Sr and Ti were affected by increasing their TF; no significant effects for TF were detected for Ga and P, while S and Ni decreased it, as summarized in Table 4. In contrast, the AM colonization increased the $BC_{(S)}$ for P, As, Cr, Ni, Rb, K and decreased for Mn, Fe, Ti and Ba. Beside the $BC_{(R)}$ for P, Cr, S and Ni increased, but for Mn, Fe, As, Zn, Ti, Cu, Rb, Sr, Al, Ba and Ca a decrease was observed (Table 4).

Table 4. Overall effect of arbuscular mycorrhizal colonization on $BC_{(S)}$, $BC_{(R)}$ and TF, compared to control tests (not mycorrhized) (+, increase; −, decrease; =, no significant effect).

Parameter	Mn	Ga	Fe	P	As	Zn	Ti	Cr	Ni	Cu	Rb	Sr	Al	Ba	K	S	Ca
BC_S	(−)	(=)	(−)	(+)	(+)	(=)	(−)	(+)	(+)	(=)	(+)	(=)	(=)	(−)	(+)	(=)	(=)
BC_R	(−)	(=)	(−)	(+)	(−)	(−)	(−)	(+)	(+)	(−)	(−)	(−)	(−)	(−)	(=)	(+)	(−)
TF	(+)	(=)	(+)	(=)	(+)	(+)	(+)	(+)	(−)	(+)	(+)	(+)	(+)	(+)	(+)	(−)	(+)

The extractive power of MAP in VDM is shown in Table 5 considering 290 plants and 1 m^3 of contaminated soil.

Table 5. Estimated elements extracting potential by MAP in a VDM containing 1 m^3 of contaminated soil. Standard Deviation is reported in brackets.

Parameter	Mn (ppm)	Fe (ppm)	Ga (ppm)	P (ppm)	As (ppm)	Zn (ppm)	Ti (ppm)	Cr (ppm)	Ni (ppm)	Cu (ppm)	Rb (ppm)	Sr (ppm)	Al (ppm)	Ba (ppm)	K (ppm)	S (ppm)	Ca (ppm)
$C_{p(S)}{}^+$	520 (42)	270,5 (20,5)	31.5 (13,4)	3803 (292)	4.05 (0.07)	1328 (53)	9.5 (0.7)	2.8 (0.3)	1.8 (0.1)	16.5 (0.7)	116 (5.66)	410 (240)	0.95 (0.07)	7.85 (0.21)	77,000 (5657)	5944.5 (227)	22,500 (2121)
$C_{p(R)}{}^+$	4550 (343)	11,888 (898)	25.5 (5,4)	1228 (50)	6.5 (0.7)	1162 (45)	446 (34)	29.5 (2.1)	38 (2.8)	17.5 (0.71)	12.5 (0.71)	79 (5.7)	5550 (3181)	98.5 (2.12)	46,500 (2121)	16,885 (941)	9850 (212)
Extracting g/VDM *	34.82	60.01	1.017	114.7	0.145	43.22	2.199	0.208	0.216	0.550	3.392	12.14	23.99	0.6515	2417	243.9	690.0

* VDM contained 1 m^3 of contaminated soil, the calculation can be also expressed as per cubic meter of contaminated soil.

4. Discussion

The MAP system at TRL-2 showed good extraction capabilities for many SRMs and CRMs in terms of BC and TF, although biomass growth observed in mining waste substrate was relatively low. Likewise, different behaviors of accumulation and translocation for each chemical element were observed in the MAP system. On the other hand, the mycorrhized soil at the end of the experiment had a significant higher concentration for Ga, Mn, Fe, As, Ni, Cu, Al than the soil without mycorrhization, while for the rest of the elements there were no significant differences (P, Ti, Sr, Zn, Cr, Rb, Ba, K, S and Ca). This could be due to the accumulation of elements in fungal structures such as mycelium and/or spores [24,25].

In regard to Ca, Zn and Cu, no differences were found in the $BC_{(S)}$ between mycorrhized and non-mycorrhized plants. However, the TF values were high in mycorrhized plants for these elements. It is known that Cu ions are involved in photosynthesis, respiration, diverse enzymes of antioxidative defense and hormone metabolism [26]. In a previous work, we found that the same MAP system was highly efficient in bioextracting Cu than other metals when the mass balance was analyzed in VDM [8]. We disclaimed that alkaline substrate could facilitate the mobility of this metal and increase the total Cu fraction available for plants. In opposition, the BP value of Cu by the MAP in the VDM containing Fe and Mn mining wastes was not notorious (Table 5). Likewise, the colonized roots of sunflower plants did not accumulate Ca, Zn and Cu but non-colonized roots did. Cornejo et al. [24] demonstrated that AM fungi usually store Cu in vesicles and spores in roots for minimizing the toxic effects on plant and fungal metabolism. Probably, no Cu accumulation in root tissues was observed because of the low intensity of AM colonization by *R. intraradices* strain GA5 after 133 days. In addition, in contrast with our previous results [8], Zn was not accumulated in mycorrhized plant roots. These discrepancies could be due to the characteristics/properties of the multi-metal mining substrate under study. The different concentrations of each metal could lead to competition for specific sites of entry into the plant and the AM fungal structures. It was reported that the extraradical mycelium and glomalin released in soil by AM fungi are capable of sequestering Ca, Zn and Cu ions limiting their entrance to the roots [25,27]. In addition, Chen et al. [28] observed a higher affinity of Cu and Zn ions in the fungal biomass rather than roots or shoots of the host plants. Otherwise, the storage of Ca ions in external hyphae might be enhancing the nutrient absorption, as we observed the higher P concentrations in shoots and roots of mycorrhized sunflower plants. The acidification of the hyphosphere (the zone of soil adjacent to hyphae) caused by the regulation of transmembrane fluxes potentially promotes the absorption of nutrients between hyphae and the soil environment under stress conditions [29]. As in Scotti et al. [8] the BP of Zn by the MAP system in the MDV reached a very good performance, reaching a value of 43.2 g/m^3, even improving the previously found value of 29.3 g/m^3. In the same way, the BP for Ca was very important (690 g/m^3).

The concentration of S was higher in roots and the TF value was lower in mycorrhized plants than non-mycorrhized; no differences in $BC_{(S)}$ between treatments were found. Allen and Shachar-Hill [30] demonstrated that AM fungi are capable of increasing root S contents by 25% in a moderate (not growth-limiting) concentration of sulfate. Additionally, Aziz et al. [31] showed that sulfur supplementation up to 80 ppm in soil increases the content of K and Ca in plant tissues, and enhances the photosynthetic rate, water content and tolerance to salinity in *H. annuus* without limiting effect; although higher sulfur concentrations were present in contaminated soils used in our study, the concomitant increase of Ca and K was observed.

Mn and Ti are essential micronutrients involved in the activation of Krebs cycle enzymes and the chlorophyll synthesis [32,33]. Additionally, Ti is considered a CRM in the fourth list of the European Commission (COM 2020/474, EC) [34]. In our work, Mn and Ti were translocated more efficiently to aerial parts when the AM fungus *R. intraradices* was present. However, non-mycorrhized plants accumulated more Mn and Ti in roots and shoots. The uptake of Mn was repeatedly found to be lower in mycorrhized plants [35]. The negative effect on Mn uptake was attributed to a shift in the composition and activity of rhizosphere microorganisms caused by inoculation with AM fungi, which decreases the abundance of Mn reducers microbes [36]. However, the BP potential of Mn for the MAP system in VDM was effective, as in Scotti et al. [8] but mainly due to percolation and recovery in the collecting chamber of the VDM. Probably, the high concentration of Mn of the mining wastes translates it into significant amounts of Mn recovering into biomass (34.82 g/m^3). Our study showed that AM colonization augmented Cr and Ni uptake in shoots and roots of sunflowers, but also contributed to enhance nutrient acquisition by high P contents in plants. It is widely known that AM fungi improve nutritional status of colonized plants, mainly by P and N uptake [9] and contribute to the enhancement of Ni and Cr in sunflower biomass [4]. This result is of great importance given that P is a valuable CRM and an essential nutrient for plant growth. As in Scotti et al. [8], there was not a good recovery in biomass of either Cr or Ni (0.21 g/m^3), but the recovery of P was very important (114.7 g/m^3).

According to bioaccumulation results, mycorrhized plants accumulated more As, Sr and Rb in shoots than in roots, and showed higher TF values than non-mycorrhized plants. Besides, the behavior of K ions was like these elements, but no differences were found in K concentrations in roots of mycorrhized and non-mycorrhized plants. K is one of the most abundant chemical elements in soil composition, but it very low availability limits plant growth. Previously, it was observed that AM symbiosis increased K acquisition in Asteraceae plants [37]. Interestingly, a strong correlation between K and P during AM symbiosis was reported [38], coinciding with our results. Otherwise, Rb is an analog for K, which is taken up along the same pathways as K and typically occurs at very low levels in soils and plant tissues [39]. We also observed the competitive behavior between K and Rb for the potential recovery in the VDM. In a previous work [40], it was found that As is predominantly deposited into leaves of sunflowers when grown in arsenic contaminated soil. The chemical similarity of P and As, combined with the mycorrhizal role in P nutrition, provides the likelihood that AM fungi may enhance As uptake [41]. Given that AM symbiosis enhanced P uptake in sunflower plants, it was possible that the AM fungus also increased the uptake of As in shoot tissues. In accordance with our previous work [8], Sr (CRM) showed high translocation to shoots and accumulation within above ground tissues. Many authors have confirmed that Sr can be accumulated in leaves, leaf trichomes and stems of several plant species [42,43]. The accumulation of toxic elements in trichomes was widely cited as a tolerance mechanism of plants subjected to stressful toxic conditions [43]. The Sr bioextracting potential was moderate (12.14 g/m^3), as was also demonstrated in Scotti et al. [8].

No differences in BC and TF values for Fe were found between mycorrhized and non-mycorrhized plants. Ferrol et al. [36] noticed that the effect of mycorrhizal colonization on Fe concentration in plant tissues is variable and inconsistent. The external mycelium and glomalin in soil immobilize Fe ions as it was demonstrated by detection of abundant crystal-like aggregates present on the mucilaginous outer hyphal wall layers of AM fungi [27]. It was demonstrated that Fe elements accumulated at the hyphosphere enhance the nutrient absorption by hyphae, as Ca ions performance [29]. In opposition, the biomass extraction in the VDM for Fe was important due to the high concentration of the metal in the mining wastes.

The Ga accumulation in shoots and roots did not differ between mycorrhized and non-mycorrhized plants. Jensen et al. [44] demonstrated that movement of Ga into the aboveground portions of perennial ryegrass (*Lolium perenne* L.) was low ($BC_{(S)}$: 0.0037), concluding that the mobility of Ga in the soil-plant system is low compared to other common trace element contaminants. While Yu et al. [45] demonstrated roots of rice seedlings were the dominant site for Ga accumulation and the total Ga accumulation rates were positively correlated to its concentrations. In our work, we agree with these researchers on the increase in Ga content in the roots when there is no mycorrhization (22 ppm in roots vs. 13.7 ppm in shoots), while when there is mycorrhization no significant differences were detected. The bio-extraction in the VDM reached an interesting value, thus increasing the commercial importance of Ga as CRM by using this biotechnology. It is known that Ga is chemically similar to Al. Others reports have indicated that AM fungi increased or did not affect Al accumulation in leaves and roots of host plants, and that glomalin and fungal hyphae modulate interactions between plants and soil Al [46]. Accordingly, Al followed the same behavior that Ga ions after observing an increase in roots when there is no AM colonization. More studies should be performed to analyze the role of AM fungal mycelium in Ga sequestration.

Finally, Ba did not show high BP (0.65 g/m^3), as previously reported by Coscione et al. [47]. In the same way, Ba control concentration in shoots was coincident with these authors. To the author's knowledge, no studies were found about Ba accumulation in sunflower plants colonized by AM fungi.

The average estimated BP encourages the subsequent recovery of SRMs and CRMs by hydrometallurgical techniques, with final purification by selective electrodeposition, as a viable and cost-effective option. For a sustainable recovery of metals from an economic and environmental point of view, the application of hydrometallurgical processes is important, because they allow, through the use of specific innovative reagents and the application of electrochemical techniques, their selective recovery at high degree of purity, which determines an appropriate commercial reuse. The fast kinetics guarantees the economy of the process because it allows reduction of the capital costs (smaller capacity reactors can be used) and the energy requirement, the latter representing an advantage also from the environmental point of view.

Assuming an integrated process circuit consisting of a leaching phase of accumulated and concentrated SRMs and CRMs, and subsequent purification by electrochemical methods, with purity levels of more than 90% will allow an industrial reuse of SRMs and CRMs, in the context of a circular economy approach [48].

5. Conclusions

In summary, the present work confirmed that it is possible to apply the MAP system in contaminated soils from mining areas classified by remote sensing as Class 2 and 4, and to carry out the bio-extraction of all the elements studied. Only S, Ni, P and Ga did not show an increase in the TF under the conditions of the experiment, the other elements under study showed a significant increase in the TF when plants were mycorrhized. Despite such promising results, it must be noticed that process conditions must be optimized to maximize plant growth and sustain a significant recovery of SRMs and CRMs by hydrometallurgical methods. It is important to monitor pH and Eh (BS ISO 11271:2002) values at various depths of soil, in order to relate the solubilization of the elements with their bioaccumulation. In addition, the simulation of MAP at TRL-6 (i.e., in a VDM) yielded novel and interesting results: the quantity of biomass extracted in the VDM is a more reliable indicator of the bio-extraction potential than the analysis of the bioaccumulation coefficient, which is directly depending on the metal concentration in soil and the possible inhibition of enzymes in metabolic pathways. Results are encouraging and the application of such multidisciplinary approach can be important to develop a circular model for sustainable exploitation of mining residues.

Author Contributions: Conceptualization, A.S., V.S. and S.M.; methodology, A.S., V.S., G.C. and D.G.; software, D.G.; validation, A.S., V.S. and A.G.; formal analysis, F.T., E.T. and D.P.; investigation, A.S. and V.S.; resources, A.S., A.G. and S.U.; data curation, A.S. and V.S.; writing—original draft preparation, A.S., V.S., S.M. and D.G.; writing—review and editing, S.M., A.G. and S.U.; visualization, A.G. and S.U.; supervision, A.S., S.M. and G.C.; project administration, A.S., S.M. and S.U.; funding acquisition, A.S., M.G. and S.U. All authors have read and agreed to the published version of the manuscript.

Funding: This research was funded by TECO Project e Technological eco-innovation for the quality control and the decontamination of polluted waters and soils-reference: "EuropeAid/135-474/DD/ACT/IN, "EU-India Research and Innovation Partnership"—and by S&T Cooperation Programme between Italian Republic and Argentina, financially supported by Italian Ministry of Foreign Affairs and International Cooperation (MAECI) and Ministry of Science, Technology and Innovation of the Argentine Republic (SGCTeIP), entitled: "Integrated multidisciplinary approach for the identification and recovery of raw materials from mining waste, by remote sensing".

Institutional Review Board Statement: Not applicable.

Informed Consent Statement: Not applicable.

Data Availability Statement: The study did not report any data.

Acknowledgments: We acknowledge to Jorge Álvarez, Paola Babay, Florencia Gonzalez, and Roxana Leguizamón from Chemistry Management, FRX Laboratory of National Atomic Energy Commission, Buenos Aires, Argentina for their technical assistance in FRX fluorescence analyses.

Conflicts of Interest: The authors declare no conflict of interest.

Appendix A

Table A1. Concentrations of elements in mycorrhized and non-mycorrhized plants grown in blank soil. Standard Deviation is reported in brackets.

Treatment	N	Mn (%)	P (ppm)	As (ppm)	Ga (ppm)	Zn (ppm)	Ti (%)	Cr (ppm)	Ni (ppm)	Cu (ppm)	Rb (ppm)	Fe (%)	Sr (ppm)	Al (%)	Ba (%)	K (%)	S (ppm)	Ca (%)
B^-	4	0.068 (0.011)	2275 (170.8)	7.050 (2.106)	5.125 (0.378)	213.8 (7.411)	0.135 (0.006)	87.25 (10.72)	43.50 (3.317)	110.5 (3.697)	206.8 (48.03)	1.475 (0.096)	1081 (114.3)	1.525 (0.250)	0.043 (0.006)	1.350 (0.100)	3450 (264.6)	5.500 (0.316)
B^+	4	0.062 (0.002)	2125 (50.00)	8.775 (0.818)	5.000 (0.683)	210.8 (18.79)	0.128 (0.005)	86.50 (18.23)	42.25 (1.500)	108.0 (5.354)	211.5 (28.21)	1.425 (0.050)	1100 (115.5)	1.363 (0.048)	0.043 (0.004)	1.325 (0.096)	3525 (221.7)	5.325 (0.126)

Table A2. Biomass per specimen (sp) of *Helliantus annuus* grown in contaminated and blank soil. Standard Deviation is reported in brackets.

Sample Type	Biomass (g/sp)	M^+	M^-
Contaminated	Shoots	0.24 ^{a} (0.03)	0.24 ^{a} (0.02)
	Roots	0.02 ^{b} (0.01)	0.02 ^{b} (0.004)
Blank	Shoots	0.35 ^{c} (0.01)	0.23 ^{a} (0.02)
	Roots	0.11 ^{d} (0.001)	0.02 ^{b} (0.002)

$^{a\text{–}d}$: different letters correspond to significant differences.

References

1. Guglietta, D.; Belardi, G.; Cappai, G.; Casentini, B.; Godeas, A.; Milia, S.; Passeri, D.; Salvatori, R.; Scotti, A.; Silvani, V.; et al. Toward a multidisciplinary strategy for the classification and reuse of iron and manganese mining wastes. *Chem. J. Mold.* **2020**, *15*, 21–30. [CrossRef]
2. Mathieux, F.; Ardente, F.; Bobba, S.; Nuss, P.; Blengini, G.; Alves Dias, P.; Blagoeva, D.; Torres De Matos, C.; Wittmer, D.; Pavel, C.; et al. *Critical Raw Materials and the Circular Economy–Background Report*; Publications Office of the European Union: Bruxelles, Belgium, 2017. [CrossRef]
3. Glick, B.R. Phytoremediation: Synergistic use of plants and bacteria to clean up the environment. *Biotechnol. Adv.* **2003**, *21*, 383–393. [CrossRef]
4. Ker, K.; Charest, C. Nickel remediation by AM colonized sunflower. *Mycorrhiza* **2010**, *20*, 399–406. [CrossRef] [PubMed]
5. Kastori, P.; Petrovic, N.; Petrovic, M. Effects of lead on water relations, proline concentration and nitrate reductase activity in sunflower plants. *Acta Agron. Hungar.* **1996**, *44*, 21–28.
6. Turgut, C.; Pepe, M.K.; Cutright, T.J. The effect of EDTA and citric acid on phytoremediation of Cd, Cr, and Ni from soil using Helianthus annuus. *Environ. Pollut.* **2004**, *131*, 147–154. [CrossRef]
7. Davies, F.T.; Puryear, J.D.; Newton, R.J.; Egilla, J.N.; Saraiva Grossi, J.A. Mycorrhizal fungi increase chromium uptake by sunflower plants: Influence on tissue mineral concentration, growth, and gas exchange. *J. Plant. Nutr.* **2002**, *25*, 2389–2407. [CrossRef]
8. Scotti, A.; Silvani, V.; Cerioni, J.; Visciglia, M.; Benavidez, M.; Godeas, A.A. Pilot testing of a bioremediation system of water and soils contaminated with heavy metals: Vegetable Depuration Module. *Int. J. Phytoremediation* **2019**, *21*, 899–907. [CrossRef]
9. Smith, S.E.; Read, D.J. *Mycorrhizal Symbiosis*, 3rd ed.; Academic Press: New York, NY, USA, 2008.
10. Lenoir, I.; Fontaine, J.; Lounès-Hadj Sahraoui, A. Arbuscular mycorrhizal fungal responses to abiotic stresses: A review. *Phytochemistry* **2016**, *123*, 4–15. [CrossRef] [PubMed]
11. Gohre, V.; Paszkowski, U. Contribution of the arbuscular mycorrhizal symbiosis to heavy metal phytoremediation. *Planta* **2006**, *223*, 1115–1122. [CrossRef] [PubMed]
12. Yang, Y.; Liang, Y.; Han, X.; Chiu, T.Y.; Ghosh, A.; Chen, H.; Tang, M. The roles of arbuscular mycorrhizal fungi (AMF) in phytoremediation and tree-herb interactions in Pb contaminated soil. *Sci. Rep.* **2016**, *6*, 20469. [CrossRef]
13. Rivelli, A.R.; De Maria, S.; Puschenreiter, M.; Gherbin, P. Accumulation of cadmium, zinc, and copper by *Helianthus annuus* L.: Impact on plant growth and uptake of nutritional elements. *Int. J. Phytoremediation* **2012**, *14*, 320–334. [CrossRef] [PubMed]
14. Scotti, A.; Godeas, A.; Silvani, V. Sistema Biorremediador Para Tratamiento de Suelo y/o Aguas Contaminadas. Argentina Patent 130100620, 22 October 2014.
15. Scotti, A.; Milia, S.; Silvani, V.; Cappai, G.; Guglietta, D.; Trapasso, F.; Belardi, G.; Salvatori, R.; Tempesta, E.; Passeri, D.; et al. Development of an integrated multidisciplinary strategy for gallium, iron and manganese recovery from mining residues in a context of circular economy. In Proceedings of the International Center of Earth Sciences Conference, Mendoza, Argentina, 23 November 2020; Martín, G., Dino, F., Eds.; CNEA: Buenos Aires City, Argentina, 2021; pp. 898–910.
16. Kadlec, R.H.; Knight, R.L.; Vymazal, J.; Brix, H.; Cooper, P. *Constructed Wetlands for Pollution Control*; IWA Publishing: London, UK, 2000.

17. Mena, J.; Rodríguez, L.; Núñez, J.; Fernández, F.J.; Villaseñor, J. Design of horizontal and vertical subsurface flow constructed wetlands treating industrial wastewaters. *WIT Trans. Ecol. Environ.* **2008**, *111*, 555–564.
18. Madera-Parra, C.A. Treatment of landfill leachate by polyculture constructed wetlands planted with native plants. *Ing. Comptet.* **2016**, *18*, 183–192.
19. Ibañez, J. Niveles de madurez de la tecnología [Technology readiness levels: TRLS]: Una introducción. *Econ. Indus.* **2014**, *393*, 165–171.
20. Silvani, V.A.; Fernández Bidondo, L.; Bompadre, M.J.; Pérgola, M.; Bompadre, A.; Fracchia, S.; Godeas, A.M. Growth dynamics of geographically different arbuscular mycorrhizal fungal isolates belonging to the '*Rhizophagus* clade' under monoxenic conditions. *Mycologia* **2014**, *106*, 963–975. [CrossRef]
21. Bompadre, M.J.; Silvani, V.A.; Fernández Bidondo, L.; Ríos de Molina, M.C.; Colombo, R.; Pardo, A.G.; Godeas, A. Arbuscular mycorrhizal fungi alleviate oxidative stress in pomegranate plants growing under different irrigation conditions. *Botany* **2014**, *92*, 187–193. [CrossRef]
22. Phillips, J.M.; Hayman, D.S. Improved procedures for clearing roots and staining parasitic and vesicular-arbuscular mycorrhizal fungi for rapid assessment of infection. *Trans. Br. Mycol. Soc.* **1970**, *55*, 158–160. [CrossRef]
23. Giovannetti, M.; Mosse, M. An evaluation of techniques for measuring vesicular arbuscular infection in roots. *New Phytol.* **1980**, *84*, 589–600. [CrossRef]
24. Cornejo, P.; Pérez-Tienda, J.; Meier, S.; Valderas, A.; Borie, F.; Azcón-Aguilar, C.; Ferrol, N. Copper compartmentalization in spores as a survival strategy of arbuscular mycorrhizal fungi in Cu-polluted environments. *Soil Biol. Biochem.* **2013**, *57*, 925–928. [CrossRef]
25. Gonzalez-Chavez, C.; D'Haen, J.; Vangronsveld, J.; Dodd, J.C. Copper sorption and accumulation by the extraradical mycelium of different *Glomus* spp. (arbuscular mycorrhizal fungi) isolated from the same polluted soil. *Plant Soil* **2020**, *240*, 287–297. [CrossRef]
26. Yruela, I. Copper in plants. *Braz. J. Plant Physiol.* **2005**, *17*, 145–156. [CrossRef]
27. Gonzalez-Chavez, M.; Carrillo-Gonzalez, R.; Wright, S.F.; Nichols, K.A. The role of glomalin, a protein produced by arbuscular mycorrhizal fungi, in sequestering potentially toxic elements. *Environ. Pollut.* **2004**, *130*, 317–323. [CrossRef] [PubMed]
28. Chen, B.; Christie, P.; Li, X. A modified glass bead compartment cultivation system for studies on nutrient and trace metal uptake by arbuscular mycorrhiza. *Chemosphere* **2001**, *42*, 185–192. [CrossRef]
29. Xu, L.J.; Hao, Z.P.; Xie, W.; Li, F.; Chen, B.D. Transmembrane H + and Ca 2+ fluxes through extraradical hyphae of arbuscular mycorrhizal fungi in response to drought stress. *Chin. J. Plant. Ecol.* **2018**, *42*, 764–773.
30. Allen, J.W.; Shachar-Hill, Y. Sulfur Transfer through an Arbuscular Mycorrhiza. *Plant. Physiol.* **2009**, *149*, 549–560. [CrossRef] [PubMed]
31. Aziz, A.; Muhammad, A.; Sultan, S.; Muhammad, A.; Naeem, A.; Sher, M.S.; Allah, W.; Ali, R.; Babar, H.B. Optimizing sulfur for improving salt tolerance of sunflower (*Helianthus annuus* L.). *Soil Environ.* **2019**, *38*, 222–233. [CrossRef]
32. Ducic, T.; Polle, A. Transport and detoxification of manganese and copper in plants. *Braz. J. Plant. Physiol.* **2005**, *17*, 103–112. [CrossRef]
33. Lyu, S.; Wei, X.; Chen, J.; Wang, C.; Wang, X.; Pan, D. Titanium as a Beneficial Element for Crop Production. *Front. Plant. Sci.* **2017**, *8*, 597. [CrossRef] [PubMed]
34. European Commission, COM 2020/474 Final. Available online: https://eur-lex.europa.eu/legal-content/EN/TXT/?uri=CELEX:52020DC0474 (accessed on 21 July 2021).
35. Correa, A.; Cruz, C.; Pérez-Tienda, J.; Ferrol, N. Shedding light onto nutrient responses of arbuscular mycorrhizal plants: Nutrient interactions may lead to unpredicted outcomes of the symbiosis. *Plant. Sci.* **2014**, *221–222*, 29–41. [CrossRef]
36. Ferrol, N.; Tamayo, E.; Vargas, P. The heavy metal paradox in arbuscular mycorrhizas: From mechanisms to biotechnological applications. *J. Experim. Bot.* **2016**, *67*, 6253–6265. [CrossRef]
37. Garcia, K.; Zimmermann, S.D. The role of mycorrhizal associations in plant potassium nutrition. *Front. Plant Sci.* **2014**, *5*, 337. [CrossRef] [PubMed]
38. Olsson, P.A.; Hammer, E.C.; Pallon, J.; van Aarle, I.M.; Wallander, H. Elemental composition in vesicles of an arbuscular mycorrhizal fungus, as revealed by PIXE analysis. *Fungal Biol.* **2011**, *115*, 643–648. [CrossRef] [PubMed]
39. Hawkes, C.V.; Casper, B.B. Lateral root function and root overlap among mycorrhizal and non mycorrhizal herbs in Florida shrubland, measured using rubidium as a nutrient analog. *Am. J. Bot.* **2002**, *89*, 1289–1294. [CrossRef] [PubMed]
40. He, X.; Lilleskov, E. Arsenic Uptake and Phytoremediation Potential by Arbuscular Mycorrhizal Fungi. In *Mycorrhizal fungi: Use in Sustainable Agriculture and Land Restoration*; Solaiman, Z., Abbot, L.K., Varma, A., Eds.; Springer Nature: Berlin/Heidelberg, Germany, 2014; Volume 41, pp. 259–275.
41. Imran, A.; Chaudhry, M.N.; Masood, K.R.; Iqbal, N. Response of sunflower (*Helianthus annuus* L.) to arsenic stress: Accumulation and portioning in different plant parts. *Soil Environ.* **2015**, *34*, 44–50.
42. Burger, A.; Lichtscheidl, I. Strontium in the environment: Review about reactions of plants towards stable and radioactive strontium isotopes. *Sci. Total Environ.* **2019**, *653*, 1458–1512. [CrossRef]
43. Hanaka, A.; Dresler, S.; Wójciak-Kosior, M.; Strzemski, M.; Kováčik, J.; Latalski, M.; Zawiślak, G.; Sowa, I. The Impact of Long- and Short-Term Strontium Treatment on Metabolites and Minerals in Glycine max. *Molecules* **2019**, *24*, 3825. [CrossRef] [PubMed]
44. Jensen, H.; Gaw, S.; Lehto, N.J.; Hassall, L.; Robinson, B.H. The mobility and plant uptake of gallium and indium, two emerging contaminants associated with electronic waste and other sources. *Chemosphere* **2018**, *209*, 675–684. [CrossRef]

45. Yu, X.Z.; Feng, X.H.; Feng, Y.X. Phytotoxicity and Transport of Gallium (Ga) in Rice Seedlings for 2-Day of Exposure. *Bull. Environ. Contam. Toxicol.* **2015**, *95*, 122–125. [CrossRef]
46. Seguel, A.; Cumming, J.R.; Klugh-Stewart, K.; Cornejo, P.; Borie, F. The role of arbuscular mycorrhizas in decreasing aluminium phytotoxicity in acidic soils: A review. *Mycorrhiza* **2013**, *23*, 167–183. [CrossRef]
47. Coscione, A.R.; Berton, R.S. Barium extraction potential by mustard, sunflower and castor bean. *Sci. Agric.* **2009**, *66*, 59–63. [CrossRef]
48. Ubaldini, S.; Guglietta, D.; Vegliò, F.; Giuliano, V. Valorization of Mining Waste by Application of Innovative Thiosulphate Leaching for Gold Recovery. *Metals* **2019**, *9*, 274. [CrossRef]

metals

MDPI

Article

Evaluation of Copper Leaching for Subsequent Recovery from the Waste Dumps of the Linares Mining District and Their Use for Construction Materials

Juan María Terrones-Saeta [1,*], Jorge Suárez-Macías [1], Ana María Castañón [2] and Francisco Antonio Corpas-Iglesias [1]

1 Department of Chemical, Environmental, and Materials Engineering, Higher Polytechnic School of Linares, University of Jaen, Scientific and Technological Campus of Linares, 23700 Linares, Spain; jsuarez@ujaen.es (J.S.-M.); facorpas@ujaen.es (F.A.C.-I.)
2 Department of Mining, Topography and Structures, University of León (ESTIM), Campus de Vegazana, s.n, 24071 León, Spain; amcasg@unileon.es
* Correspondence: terrones@ujaen.es

Abstract: The development of the population's well-being involves the use of different raw materials. However, metallic elements such as copper are currently scarce due to their intensive use in different sectors. Therefore, new sources of raw materials that provide these elements, are of lower cost, and use waste for their extraction must be sought. For this reason, in this research, different waste dumps of the mining district of Linares (Spain) are studied to evaluate the existence of recoverable copper by hydrometallurgical techniques. The material from the waste dump selected as potentially viable is leached with different sulfuric acid solutions (0.25, 0.5, 1, and 2 mol) and at different times, obtaining copper concentrations usable for subsequent hydrometallurgical processes. In addition, in order to develop an environmental hydrometallurgy, the leach waste is characterized, and bituminous mixtures are made with it. The results of the present investigation showed that it was possible to recover 80% of the copper in the waste dumps of the Linares mining district with 1 and 2 mol solutions of sulfuric acid. At the same time, the waste from the leaching process was found to be suitable for use as an aggregate in bituminous mixtures. Therefore, bituminous mixtures were conformed, and it was obtained that the optimum percentage of bituminous emulsion was 6.95% for the proposed granulometry. This emulsion percentage, which corresponds to a residual bitumen percentage of 4.17%, showed particle loss test results of 14% and 18% after immersion. In addition, the stability test values for the Marshall test with the above-mentioned bitumen emulsion percentage and leaching waste showed a stability of 8.99 KN. This fact demonstrates the quality of the bituminous mixture made with the leaching waste for use in bituminous mixtures. Consequently, it can be affirmed that in the present investigation, a significant percentage of copper has been extracted from the waste dumps of the mining district of Linares (Spain) and that the waste after processing can be used in bituminous mixtures, there being a closed cycle of materials in which no waste is produced.

Keywords: hydrometallurgy; metallic elements recovery; mining waste; bituminous mixture; circular economy; sustainability; construction materials

Citation: Terrones-Saeta, J.M.; Suárez-Macías, J.; Castañón, A.M.; Corpas-Iglesias, F.A. Evaluation of Copper Leaching for Subsequent Recovery from the Waste Dumps of the Linares Mining District and Their Use for Construction Materials. *Metals* **2021**, *11*, 1328. https://doi.org/10.3390/met11081328

Academic Editor: Anna H. Kaksonen

Received: 6 July 2021
Accepted: 20 August 2021
Published: 23 August 2021

1. Introduction

Mining is an essential activity for maintaining the well-being of the population [1]. It is through this industry that resources essential to daily life are obtained [2]. Consequently, the disappearance of this activity cannot be envisaged. However, mining operations produce a number of environmental impacts that need to be taken into account and reduced as much as possible [3,4]. One of the most important environmental impacts of mining activity is the production of waste [5]. This waste derives mainly from the extraction of the material and the concentration of the elements of interest, whereby there is extracted material that is separated and deposited in landfills. This material, which is currently of no use, has, in some cases, contaminating elements and chemical compounds that can cause a significant impact on the environment if not properly treated [6]. Consequently, it must be studied and evaluated how to make use of these mining wastes and incorporate them into new materials to avoid environmental pollution [7–9].

It should be noted that environmental regulations regarding mining are becoming increasingly restrictive [10]. However, in the past, there were no such environmental regulations, and waste was deposited without precaution. These waste deposit derivatives from former mining operations now cause significant pollution of surface water, groundwater, flora, fauna, landscape, etc. [11,12]. Specifically, in Linares, the area on which this research focuses, several authors have corroborated the contamination produced by mining waste [13]. At the same time, it is important to highlight that the mining waste produced in other times with less efficient equipment still contains important percentages of the main extraction element of the mine since it could not be concentrated by the old techniques. Moreover, because the demand for different minerals has been changing according to the development of the population, these wastes contain in some cases minerals that are economically viable to extract nowadays [14]. This is the case of the Linares lead mines, where the lead sulfides extracted in the past are associated mainly with copper sulfides, as well as other elements that, because they are of no economic interest or because they are difficult to extract using the old techniques, are found in the mining waste [15].

Based on the above, the purpose of this article is to study different waste dumps in the Linares mining district to evaluate the existence of copper, as well as to select the waste dump with the highest copper content for copper extraction by hydrometallurgical methods.

It should be noted that copper is one of the most demanded elements nowadays. This fact is due to its properties, being mainly its excellent electrical conductivity, good thermal conductivity, its malleability, resistance to corrosion, etc. This fact causes copper to be used in several areas such as energy, construction, military, naval, etc. [16]. Therefore, it is an essential element currently that has a high price due to the reduction of its reserves [17].

Worldwide, 80% of copper is extracted by pyrometallurgical techniques and 20% by hydrometallurgical processes [18]. However, the decrease in reserves, the existence of reserves with low percentages of copper, and the existence of this mineral, mainly copper sulfide, associated with other minerals, have made hydrometallurgical techniques economically unviable [19,20]. It is, therefore, necessary to develop economically profitable hydrometallurgical techniques that extract copper from materials with low copper content, such as mining waste [21]. This economic profitability is achieved through the use of robust techniques capable of adapting to variation in the chemical composition of the raw materials, slow recovery, and very low processing cost [22].

The hydrometallurgical techniques suitable for this system are currently carried out with acidic solutions and atmospheric pressure because the cost of this process is very low [23]. However, as mentioned above, the process is much slower and can take up to several months.

It also should be noted that the extraction of copper by hydrometallurgical techniques also produces a residue from the process [24]. This residue, based on a new circular economy, must be used as raw material for new products [25]. In this case, and due to the fact that the construction sector is one of the most demanding of raw materials and, in turn, one of the most polluting [26], the residue from the leaching process of the Linares mine waste dumps was used as aggregate for bituminous mixtures for roads. This waste, mainly granite, was conformed with bitumen emulsion to create a bituminous mix of discontinuous grading capable of withstanding the traffic loads of medium to low traffic roads in the area. The incorporation of leaching residues into bituminous mixtures was chosen as the best option because, as mentioned above, road construction requires huge quantities of raw materials [27]. Furthermore, the selection of this type of mix is due to the fact that it is a sustainable mix, incorporating bitumen emulsion as a binder, [28] and it possesses suitable mechanical and physical properties [29].

In short, this research studies the recovery of copper by hydrometallurgical techniques through the leaching with sulfuric acid solutions of materials from waste dumps, or dumps, from lead mines in the mining district of Linares. For this purpose, several waste dumps from the district were analyzed, the one with the highest percentage of copper was selected, and the waste was subjected to leaching with acidic media, measuring the pH, the electrical conductivity, and the concentration of copper in the recovered leachate continuously during the test time. After the leaching process, the residue was used as an aggregate for bituminous mixtures.

These bituminous mixtures would provide a suitable use of the leach residue and could create new bituminous mixtures for various roads in order to avoid the extraction of new raw materials and to avoid landfilling of the leach residue. Specifically, the conformed bituminous mixture was a discontinuous grading with leach residue and bitumen emulsion. This sustainable bituminous mixture was tested to determine its physical and mechanical properties by means of maximum density, bulk density, void content, simple compressive strength, and Marshall test.

2. Materials and Methods

2.1. Materials

The materials used in this research are mainly from the waste dumps of the Linares mining district and commercial products. The commercial materials used in the methodology are distilled water, sulfuric acid, and bitumen emulsion. The first two materials mentioned were used for the first stage of the work, in which the material from the selected dump was leached with different acid solutions at atmospheric pressure and ambient temperature. The bitumen emulsion was used in the second stage of this research to conform bituminous mixtures using the leaching residue. These materials are described in detail in the following sections.

2.1.1. Mining Waste Dumps

The residues assessed come from the waste dumps belonging to the mining district of Linares, Spain. This area was intensively mined mainly for lead. At the same time, and to a much lesser extent, silver was extracted as a secondary element of the argentiferous galena.

The mines in the Linares district were exploited for hundreds of years, until the end of the 20th century, when they ceased to operate. The cessation of their activity was mainly due to the enormous drop in lead prices, as it was no longer such an essential element for society, as well as the lack of research into new lead seams. As a result, there are several abandoned infrastructures and enormous quantities of mining waste derived from the exploitation of these mines all over the Linares area. Therefore, mining waste is currently a major environmental problem in the area.

In this mining waste, as is usual in lead sulfide formations, there are other chemical compounds associated with these mining wastes. Among the main chemical elements in the residues are copper, zinc, and even arsenic. More specifically, residue from waste dumps is rather random in size. Mainly, the particle size is 10 to 15 cm. These waste dump particles are essentially composed of granite, as this is the country rock of the exploited lead seam. This granite is fissured, presenting in its cracks other types of hydrothermal chemical compounds. This fact is easily observed in Figure 1.

Consequently, the milling of the waste rock will occur mainly through these cracks, as they are the weakest points of the joint. Therefore, this milling exposes the hydrothermal chemical compounds present in the cracks for leaching under acidic media, atmospheric pressure, and ambient temperature. Finally, the leaching residue, being granite, will be used as aggregate for bituminous mixes destined for roads.

Figure 1. Sample of waste dump showing cracks filled with hydrothermal chemical compounds.

Among all the waste dumps existing in the mining district of Linares, the waste dumps selected for the present study correspond to the mine workings of San Andrés, Arrayanes, and La Cruz.

2.1.2. Sulfuric Acid and Water

The sulfuric acid used in this investigation had a purity of 96% with a molar mass of 98.08 g/mol. This acid was diluted with different concentrations of distilled water with the aim to obtain sulfuric acid solutions of different molarities. In this way, it was possible to check the concentration of copper produced with different acid solutions at atmospheric pressure and at different periods of time. Therefore, the use of distilled water was necessary to obtain the different sulfuric acid solutions.

The distilled water used had a colorless liquid appearance, a density at 20 °C of 0.995 g/cm^3, a pH of 7, a conductivity in µS/cm of 2.5, a silica percentage of 0.003 mg/L, and a hardness measured in mg/L $CaCO_3$ of 0.3.

2.1.3. Bitumen Emulsion

In this research, with the aim of carrying out an environmentally sustainable project in which the waste produced was zero and usable materials were obtained for society, the by-products of the leaching of the waste dumps, after the extraction of the interesting metallic elements, were used as raw material for the conformation of bituminous mixtures. It is therefore essential to define the main elements used in the conformation of these bituminous mixtures, namely, the leaching residue (aggregate) and the bitumen emulsion (binder).

The bitumen emulsion is the binder of the bituminous mix of discontinuous grading conformed with the leaching residues. In other words, the emulsion is the element that binds the aggregates (leaching residues) together to withstand the tensile loads of traffic. The advantage of using bitumen emulsions over traditional bitumen lies in the fact that they can be conformed at ambient temperature, as well as extended and compacted. Consequently, the CO2 emissions from the process are much lower, as it is not necessary to heat the entire bituminous mix to a temperature of 180 °C. However, the bitumen emulsion must be compatible with the aggregate used, in this case the leaching residue. Therefore, a cationic emulsion with medium breaking time was selected, more specifically, the bitumen emulsion C60BF3 MBA. The technical data sheet of the bitumen emulsion C60BF3 MBA used is detailed in Table 1.

Table 1. Technical details of the bitumen emulsion C60BF3 MBA used.

Characteristics	Unit	Standard	Min.	Max.
Original Emulsion				
Particle polarity	-	UNE EN 1430	Positive	-
Breaking value (Forshammer filler)	g	UNE EN 13075-1	70	155
Binder content (per water content)	%	UNE EN 12846-1	58	62
Oil distillate content	%	UNE EN 1431	-	5
Efflux time (2 mm. 40 °C)	s	UNE EN 12846	40	130
Residue on sieving (0.5 mm)	%	UNE EN 1429	-	0.10
Setting tendency (7 days storage)	%	UNE EN 12847	-	5
Water effect of binder adhesion	%	UNE EN 13614	90	-
Binder after distillation (UNE EN 1431)				
Penetration (25 °C; 100 g; 5 s)	0.1 mm	UNE EN 1426	-	220
Softening point	°C	UNE EN 1427	35	-
Evaporation residue (UNE EN 13074-1)				
Penetration (25°C; 100 g; 5 s)	0.1 mm	UNE EN 1426	-	330
Softening point	°C	UNE EN 1427	35	-
Stabilizing residue (UNE EN 13074-2)				
Penetration (25 °C; 100 g; 5 s)	0.1 mm	UNE EN 1426	-	220
Softening point	°C	UNE EN 1427	35	-

2.2. Methodology

First, the mining district of Linares was studied in order to identify those waste dumps belonging to different mine workings that could present viable percentages of interesting metallic elements such as copper. The selected waste dumps were visited to take samples of representative materials, taking into account the most abundant formations therein and collecting this type of material. Subsequently, the material from the different samples was treated for subsequent analysis with micro-X-ray fluorescence tests, identifying the existence of copper in the dumps. Based on the results obtained, the dump with the highest potential percentage of copper was selected for extraction by hydrometallurgical techniques.

The selected waste dump was analyzed in more detail by taking samples from different points of the waste dump in order to quantify in more detail the chemical elements present, the percentage of copper present, as well as the variability of the results. For this purpose, samples were taken and, subsequently, analyzed chemically. Once the chemical composition of the selected waste dump was evaluated, the material was milled and leached in acidic media. These acidic media consisted of sulfuric acid solutions of different molarities (0.25 mol, 0.50 mol, 1.00 mol, and 2.00 mol), measuring the pH, electrical conductivity, and copper concentration at different times. With these data, it was possible to obtain the appropriate leaching solution and the time required for the recovery of an economic percentage of copper.

The leaching residue was subjected to a leachate test in order to determine its suitability for the conformation of bituminous mixtures. This process was carried out in order to assess that the leachate from the aggregate did not cause significant environmental contamination. Subsequently, the leaching residue was subjected to different physical and mechanical tests to determine its suitability as an aggregate for bituminous mixtures.

Finally, and having determined the suitability of the leaching residue for conformation bituminous mixtures, different discontinuous grading and bitumen emulsion bituminous mixtures were conformed. These bituminous mixtures were conformed with increasing percentages of binder and were subsequently subjected to physical and mechanical tests in order to establish the optimum percentage of bitumen emulsion that developed the best properties.

The following sections describe the tests carried out in greater detail in the different blocks into which this research is divided.

2.2.1. Analysis of the Different Waste Dumps in the Mining District of Linares

The waste dumps selected were those known as San Andrés, Arrayanes, and La Cruz. These waste dumps correspond to the deposition of waste from mining operations over decades, so that at the three study points, they have a volume of thousands of cubic meters. As detailed above, these waste dumps were considered in this investigation due to the existence of abundant bibliographical references that testified the existence of polymetallic sulfides, essentially copper.

The samples were taken at different points of the waste dumps. For this purpose, the UNE-EN 932-1 standard was applied for taking samples in large depositions of material. In this way, it was possible to obtain a final sample of 20 kg as representative as possible of the waste dumps analyzed.

The samples from the three waste dumps were dried at a temperature of 105 ± 2 °C and, subsequently, reduced in size according to UNE-EN 932-2. The subsample obtained was milled to a particle size of less than 100 micrometers for chemical characterization. These samples from each waste dump were analyzed by micro-X-ray fluorescence. The equipment used was the Bruker M4 Tornado energy dispersive micro-X-ray fluorescence spectrometer (M4 Tornado, Bruker, Billerica, MA, USA). The analysis of the samples with the microfluorescence equipment provided a rapid elemental composition of each waste dump, thus, allowing us to select the potentially viable waste dump for further leaching and copper extraction.

2.2.2. Chemical Characterization of the Selected Waste Dump

Once the three samples from the three waste dumps had been analyzed, the waste dump of greatest interest for copper element recovery was selected. The selected waste dump sample was further reduced according to UNE-EN 932-2 to obtain sufficient material with a particle size of less than 100 micrometers for detailed chemical characterization.

The first of the chemical tests carried out was elemental analysis. This test makes it possible to detect and quantify the percentages of carbon, nitrogen, hydrogen, and sulfur present in the sample. For this, the sample is calcined at a temperature of 1000 ± 5 °C, and the gases are analyzed with LECO's commercial TruSpec Micro equipment (TruSpec Micro, LECO, St. Joseph, MI, USA).

In order to provide further information on the sample, the loss-on-ignition test was also performed. This test shows the difference in mass in percentage that exists between the sample before and after calcination at 1000 ± 5 °C. This loss of mass of the sample corresponds mainly to the organic matter in the sample, the loss of volatile elements, or the transformation of hydrated or carbonated chemical compounds.

Finally, and because the nature of the sample is mainly inorganic, the X-ray fluorescence test was performed. This test, performed with the commercial ADVANT'XP+ equipment (ADVANT'XP+, Thermo Fisher, Waltham, MA, USA) was also performed to determine the percentage of copper in the sample. The determination of the percentage of copper is essential to assess the percentage recovery occurring in the leachate, as well as the feasibility of mining.

The chemical characterization of the selected waste dump sample not only allows the accurate determination of the percentage of copper in the sample, but also identifies critical points where special attention should be paid.

2.2.3. Leaching in Acidic Media and Atmospheric Pressure

Once the chemical composition of the selected waste dump sample was analyzed, we proceeded to study the leaching of copper with different sulfuric acid solutions at different leaching times and atmospheric pressure.

First, the sample of the selected waste dump was milled, and different representative sub-samples were obtained according to the UNE-EN 932-2 standard. The grading curve of the crushed waste dump sample was determined using the UNE-EN 933-1 grading curve analysis test. The milling was carried out with a double purpose: on the one hand, to reduce the sample size so that it was easier to handle in the industrial process; on the other hand, and due to the fact that the granite of the waste dump was pre-cracked and in these cracks there were copper hydrothermal chemical compounds, the milling produced the fracture of the rock through these cracks obtaining, consequently, the exposure of the interesting chemical compounds to the acid dissolution.

Once the sample was crushed, the leaching test was carried out in acidic media and with different solutions. The samples selected for leaching had a mass of 300 ± 0.1 g, after drying to remove humidity. The sulfuric acid solutions, 3000 ± 10 mL volume per process, were 0.25 mol, 0.50 mol, 1.00 mol, and 2.00 mol. The use of these low molarity solutions is motivated by different reasons: on the one hand, the use of low molarity solutions of sulfuric acid develops a leaching residue that is easier to treat afterwards; on the other hand, a lower concentration of sulfuric acid makes it possible to increase the working life of the equipment; and finally, higher molarities of sulfuric acid would create a higher percentage of sludge that is difficult to treat. Furthermore, it should be noted that the use of the aforementioned solid/liquid ratio is to avoid the need to carry out subsequent concentration processes at a higher economic cost, as would be the case if higher temperatures than ambient were used.

The leaching process was carried out in equipment specially designed for the research. This equipment consisted mainly of sieves that retained the sample and an upper pulverizer. The sprayer sprayed the sample remaining on the sieves, depositing the fluid by gravity in a leaching tank. The fluid from the leach tank was continuously pumped into the pulverizer for a continuous recirculation process. This process was maintained constant during the 35 days of testing without interruption, controlling the temperature at 25 ± 1 °C. It should be noted that the operation of the equipment is very similar to that of the industrial process, in which large quantities of material are irrigated with acid solutions that are continuously recirculated. The image of the equipment can be seen in Figure 2.

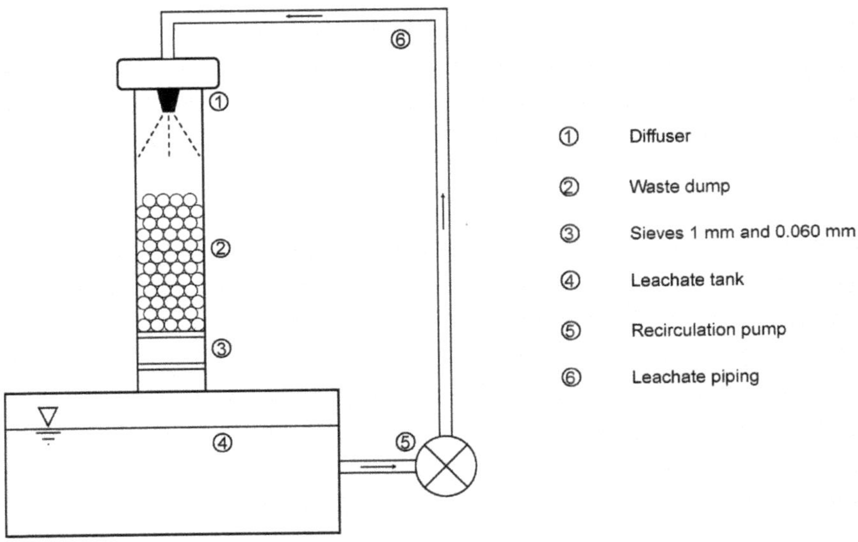

Figure 2. Leachate equipment for leaching of mining waste in acidic media, ambient temperature, and atmospheric pressure.

This process was carried out for the four detailed sulfuric acid solutions, measuring the pH and electrical conductivity of the solution every 24 h. In addition, samples were taken every 24 h for subsequent analysis by inductively coupled plasma mass spectrometry (7900, Agilent, Santa Clara, CA, USA). For the chemical analysis with the detailed equipment, standardized standards of the chemical element copper were used, repeating each measurement three times to obtain statistically reliable results. In this way, the percentage of copper leached in the solution used and at different times could be verified, as well as the percentage of recovery that occurred with respect to the total copper in the sample.

With the results of the variation of the pH of the solution, the variation of the electrical conductivity, as well as the concentration of copper in the different solutions for different times, the optimum solution was determined, as well as the time necessary for the leaching of copper at an economically viable recovery percentage.

2.2.4. Characterization of the Leaching Residue

Once the solution that developed a leaching of copper from the waste dump was selected in a viable time, a significant amount of material was leached in order to study the leaching residue produced after the leaching process.

The leaching residue, after the leaching process with the conditions set, was evaluated chemically, physically, and mechanically. This characterization was carried out with the aim of determining the quality of the residue after the copper extraction process for use in bituminous mixtures, thus, developing a sustainable process in which the residues produced are zero and feed different industries of essential raw materials, such as copper and aggregates for bituminous mixtures.

First, and due to the fact that the leaching residues come from leaching processes with sulfuric acid, as well as from materials in which heavy metals are usually found, we proceeded to analyze the leachate from the residue in order to determine that its use did not entail significant environmental problems. This leaching residue test is essential for the use of aggregates in bituminous mixtures for road infrastructures. To this end, the leaching test was carried out in accordance with the UNE-EN 12457-3 standard, detecting the presence of contaminating elements in high proportion, comparing them with the limits set by Spanish regulations, and limiting their use for the proposed purpose.

Once the feasibility of using the leaching residue had been chemically determined, a series of physical and mechanical tests were carried out to evaluate its possible use in bituminous mixtures.

The leaching residue was tested for: particle density according to the UNE-EN 1097-7 standard in order to determine if volumetric corrections were necessary due to a density different from that of a conventional aggregate, 2.65 t/m^3; the sand equivalent test according to UNE-EN 933-8 to determine the presence of colloidal particles; the percentage of crushed and broken surfaces test according to UNE-EN 933-5; and the flakiness index test according to UNE-EN 933-3 to determine the shape of the particles and their suitability for use in bituminous mixtures. In order to determine the resistance of the leaching residue, resistance to fragmentation tests were carried out in accordance with the UNE-EN 1097-2 standard to qualify the strength of the material, as well as tests for resistance to freezing and thawing cycles (UNE-EN 1367-1 standard) to evaluate the resistance to thermal fatigue of the aggregate and determination of the polished stone value (UNE-EN 1097-8 standard) to quantify the resistance to the continuous friction of the tire.

These chemical, physical, and mechanical tests allowed an objective assessment of the quality of the leaching residue for use in bituminous mixtures.

2.2.5. Bituminous Mixtures Conforming and Testing

Once the chemical, physical, and mechanical suitability of the leaching residue had been analyzed, the bituminous mix used was formulated and named a discontinuous grading mixture, hereinafter referred to as AF12. This type of bituminous mix was developed with the fundamental aim of obtaining a sustainable, low-cost mix with a high use of waste.

It should be noted that in this work, the best practice guide provided by the Technical Association for Bitumen Emulsions (hereinafter ATEB) will be used for the formulation of bituminous mixes. This guide was written by experts in the field and sets out the necessary steps for assessing the suitability of a mixture of this type.

The grading curve of the leaching residue was that established by the envelope grading curve detailed in the aforementioned standard. This grading curve, shown in Figure 3, was intermediate between the upper and lower limit of the grading envelope in order not to induce more variables.

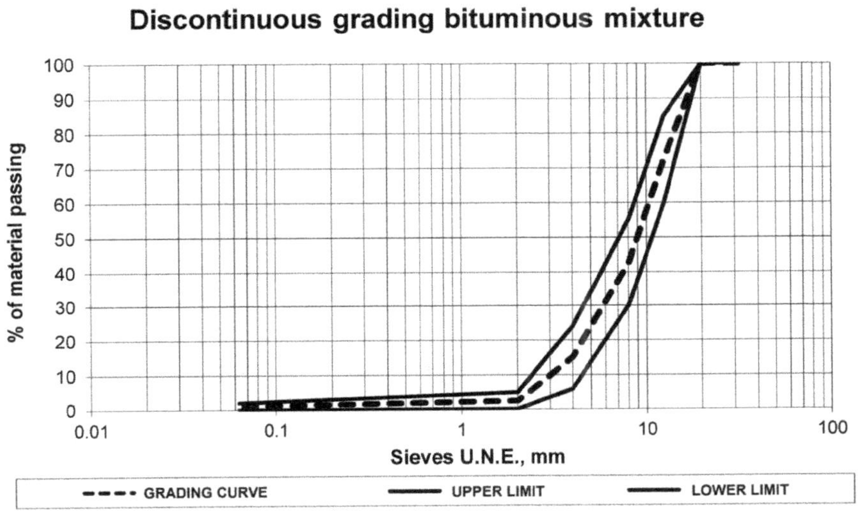

Figure 3. Grading curve of the bituminous mixture conformed with leaching residue and bitumen emulsion.

Once the grading curve had been defined, the analysis of the compatibility of the aggregate with the emulsion to be used was carried out. For this purpose, the coating test was carried out in accordance with the NLT-196/84 standard. This test evaluates the quality of the aggregate's coating by the emulsion and, consequently, the quality of the final mix to be developed.

First, the minimum percentage of emulsion, residual bitumen, that the bituminous mix could contain was calculated. To do this, the percentage of emulsion to be added to the bituminous mix was calculated according to the mathematical formula described by ATEB in its good practice guide. It should be noted that the bitumen emulsion is mainly composed of bitumen and water, and that after the mixture is spread, the water evaporates by natural processes. Therefore, the calculation of the percentage of residual bitumen from the percentage of bitumen emulsion is immediate if the technical data sheet of the bitumen emulsion is known. In this case the bitumen emulsion used had a residual bitumen percentage of 60%. The formula provided by ATEB for this type of bituminous mixture is based on the specific surface area of a conventional aggregate and is detailed below.

$$BR = (K/100) \cdot (1.5A + 2.5B + 4C + 6D + 9E + 12F) \quad (1)$$

where:

BR = Proportion of residual bitumen on the dry mass of the aggregates.

K = Coefficient of enrichment, the value of which is 1 in the wearing course and 0.9 in the lower course.

A = Proportion of aggregates retained by the sieve UNE 20 mm.

B = Proportion of aggregates passing through the sieve UNE 20 mm and retained by the sieve UNE 8 mm.

C = Proportion of aggregates passing through the sieve UNE 8 mm and retained by the sieve UNE 4 mm.

D = Proportion of aggregates passing through the sieve UNE 4 mm and retained by the sieve UNE 2 mm.

E = Proportion of aggregates passing through the sieve UNE 2 mm and retained by the sieve UNE 0.063 mm.

F = Proportion of aggregates passing through the sieve UNE 0.063 mm.

Consequently, and according to the grading defined above, the ATEB formula determined a residual bitumen content of the aggregate of 3.5%, corresponding to a percentage of bitumen emulsion of the aggregate of 5.8%. Once the optimum point was calculated empirically using the formula provided by ATEB, different families of test samples were made with increasing percentages of bitumen from 0.25% onwards. As mentioned above, when conforming, it must be taken into account that the water in the emulsion will be evaporated later, so the reference values are the bitumen that will remain in the bituminous mix after evaporation.

The test samples of each family of specimen with increasing percentages of bitumen were conformed and compacted according to UNE-EN 12697-30. Subsequently, all the test samples were subjected to a curing process to eliminate the water from the emulsion. For this purpose, the Marshall type test samples manufactured were leveled with the mold so that they could be placed on sieves with an opening of less than 2 mm. The subsequent curing process consisted of 2 days at a temperature of 75 $\pm$ 2 °C and 5 more days at a temperature of 90 $\pm$ 2 °C (the curing processes are detailed in ATEB). During the curing process, binder drainage must be continuously observed; for this purpose, a filter paper was placed at the bottom of the test samples during the curing process. The test sample family that caused binder drainage was considered as null, and consequently, the percentage of bitumen or emulsion in this family was the maximum permissible. Therefore, the bituminous mixes viable for use were those incorporating a percentage of emulsion from the minimum set by ATEB to the maximum produced by binder drainage.

After curing, the test samples were stripped and subjected to a series of physical tests to determine the general physical properties. The first test samples were tested for maximum density, according to UNE-EN 12697-5; bulk density, according to UNE-EN 12697-6; and void characteristics in the mix, according to UNE-EN 12697-8.

Subsequently, and in order to determine the resistance of the bituminous mixtures to fracture and aggregate pull-out, the particle loss test was carried out in accordance with the UNE-EN 12697-17 standard for all families of mixtures. In turn, in order to evaluate the influence of water on the bituminous mix and, consequently, to determine the compatibility between the aggregate and the emulsion, the same type of test was carried out with immersion in water at 45 $\pm$ 1 °C for 24 h, in accordance with standard NLT-362/92.

In this way, the quality of the mastic conformed with the filler and bitumen in the emulsion could be determined. However, due to the incorporation of high percentages of bitumen, the problem of plastic deformation in the infrastructure can occur. To avoid this and to quantify the resistance of the mix at high temperatures, the Marshall test was carried out in accordance with the UNE-EN 12697-34 standard.

Finally, with the results of the aforementioned tests for the different groups of samples, it was possible to observe, graphically and mathematically, the variation of the physical and mechanical properties of the bituminous mixes with increasing bitumen percentage. In this way, the study of the graphical and mathematical models provided an optimum percentage of bitumen emulsion that developed the best physical and mechanical characteristics. This combination is the one that was called optimum, with which all the above-mentioned tests were carried out again to corroborate the mathematically determined properties. These tests for determining the physical properties of the final bituminous mix are: maximum density UNE-EN 12697-5, bulk density UNE-EN 12697-6, and void characteristics UNE-EN

12697-8; and for the mechanical properties: loss of particles without immersion and after immersion UNE-EN 12697-17 and the Marshall test UNE-EN 12697-34.

3. Results

3.1. Analysis of the Different Waste Dumps in the Mining District of Linares

The results of the micro-X-ray fluorescence test of the three waste dumps (San Andrés, Arrayanes, and La Cruz) are shown in Table 2.

Table 2. Micro-X-ray fluorescence of the waste dumps in study.

Element	San Andrés, Wt %	Arrayanes, Wt %	La Cruz, Wt %
Si	20.59	19.54	18.47
Al	10.37	1.27	8.54
Fe	7.56	36.28	15.16
Sx	1.34	0.06	4.52
Cu	4.54	1.35	0.45
Mg	2.76	0.18	3.14
K	2.01	0.03	3.57
Ca	0.73	0.06	2.45
Pb	1.48	0.97	2.33

The micro-X-ray fluorescence results show that the waste dump with the highest percentage of potentially extractable copper is the San Andrés waste dump, obtaining results of 4.54% copper according to this test. However, since this test was used for a rapid detection of the elements present in the samples without an exhaustive preparation of the sample, the values obtained could be influenced with high probability by the shape of the particles, the excitation of the chemical elements present, as well as by the time of analysis. However, all other things being equal, the sample with the highest percentage of copper could be evaluated, this being the one from the San Andrés waste dump. This sample was the one selected, being analyzed later in greater detail for the evaluation of its chemical composition.

3.2. Chemical Characterization of the Selected Waste Dump

The chemical composition of the waste dump residue determines the percentage of copper in the sample and quantifies the existence of other chemical elements that may impair the leaching process. It is, therefore. essential to perform.

The first of the tests carried out was elemental analysis, and the values are shown in Table 3.

Table 3. Elemental analysis of the chemical elements nitrogen, carbon, hydrogen, and sulfur in the sample from the San Andrés waste dump (ESA).

Sample	Nitrogen, %	Carbon, %	Hydrogen, %	Sulfur, %
ESA	0.000 ± 0.001	0.843 ± 0.019	0.410 ± 0.015	3.122 ± 0.073

Elemental analysis of the mining waste sample shows a very low percentage of carbon. This low percentage of carbon confirms the low proportion of organic matter and carbonates. The low percentage of hydrogen may be due to the transformation of the hydrated compounds or to unavoidable traces of humidity from the testing process. However, as in the case of carbon, this value is quite low. Finally, it can be observed that there is a percentage of sulfur. The existence of this percentage is obvious, since the sample contains copper sulfides, iron sulfides, lead sulfides, and to a much lesser extent, zinc sulfides. However, it should be noted that the percentage of sulfur determined by elemental analysis is not the total sulfur in the sample, as this test was performed by analyzing the gases produced in the combustion of the sample at 1000 ± 5 °C. Subsequently, the percentage of remaining sulfur was evaluated by the X-ray fluorescence method.

The loss-on-ignition test reflects the mass variation in the sample when the sample is subjected to the temperature of 1000 ± 5 °C. The loss on ignition of the sample of the mining waste under study is 4.35 ± 0.10%. This very low loss on ignition is mainly due to the transformation of the chemical compounds. This loss on ignition value is usual and similar to those obtained in tests of this type for mining waste.

X-ray fluorescence determined the percentage of the chemical elements with the highest atomic weight. The elemental composition of the mining waste obtained by X-ray fluorescence is shown in Table 4.

Table 4. X-ray fluorescence of the San Andrés waste dump.

Element	Wt %	Est. Error
Si	22.45	0.12
Al	8.02	0.09
Fe	5.85	0.10
Sx	2.29	0.05
Cu	3.32	0.08
Mg	2.49	0.06
K	3.32	0.08
Ca	2.82	0.07
Na	0.522	0.026
Ti	0.379	0.019
Pb	0.257	0.013
Mn	0.115	0.0058
Px	0.0513	0.0028
Sn	0.0477	0.0061
Zn	0.0345	0.0019
Cr	0.0273	0.0014
Cl	0.0318	0.0031
V	0.0137	0.0012
Ni	0.0178	0.0012
Zr	0.0119	0.0024
Oxygen	43.54	1.11

Table 4 shows the percentage of the different chemical elements analyzed in the sample. This is, therefore, the most reliable method for determining the actual percentage of each of the chemical elements existing in the sample, regardless of their combination in different chemical compounds.

From the results, it can be seen that the percentage of copper, the main element of this study, was 3.32 ± 0.08%. This copper came from the ores of the waste dump, which is the one that wass intended to be extracted by leaching in acidic media at atmospheric pressure and ambient temperature. The percentage of copper is contemplated as a viable percentage for recovery, i.e., it is not a small percentage, but it would present problems for extraction by pyrometallurgical means. This is due to the diversity of existing minerals, their combination, and the existence of some elements that are detrimental to the pyrometallurgical process, which will be detailed later. In short, the percentage of copper present for the development of this new hydrometallurgical technique is correct and to be expected for the material to which it corresponds.

It should be noted that there was a percentage of sulfur in the elemental composition. This is indicative of the existence of some sulfides, probably copper. Therefore, with the use of this low molarity leaching technique, and without the development of a previous treatment, it was difficult to achieve a complete recovery of the existing copper in the waste dump.

In turn, calcium represented an important percentage in the composition, corresponding to the feldspars and mica of the granite, as well as aluminum. It should be remembered that granite is the main rock in which the metal sulfides deposits have formed, so its presence in the waste dump is important. However, calcium and aluminum are not chemical elements that are detrimental to the hydrometallurgical leaching process, so they are not a problem. Similarly, silicon is present in significant percentages as granite is a siliceous rock.

Magnesium and potassium, as in the previous cases, come directly from the granite and are usually found in the percentages detailed in this rock.

The lead comes from galena, a mineral mainly extracted in the mine working of the vein. The percentage of lead is reduced as the waste dumps belong to mine tailings, so the mining process drastically reduced the percentage of lead in the waste.

Zinc, belonging to the blende ore, appeared in low proportion in the composition. Therefore, it was not the main element of extraction by leaching in an acidic media, but it could be extracted at a lower recovery rate after obtaining copper, being also a sought-after element by the industry.

The other chemical elements were found in very low proportions, and wet test methods are needed to quantify them under higher pressure. However, they do not fall within the scope of this study and will not be detrimental to the industrial process. The quality of the waste in terms of arsenic percentages is noteworthy. It is usual to find arsenic sulfides in this type of mining formations that seriously damage the pyrometallurgical process; however, in hydrometallurgy, its influence is much less and the percentage it represents in the mining waste is very small.

In short, and after the analysis of the chemical composition, it can be confirmed that the chemical study of the San Andrés waste dump carried out previously shows a viable copper percentage and reflects the inexistence of chemical elements in high percentages that could damage the process, making possible the subsequent leaching section of the sample in acidic media at atmospheric pressure and ambient temperature, according to hydrometallurgical techniques.

3.3. Leaching in Acidic Media and Atmospheric Pressure

The dry sample of the waste dump was milled in the jaw crusher in order to free the copper compounds in the waste dump for direct contact with the solution and, consequently, its leaching. Figure 4 shows the grading curve of the waste dump after the milling process.

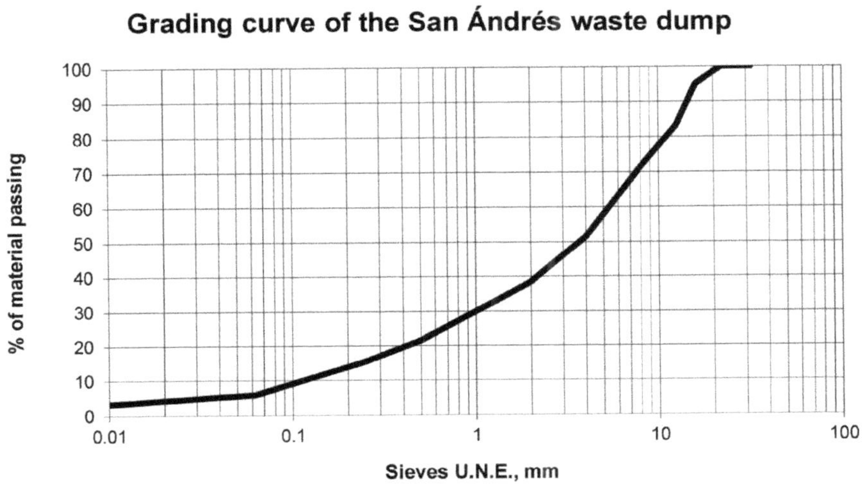

Figure 4. Grading curve of the San Andrés waste dump after the milling process.

The grading of the waste dump sample selected coincides with a continuous grading in which the maximum aggregate size is less than 20 mm. It should be noted that this milling process is simple and economical, so it would not increase the cost of treatment and a leaching residue could be obtained that could be used for bituminous mixtures.

Subsequently, the milled mining waste sample was divided into several sub-samples for the leaching process. This process was carried out at ambient temperature (25 ± 1 °C), atmospheric pressure, for 35 days, and with different sulfuric acid solutions: 0.25 molar, 0.50 molar, 1.00 molar, and 2.00 molar.

First, Figure 5 shows the variation of the pH of the leachate as a function of time and according to the concentration of sulfuric acid used.

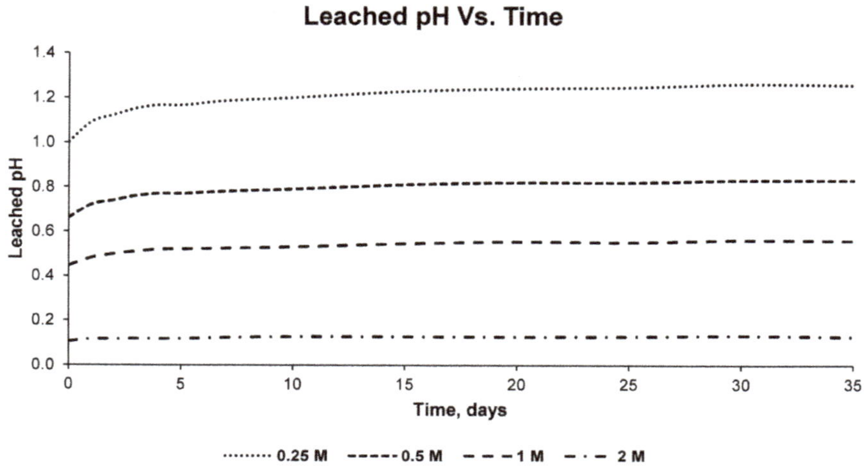

Figure 5. Graph of the pH of the leachate, as a function of time, for the 0.25 molar, 0.50 molar, 1.00 molar, and 2.00 molar sulfuric acid solutions.

As can be seen in Figure 5, the pH of the solution was lower the higher the molarity of sulfuric acid, which is to be expected if the activity of this type of acid is taken into account. At the same time, it was observed that, in the first 24 h in all four solutions, there is a higher pH increase than in the following days due to the direct contact of the solution with the mining waste from the landfill.

The 2.0 mol sulfuric acid solution showed a rapid stabilization of the pH after 24 h of the process, producing something similar with the 1.00 mol solution but with a longer leaching time. In contrast, the 0.50 mol and 0.25 mol solutions did not seem to show a stabilization of the pH over the time under study, so it is highly probable that they could represent the need for a longer leaching time to obtain copper concentrations similar to those obtained with higher molarity solutions.

On the other hand, the electrical conductivity of the leachates of the four solutions under study and for different times was measured during the 35 days of testing, reflecting the results shown in Figure 6.

The electrical conductivity of the different solutions at different leaching times clearly showed that the solutions with higher molarity had higher electrical conductivity. This conductivity decreased in contact with the waste dump waste, mainly in the first days. However, the electrical conductivity of the solutions did not seem to stabilize over the days, unlike the pH. This is because this more accurate measure reflects the dissolution of elements in the leachate not only of copper but also of other potentially leachable and less economically interesting elements in the sample. Therefore, this test demonstrates the need to select not only the correct dilution, as in the pH test, but also the correct leaching times. If the correct leaching times were not selected, concentrations of other elements of less economic interest would be produced in the leachate, which would also hinder subsequent hydrometallurgical processes to obtain copper.

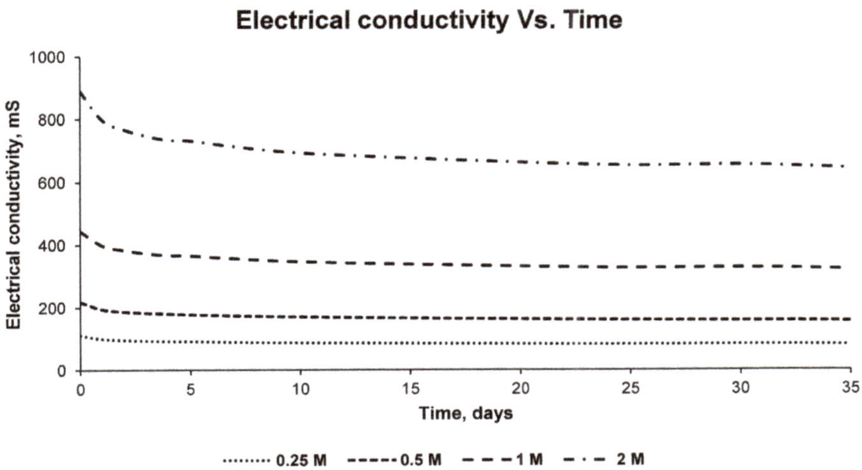

Figure 6. Graph of the electrical conductivity of the leachate, as a function of time, for the 0.25 molar, 0.50 mol, 1.00 mol, and 2.00 mol sulfuric acid solutions.

Finally, Figure 7 shows the variation of copper concentration as a function of time and the solution used. The units used are milligrams per gram of sample, with a maximum leachable copper concentration of 33.2 mg/g, according to the X-ray fluorescence test.

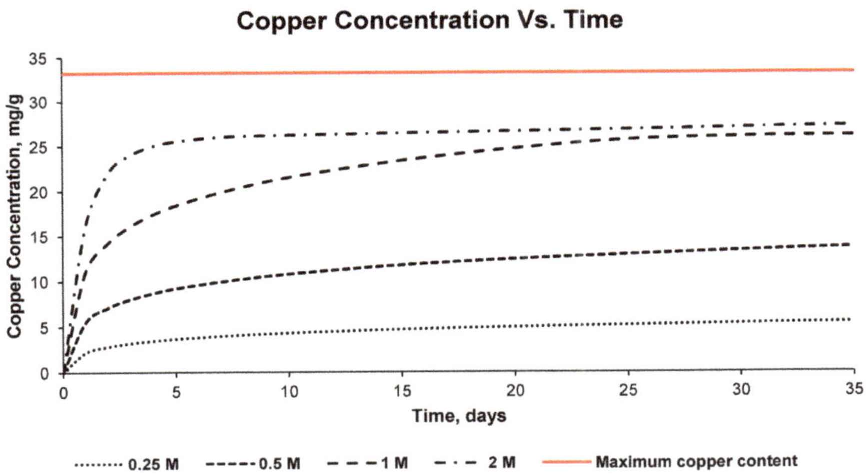

Figure 7. Graph of leachate copper concentration as a function of time for 0.25 mol, 0.50 mol, 1.00 mol, and 2.00 mol sulfuric acid solutions.

The graph of copper concentration in the leachate of the four solutions shows how the 0.25 mol sulfuric acid solution produces a reduced concentration of copper in the leachate, which makes its use undesirable. In turn, the 0.5 mol solution shows how the copper concentration increased gradually with time, not reaching a significant copper recovery percentage in the time evaluated and, consequently, needing more time for its recovery. For these detailed reasons, the 0.5 mol solution was also discarded as a possible solution for copper leaching, as the time required would be too long.

The 1.0 molar and 2.0 mol sulfuric acid solutions show that the maximum percentage of copper recovery was approximately 80%, as the 2.0 mol solution reached this maximum and stabilized, not producing higher copper concentrations over time. This is because, as discussed in the chemical analysis of the San Andrés waste dump, there are still copper sulfides in the material that have not been oxidized and are, therefore, very difficult to leach with this hydrometallurgical technique. Therefore, complete leaching of the copper in the sample will be achieved by pre-treating the waste dump material and transforming the copper sulfides into oxides. Since the interest of this research is to obtain an environmentally and economically sustainable process, this pre-treatment was not carried out, and a recovery rate of 80% of the copper in the sample was considered workable.

There are important differences between the 1.0 mol and 2.0 mol sulfuric acid solutions, as the 2.0 mol solution reached stabilization of the copper concentration in the leachate practically 5 days after the start of the process. On the other hand, the 1.0 mol sulfuric acid solution reached stabilization of the copper concentration after 25 days, obtaining similar results to those provided by the 2.0 mol solution. Based on the molarities of the solution, the environmental damage that it can cause, the wear and tear or deterioration of the equipment, as well as the previous activities for the execution of the leaching, in this research, the 1.0 mol solution of sulfuric acid has been taken as the optimum solution for copper recovery, setting a leaching time of 25 days.

3.4. Characterization of the Leaching Residue

The leaching residue obtained after the leaching process at the San Andrés waste dump was recovered for chemical and physical characterization. In this way, it was possible to assess its viability for use in bituminous mixtures.

The test for the evaluation of the viability of using the leaching residue in bituminous mixtures was the leachate test, according to the standard defined in the methodology. This test quantifies the concentrations of the contaminating elements of the leaching residue in the leachate, assessing its suitability for use. The results of the leachate test are shown in Table 5.

Table 5. Concentrations of contaminants in the leachate of the leaching residue compared to regulatory maximum limits.

Element	Leaching Residue, mg/kg	Maximum Limits, mg/kg
Ba	0.003 ± 0.001	17
Cd	0.001 ± 0.001	0.009
Cr	0.003 ± 0.001	0.5
Mo	0.002 ± 0.001	0.5
Ni	0.007 ± 0.001	0.4
Pb	0.165 ± 0.004	0.5
Se	0.001 ± 0.001	0.1
V	0.026 ± 0.001	1.3
Zn	0.017 ± 0.001	1.2
As	0.003 ± 0.001	0.5
Cu	0.756 ± 0.019	2
Hg	-	0.01
Sb	0.001 ± 0.001	0.06
Chlorides	87 ± 2	800
Sulfate	267 ± 6	377

The leachates from the leaching residue obtained in the hydrometallurgical process for the extraction of copper comply with the limits set by the regulations. The low concentration of heavy metals is to be highlighted, as well as the acceptable result of sulfates in the leachate. This percentage of sulfates was a critical point for its reuse, as the waste dumps had sulfur and the hydrometallurgical process was carried out with sulfuric acid.

Once the chemical suitability of the leaching residue had been analyzed, the physical properties were determined. These tests are defined in Table 6.

Table 6. Physical properties of the leaching residue.

Test	Standard	Value/Unit
Particle density	UNE-EN 1097-7	2.71 ± 0.06 t/m^3
Sand Equivalent, %	UNE-EN 933-8	77 ± 2
Broken surfaces, %	UNE-EN 933-5	95 ± 1
Flakiness index, %	UNE-EN 933-3	8 ± 1

The physical tests of the leaching residue showed that the density of this material is similar to that of commercial or conventional aggregates, 2.65 t/m^3. In turn, the sand equivalent test showed that the percentage of colloidal particles in the residue is adequate, so there should be no subsequent expansivity problems due to clayey particles. The percentage of broken surfaces in the coarse aggregate of the leaching residue, as well as the flakiness index, showed the quality of this material for use in bituminous mixtures, even more so in this type of mixture with discontinuous grading, in which the compressive loads are mainly supported by friction between aggregates.

On the other hand, and due to the fact that the aggregate in the bituminous mix is an essential element for resisting the tensile loads of traffic, the different tests established by the regulations for the quantification of the mechanical properties of the leaching residue were carried out. The mechanical properties tests are listed in Table 7.

Table 7. Mechanical properties of the leaching residue.

Test	Standard	Value/Unit
Resistance to fragmentation, %	UNE-EN 1097-2	18 ± 1
Resistance to freezing and thawing, %	UNE-EN 1367-1	1.121 ± 0.016
Polished stone value	UNE-EN 1097-8	57 ± 1

The resistance to fragmentation test of the leaching residue showed excellent mechanical properties. Therefore, this result ensures that there will be no subsequent breakage of the material in the bituminous mix and, consequently, no undue compaction due to the constant passage of vehicles. On the other hand, the test of resistance to freezing and thawing cycles has shown an excellent resistance of the leaching residue to thermal fatigue, comparable to the results obtained for a high-quality siliceous aggregate. Finally, it should be remembered that the irregular surface of the aggregate must be maintained over time in order to provide the road with sufficient roughness for friction between the tire and the pavement. This characteristic was evaluated through the polished stone value test. This test showed that the leaching residue has a high resistance to the continuous passage of vehicles and that its microtexture characteristics are preserved over time.

Consequently, based on the results of the chemical, physical, and mechanical characterization of the leaching residue, it can be stated that the material has suitable properties for use in bituminous mixtures, even for important traffic roads. This is directly due to the nature of the leaching residue as it is directly derived from a siliceous rock.

3.5. Bituminous Mixtures Conforming and Testing

Once the suitability of the leaching residue for use in bituminous mixtures had been verified, we proceeded to the development of discontinuous grading mixtures with bitumen emulsion. To this end, in the methodology, we defined the grading curve of the bituminous mix, this being the intermediate of the grading envelope detailed in the standards.

First, the compatibility of the medium cationic emulsion (C60BF3 MBA) with the leaching residue was evaluated. For this purpose, a coating or adhesion test was carried out, in which the aggregates with the fixed grading were mixed with the corresponding emulsion percentages. The images of the coating test for the leaching residue in combination with the bitumen emulsion are shown in Figure 8.

Figure 8. Coating test with the leaching residue and the C60BF3 MBA emulsion. (**a**) Dry coating test and (**b**) coating test after water irrigation.

The coating test showed that the coverage of the leaching residue particles by the bitumen emulsion was quite good in all cases. Therefore, it can be stated that the used bitumen emulsion C60BF3 MBA is compatible with the leaching residue and can be used for conforming bituminous mixtures.

After the coating test was performed and the material compatibility was conformed, 12 test samples were manufactured for each emulsion point (5.83%, 6.25%, 6.67%, 7.08%, 7.50%, and 7.92%). The test samples were conformed according to the procedure detailed in the methodology, subsequently, undergoing the curing process and continuously observing that no binder drainage occurred during this time.

In this case, the bituminous mix with an emulsion percentage of 7.92% on aggregate produced binder drainage during the curing process. Consequently, this bituminous mix was rejected for use as this binder drainage manifested the inability of the aggregate to absorb the percentage of emulsion added.

With the test samples of the families of bituminous mixtures manufactured with the viable emulsion percentages (5.83%, 6.25%, 6.67%, 7.08%, and 7.50%), we proceeded to determine the physical properties of the bituminous mixtures. The first of the physical tests was the maximum density of the bituminous mixtures, with the results shown in Figure 9.

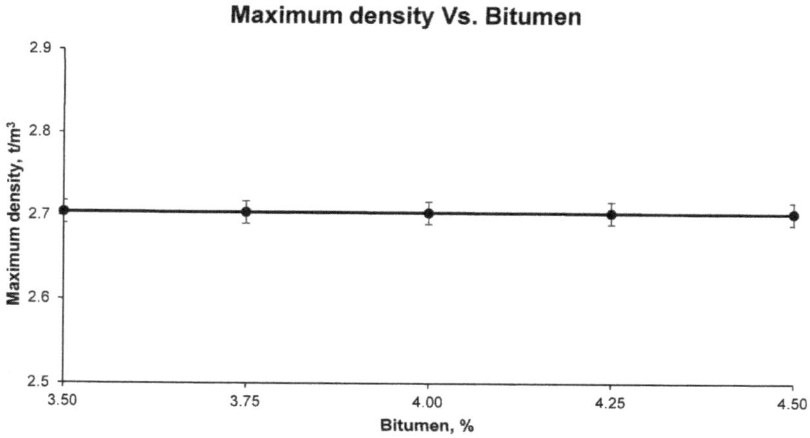

Figure 9. Maximum density of bituminous mixtures conforming with leaching residue and increasing percentages of bitumen emulsion.

The maximum density of the different families of bituminous mixes with increasing percentages of bitumen emulsion is approximately similar. However, it can be seen that increasing the percentage of emulsion makes it possible to decrease the maximum density. This is mainly due to the fact that the incorporated binder has a much lower density than the aggregate.

In turn, the bulk density of the conformed bituminous mixtures was calculated according to the standard detailed in the methodology. The results of this test are shown in Figure 10.

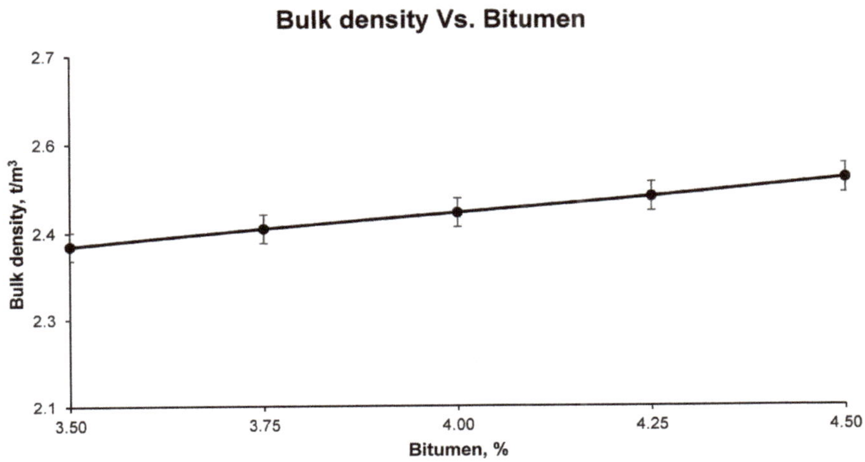

Figure 10. Bulk density of bituminous mixtures conforming with leaching residue and increasing percentages of bitumen emulsion.

The bulk density of the different bituminous mixtures, and in contrast to the maximum density, increased as the percentage of bitumen emulsion increased. It should be noted that the bulk density is the density of the bituminous mixture taking into account the internal air void content. Therefore, an increase in bulk density reflects a decrease in voids in the mix, which is to be expected if one considers that a higher percentage of emulsion will provide a higher compactability of the mix.

The last of the physical properties measured in the bituminous mixtures was the voids content. This physical variable, which is essential for the characterization of a bituminous mix, depends directly on the maximum density and bulk density. The void contents of the different families of bituminous mixtures are shown in Figure 11.

The void content in the bituminous mixes, as expected from the results of the bulk density, decreased as the percentage of bitumen emulsion increased. However, in all families of bituminous mixes, a fairly high void content was obtained in the mix, which was capable of providing the bituminous mixes with very interesting properties. In particular, a high void content in the mix is capable of producing good drainage of rainwater from the pavement, thus, preventing vehicles from slipping; it ensures a good macrotexture of the pavement, increasing safety when braking; and it even provides absorption of the noise caused by contact between the tire and the pavement, providing a comfortable and safe road surface.

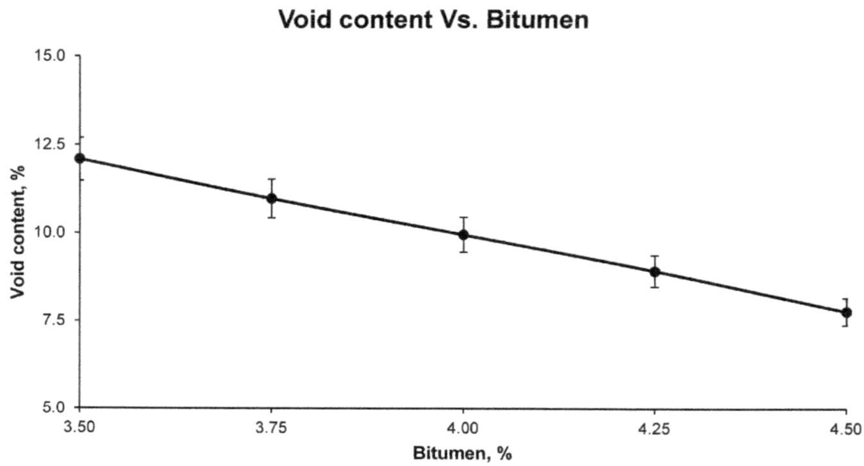

Figure 11. Void content of bituminous mixtures conforming with leaching residue and increasing percentages of bitumen emulsion.

Once the physical properties of the bituminous mixes had been analyzed, the mechanical characteristics of the mixes were determined. It should be borne in mind that a high void content in the mix favors beneficial characteristics for the pavement, however, the mastic that surrounds the aggregates must be of sufficient quality to prevent the bituminous mix from fracturing due to tensile loads or even the aggregates being torn out. Therefore, the test that best assesses the behavior of the bituminous mix and the quality of the mastic to hold the aggregates together is the particle loss test. This test was carried out in this research with and without immersion in water in order to evaluate the effect of water on the bituminous mixes, with the results shown in Figure 12.

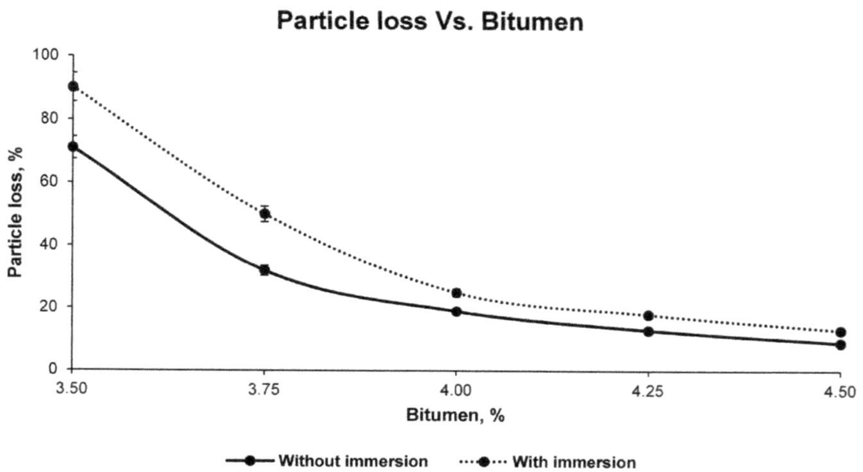

Figure 12. Particle loss test with and without water immersion of bituminous mixtures conforming with leaching residue and increasing percentages of bitumen emulsion.

The particle loss test showed a higher wear of the bituminous mixes with water immersion, which is to be expected from developing under extreme conditions. However, the difference between the wear of the mixes with and without immersion decreased as the percentage of emulsion increased. This reflects the fact that a higher quality mastic is being formed as the percentage of residual bitumen increases. It should also be noted that Spanish regulations set a maximum percentage of particle loss of 25% without immersion. Therefore, bituminous mixtures with residual bitumen percentages of 3.5% and 3.75% do not comply with the established limitations and are considered to be rejectable. The rest of the families comply with the requirements of the standard, even for test samples with water immersion.

It should be noted that a high percentage of emulsion will create a high percentage of residual bitumen after curing, therefore the mastic developed will be able to better wrap the aggregates and hold them together. However, a high percentage of bitumen can develop plastic deformations at high temperatures, i.e., variations in the physical properties of the bituminous mixes. This defect renders the infrastructure completely unusable for use, as subsidence, bleeding, etc. can occur on the pavement. It must, therefore, be monitored and controlled to ensure that this high percentage of bitumen, which is so necessary for maintaining the joint between the particles, does not develop plastic deformations. The evaluation of this problem was carried out with the Marshall test. The results of the Marshall test for conformed bituminous mixtures are shown in Figure 13.

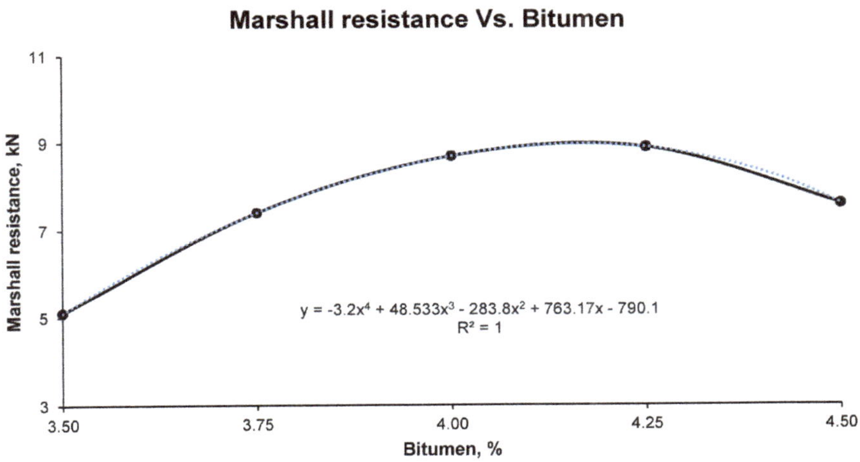

Figure 13. Marshall resistance of bituminous mixtures conforming with leaching residue and increasing percentages of bitumen emulsion.

The Marshall resistance as shown in the figure above increased up to a maximum and then decreased for the maximum percentage of residual bitumen, 4.5%. Spanish regulations stipulate that Marshall resistance must be higher than 7.5 kN; therefore, bituminous mixes with residual bitumen percentages of 4%, 4.25%, and 4.5% are acceptable, the others being rejected by this test.

It should be noted that the Marshall test results clearly show a maximum between the residual bitumen percentage of 4% and 4.25%. This maximum percentage of Marshall resistance was calculated mathematically through the expression shown in the figure, obtaining the percentage of residual bitumen and bitumen emulsion that developed this maximum stability. This percentage of emulsion will be considered as optimum, provided that the results of the other tests mentioned are acceptable according to the regulations.

In order to verify the properties and characteristics of the optimum bitumen emulsion corresponding to the maximum Marshall stability, test samples with the determined grading and this percentage of emulsion were carried out again, with the results shown in Table 8.

Table 8. Results of the physical and mechanical tests of the test samples conforming with the optimum percentage of bitumen emulsion.

Test	Standard	Value/Unit
Residual bitumen, %	-	4.17
Bitumen emulsion, %	-	6.95
Maximum density, t/m^3	UNE-EN 12697-5	2.703 ± 0.069
Bulk density, t/m^3	UNE-EN 12697-6	2.452 ± 0.062
Void content, %	UNE-EN 12697-8	9.27 ± 0.23
Particle loss without immersion, %	UNE-EN 12697-17	14 ± 1
Particle loss with immersion, %	UNE-EN 12697-17 / NLT-362/92	18 ± 1
Marshall stability, kN	UNE-EN 12697-34	8.99 ± 0.23

The results of the AF12 bituminous mix with discontinuous grading with bitumen emulsion conformed with leaching residue and with the optimum percentage of emulsion according to the Marshall test showed very suitable properties for medium traffic roads. The high void content in the mixture will be able to provide a comfortable and safe road surface, and the particle loss and Marshall resistance values show the mechanical strength of the mixture conformed with a high percentage of waste.

4. Conclusions

The tests mentioned in the methodology allowed us to obtain a series of partial conclusions that lead to the verification of the initial hypothesis of this research.

On the one hand, it should be pointed out that the element copper was present in the three waste dumps analyzed, the San Andrés waste dump being the one with the highest percentage. The detailed waste dump was leached with different sulphuric acid solutions after milling, demonstrating that the highest percentage of copper recovery produced was 80%, with the 1 and 2 mol sulphuric acid solutions. At the same time, the leaching waste was chemically characterized, determining its usefulness for use in bituminous mixtures according to the regulations. The bituminous mixture formed with the leaching waste and a bitumen emulsion percentage of 6.95% was the one with the best mechanical properties, with a residual bitumen percentage of 4.17%. This bituminous mix had a Marshall stability of 8.99 KN and a particle loss before and after immersion of 14% and 18%. Therefore, this bituminous mixture has physical and mechanical properties suitable for use on medium and low traffic roads.

Consequently, and according to the detailed partial conclusions, it can be stated that the San Andrés waste dump has an interesting material for the extraction of copper by means of hydrometallurgical techniques carried out at atmospheric pressure, ambient temperature, and acidic media, recovering a high percentage of the copper existing in it. In addition, the leaching residue after the hydrometallurgical process showed physical, chemical, and mechanical properties suitable for use as an aggregate in bituminous mixtures of discontinuous grading and bitumen emulsion. In short, this research has developed a new industrial process from a waste product with no use or value at present, in which economically interesting elements such as copper and aggregates for construction are obtained. In this way, a new environmental hydrometallurgy is developed for the rehabilitation, strengthening, and development of the city of Linares, based on circular engineering.

Author Contributions: Conceptualization, F.A.C.-I., A.M.C., J.M.T.-S. and J.S.-M.; methodology, F.A.C.-I., A.M.C., J.M.T.-S. and J.S.-M.; software, J.M.T.-S. and J.S.-M.; validation, F.A.C.-I. and A.M.C.; formal analysis, F.A.C.-I. and A.M.C.; investigation, J.M.T.-S. and J.S.-M.; resources, F.A.C.-I.; data curation, A.M.C.; writing—original draft preparation, J.S.-M.; writing—review and editing, J.M.T.-S.; visualization, J.M.T.-S.; supervision, F.A.C.-I.; project administration, J.S.-M.; funding acquisition, F.A.C.-I. All authors have read and agreed to the published version of the manuscript.

Funding: This research received no external funding.

Institutional Review Board Statement: Not applicable.

Informed Consent Statement: Not applicable.

Data Availability Statement: Data are contained within the article.

Acknowledgments: In Technical and human support provided by CICT of Universidad de Jaén (UJA, MINECO, Junta de Andalucía, FEDER) is gratefully acknowledged.

Conflicts of Interest: The authors declare no conflict of interest.

References

1. Keenan, J.; Holcombe, S. Mining as a temporary land use: A global stocktake of post-mining transitions and repurposing. *Extr. Ind. Soc.* **2021**, *8*, 100924.
2. Beylot, A.; Villeneuve, J. Assessing the national economic importance of metals: An Input-Output approach to the case of copper in France. *Resour. Policy* **2015**, *44*, 161–165. [CrossRef]
3. Souza-Filho, P.W.M.; Cavalcante, R.B.L.; Nascimento, W.R.; Nunes, S.; Gastauer, M.; Santos, D.C.; Silva, R.O.; Sahoo, P.K.; Salomão, G.; Silva, M.S.; et al. The sustainability index of the physical mining Environment in protected areas, Eastern Amazon. *Environ. Sustain. Indic.* **2020**, *8*, 100074. [CrossRef]
4. Izatt, R.M.; Izatt, S.R.; Bruening, R.L.; Izatt, N.E.; Moyer, B.A. Challenges to achievement of metal sustainability in our high-tech society. *Chem. Soc. Rev.* **2014**, *43*, 2451–2475. [CrossRef]
5. Yıldız, T.D. Waste management costs (WMC) of mining companies in Turkey: Can waste recovery help meeting these costs? *Resour. Policy* **2020**, *68*, 101706. [CrossRef]
6. Guzmán-Martínez, F.; Arranz-González, J.C.; Ortega, M.F.; García-Martínez, M.J.; Rodríguez-Gómez, V. A new ranking scale for assessing leaching potential pollution from abandoned mining wastes based on the Mexican official leaching test. *J. Environ. Manag.* **2020**, *273*, 111139. [CrossRef]
7. Singh, S.; Sukla, L.B.; Goyal, S.K. Mine waste & circular economy. *Mater. Today Proc.* **2020**, *30*, 332–339.
8. Hefni, M.; Ahmed, H.A.M.; Omar, E.S.; Ali, M.A. The potential re-use of saudi mine tailings in mine backfill: A path towards sustainable mining in saudi arabia. *Sustainability* **2021**, *13*, 6204. [CrossRef]
9. Gautam, P.K.; Kalla, P.; Nagar, R.; Agrawal, R.; Jethoo, A.S. Laboratory investigations on hot mix asphalt containing mining waste as aggregates. *Constr. Build. Mater.* **2018**, *168*, 143–152. [CrossRef]
10. Orihuela, J.C. The environmentalization of mining in Colombia, Chile, and Peru: A comparative analysis of green state formation. *Extr. Ind. Soc.* **2020**. [CrossRef]
11. Romero, A.; González, I.; Galán, E. Estimation of potential pollution of waste mining dumps at Peña del Hierro (Pyrite Belt, SW Spain) as a base for future mitigation actions. *Appl. Geochem.* **2006**, *21*, 1093–1108. [CrossRef]
12. Azapagic, A. Developing a framework for sustainable development indicators for the mining and minerals industry. *J. Clean. Prod.* **2004**, *12*, 639–662. [CrossRef]
13. Rey, J.; Martínez, J.; Hidalgo, M.C.; Rojas, D. Heavy metal pollution in the Quaternary Garza basin: A multidisciplinary study of the environmental risks posed by mining (Linares, southern Spain). *Catena* **2013**, *110*, 234–242. [CrossRef]
14. Álvarez, M.L.; Méndez, A.; Rodríguez-Pacheco, R.; Paz-Ferreiro, J.; Gascó, G. Recovery of zinc and copper from mine tailings by acid leaching solutions combined with carbon-based materials. *Appl. Sci.* **2021**, *11*, 5166. [CrossRef]
15. Terrones-Saeta, J.M.; Suárez-Macías, J.; Linares Del Río, F.J.; Corpas-Iglesias, F.A. Study of copper leaching from mining waste in acidic media, at ambient temperature and atmospheric pressure. *Minerals* **2020**, *10*, 873. [CrossRef]
16. Li, L.; Pan, D.; Li, B.; Wu, Y.; Wang, H.; Gu, Y.; Zuo, T. Patterns and challenges in the copper industry in China. *Resour. Conserv. Recycl.* **2017**, *127*, 1–7. [CrossRef]
17. Wang, M.; Chen, W.; Zhou, Y.; Li, X. Assessment of potential copper scrap in China and policy recommendation. *Resour. Policy* **2017**, *52*, 235–244. [CrossRef]
18. Copper: World Mine Production, By Country. Available online: https://www.indexmundi.com/en/commodities/minerals/copper/copper_t20.html (accessed on 19 May 2020).
19. Kowalczuk, P.B.; Bouzahzah, H.; Kleiv, R.A.; Aasly, K. Simultaneous leaching of seafloor massive sulfides and polymetallic nodules. *Minerals* **2019**, *9*, 482. [CrossRef]
20. Watling, H.R. Chalcopyrite hydrometallurgy at atmospheric pressure: 1. Review of acidic sulfate, sulfate-chloride and sulfate-nitrate process options. *Hydrometallurgy* **2013**, *140*, 163–180. [CrossRef]
21. Meshram, P.; Prakash, U.; Bhagat, L.; Abhilash; Zhao, H.; van Hullebusch, E.D. Processing of Waste Copper Converter Slag Using Organic Acids for Extraction of Copper, Nickel, and Cobalt. *Minerals* **2020**, *10*, 290. [CrossRef]
22. Jones, D.L. CESL copper process. In Proceedings of the ALTA 1996 Copper Hydrometallurgy Forum, Brisbane, Australia, 14–15 October 1996; Volume 24.
23. Chen, J.; Wang, Z.; Wu, Y.; Li, L.; Li, B.; Pan, D.; Zuo, T. Environmental benefits of secondary copper from primary copper based on life cycle assessment in China. *Resour. Conserv. Recycl.* **2019**, *146*, 35–44. [CrossRef]

24. Pelino, M. Recycling of zinc-hydrometallurgy wastes in glass and glass ceramic materials. *Waste Manag.* **2000**, *20*, 561–568. [CrossRef]
25. Avilés-Palacios, C.; Rodríguez-Olalla, A. The Sustainability of Waste Management Models in Circular Economies. *Sustainability* **2021**, *13*, 7105. [CrossRef]
26. Pomponi, F.; Stephan, A. Water, energy, and carbon dioxide footprints of the construction sector: A case study on developed and developing economies. *Water Res.* **2021**, *194*, 116935. [CrossRef] [PubMed]
27. Lee, J.; Madanat, S. Optimal policies for greenhouse gas emission minimization under multiple agency budget constraints in pavement management. *Transp. Res. Part D Transp. Environ.* **2017**, *55*, 39–50. [CrossRef]
28. Terrones-Saeta, J.M.; Iglesias-Godino, F.J.; Corpas-Iglesias, F.A.; Martínez-García, C. Study of the Incorporation of Ladle Furnace Slag in the Manufacture of Cold In-Place Recycling with Bitumen Emulsion. *Materials* **2020**, *13*, 4765. [CrossRef]
29. Terrones-Saeta, J.M.; Suárez-Macías, J.; Iglesias-Godino, F.J.; Corpas-Iglesias, F.A. Development of Porous Asphalt with Bi-tumen Emulsion, Electric arc Furnace Slag and Cellulose Fibers for Medium Traffic Roads. *Minerals* **2020**, *10*, 872. [CrossRef]

metals

MDPI

Article

Leaching of Zinc for Subsequent Recovery by Hydrometallurgical Techniques from Electric Arc Furnace Dusts and Utilisation of the Leaching Process Residues for Ceramic Materials for Construction Purposes

Juan María Terrones-Saeta *, Jorge Suárez-Macías, Evaristo Rafael Moreno-López and Francisco Antonio Corpas-Iglesias

Research Group TEP-222 "Materials and Mining Engineering", Higher Polytechnic School of Linares, Scientific and Technological Campus of Linares, University of Jaen, 23700 Linares, Spain; jsuarez@ujaen.es (J.S.-M.); erml0001@red.ujaen.es (E.R.M.-L.); facorpas@ujaen.es (F.A.C.-I.)

* Correspondence: terrones@ujaen.es

Abstract: Steel is one of the most widely used materials in the past and today. Various techniques are used to recycle this material, including the electric arc furnace. This process has several advantages, but it also has a major disadvantage, namely, the generation of waste such as electric arc furnace dusts. Electric arc furnace dusts are classified as hazardous waste due to their high percentage of heavy metals, including zinc. Consequently, in the present research, the leaching of zinc for recovery with sulfuric acid solutions at ambient temperature and atmospheric pressure is evaluated, as well as the reuse of the leaching process residue as a raw material for ceramic materials. The sulfuric acid solutions were 0.125, 0.25, 0.5, and 1 molar, using clay for ceramic conforming and percentages of the leaching residue from 0–50%. The results showed that the optimum solution was 1 molar sulfuric acid, recovering all the zinc in the sample in 36 h. Furthermore, it was found that the clay-conformed ceramics with less than 40% leaching residue showed acceptable physical and mechanical properties according to standards. Therefore, this research develops a new environmental hydrometallurgy in which metallic elements of interest are valorized and the production of waste is avoided, reducing the deposition of hazardous waste in landfills and the extraction of raw materials for the manufacture of construction materials.

Keywords: hydrometallurgy; metallic elements recovery; mining waste; ceramic; construction materials; circular economy; sustainability

Citation: Terrones-Saeta, J.M.; Suárez-Macías, J.; Moreno-López, E.R.; Corpas-Iglesias, F.A. Leaching of Zinc for Subsequent Recovery by Hydrometallurgical Techniques from Electric Arc Furnace Dusts and Utilisation of the Leaching Process Residues for Ceramic Materials for Construction Purposes. *Metals* **2021**, *11*, 1603. https://doi.org/10.3390/met11101603

Academic Editors: Jean François Blais and Anna H. Kaksonen

Received: 22 August 2021
Accepted: 30 September 2021
Published: 9 October 2021

1. Introduction

Steel is a material that was widely used in the past and in the present, as its economy and resistance make it difficult to replace, even today [1]. This material also has the quality of being totally recyclable [2], being able, through different processes, to treat the scrap to obtain a new steel [3]. Among the processes developed for this purpose, the most widely used is the so-called electric arc furnace [4,5].

The electric arc furnace established itself as a process for obtaining steel from scrap due to its simplicity, novelty and optimization [6]. However, it has a series of disadvantages that must also be considered, one of the main ones being the generation of waste [7]. Among these wastes, and due to the fact that the electric arc furnace process is carried out at high temperatures, are electric arc furnace dusts [8]. This waste derives from the capture of particles by means of electrostatic precipitators, baghouses, and cyclones, from the gases produced in the electric arc furnace [9]. The particles retained in the gases reach 99.9% [10], deriving from the most volatile elements found in the scrap, such as zinc. These electric arc furnace dusts are considered by the European Waste Catalogue as a hazardous waste [11], producing about 15–20 kg per ton of final steel [12]. Consequently, it can be stated that this waste poses a significant environmental problem if it is not properly treated. The main elements of these electric arc furnace dusts are zinc, iron, copper, lead, chromium, manganese, silicon, cadmium, nickel, and calcium [13,14]. Obviously, due to the high temperatures of the electric arc furnace, all of the above elements are in the form of oxides.

Among all the chemical elements mentioned, zinc is the one found in the highest proportion. This chemical element is currently highly valued. So much so that China's consumption in 2000 was 1.4 million tons, exceeding the consumption of the United States, and in 2014 this figure quadrupled, with 6.42 million tons being consumed in China [15]. Therefore, zinc is an element that is currently being revalued, requiring new sources of supply for its procurement and consumption.

Different processes, either pyrometallurgical or hydrometallurgical [16–19], can be used to concentrate and obtain zinc from waste or ores. Hydrometallurgy, based on the leaching of zinc in solutions for its subsequent concentration, has advantages over pyrometallurgical techniques. One of the advantages is that it requires less investment in equipment, has lower energy consumption, is more adaptable to variations in the chemical composition of the raw material being treated, and the efficiency of the process is less dependent on other chemical elements [20]. In addition, leaching of zinc oxides in acidic solutions of relatively low molarity is usually an efficient and fast process. Consequently, the leaching of electric arc furnace dusts by hydrometallurgical methods in acidic solutions at atmospheric pressure and ambient temperature appears to be an attractive solution to the problem of electric arc furnace dust production [21].

However, leaching for hydrometallurgical techniques also has a number of disadvantages. Among them is the production of wastes that can be quite harmful to the environment [22]. Therefore, if environmental hydrometallurgy is desired, not only should zinc be extracted from electric arc furnace dusts, but also the waste from the leaching process should be used as a raw material for new materials. This avoids the extraction of raw materials, avoids the pollution caused by landfill site the leaching residues, and achieves a retention of the polluting elements of the leaching residue in the new material.

In this research, after the zinc leaching process of electric arc furnace dusts in acidic media, the use of the leaching residue in ceramic materials for bricks was proposed. The construction sector is one of the most polluting sectors in existence today [23]. This is mainly due to the fact that large quantities of materials are produced, thus consuming a large amount of raw materials [24]. Furthermore, the ceramic element is capable of retaining, as was corroborated by different authors, the potentially polluting chemical elements of the leaching residue in its ceramic matrix [25,26]. Consequently, ceramic is one of the best materials for the use of the leaching residue without the subsequent production of environmental problems.

In short, in the present research, the electric arc furnace dusts are chemically characterized to evaluate their suitability for leaching, and subsequently, these dusts are leached in sulfuric acid solutions with different molarities and at different times, obtaining the optimum solution and time for the extraction of the highest percentage of zinc. With the leaching residue, different families of ceramics are conformed and the physical and mechanical properties of these families are evaluated. In this way, a ceramic is obtained that retains the leaching of polluting elements from the leaching residue. This research is therefore based on the new circular economy, since the elements of interest are extracted from a waste produced by the industry and the leaching residue is reused for new ceramic materials, creating a process in which the waste is null and reduces the environment affection.

2. Materials and Methods

This section describes the main materials used in this research, as well as the tests carried out to obtain a clear and scientific methodology to assess the possible leaching of zinc from electric arc furnace dusts and its subsequent use in ceramics.

2.1. Materials

The materials used in this research are mainly electric arc furnace dusts, distilled water and sulfuric acid for zinc leaching, as well as clay for conforming ceramics with the leaching residue. These materials are described in detail below.

2.1.1. Electric Arc Furnace Dust

The steel dust used in this research, as mentioned above, corresponds to the material that is deposited in the fume filtering installations during the melting process in the electric arc furnace, and the sample of this waste was taken directly from the producing steel company located in Andalucía, Spain.

The electric arc furnace dust samples were taken in a representative manner according to the regulations, sampling for months to verify that the physical and chemical characteristics of the waste did not vary over time with different production batches.

The samples after collection were dried at a temperature of 105 ± 5 °C to remove any humidity they might contain. In this way, unnecessary variables that could cloud the results of the tests mentioned in the methodology were avoided. Due to the small grading of the electric arc furnace dusts, this material was used directly for the characterization and leaching tests, without any intermediate milling process.

2.1.2. Sulfuric Acid and Water

The sulfuric acid used in this investigation has a purity of 96%, with a molar mass of 98.08 g/mol. This acid was diluted with different concentrations of distilled water to obtain sulfuric acid solutions of different molarities. In this way, it was possible to study the concentration of zinc produced with different acid solutions at atmospheric pressure and in different periods of time. The distilled water used has a colorless liquid appearance, a density at 20 °C of 0.995 g/cm^3, a pH of 7, a conductivity in µS/cm of 2.5, a silica percentage of 0.003 mg/L, and a hardness measured in mg/L $CaCO_3$ of 0.3.

2.1.3. Clay

The waste obtained after the leaching process of the electric arc furnace dusts was used as an additive or raw material for the production of brick ceramics. Consequently, the use of clay is necessary for the conforming of the different families of ceramics.

The clay used corresponds to that extracted in the province of Jaén, Spain. This type of red clay was used for decades for the manufacture of bricks for construction, and there are currently several production industries in the area that produce and market this material.

The red clay used has a very small particle size, similar to electric arc furnace dust. This clay was dried at a temperature of 105 ± 5 °C, as were the steel dust, for its subsequent characterization and use in ceramics. No milling process was carried out on the clay, but the sample of this material was sieved through a 0.25 mm sieve.

2.2. Methodology

The methodology followed in this research is based on a series of logically ordered tests to verify the main objective of this research, leaching zinc from electric arc furnace dusts with sulfuric acid solutions at ambient temperature and atmospheric pressure, as well as reusing the leaching residue for the shaping of ceramics that avoid environmental pollution from the leaching process waste.

To do this, first of all, the electric arc furnace dusts were chemically characterized. In this way, it was possible to identify the percentage of zinc in the sample, as well as the existence of other elements hazardous to the leaching process. Subsequently, the leaching process was carried out with different sulfuric acid solutions, measuring the fundamental leaching parameters every 24 h. The results obtained were used to select the optimum solution and leaching times for the recovery of the highest percentage of zinc, and the leaching residue was chemically analyzed after filtering. With the leaching residue, different families of ceramics with different percentages of waste and clay were conformed. These ceramics were physically and mechanically characterized, determining the maximum percentage of incorporation of the leaching residue in ceramics.

In the following sections, the tests corresponding to each phase of this research are detailed in more detail.

2.2.1. Chemical Characterization of Electric Arc Furnace Dusts

The electric arc furnace dusts after drying were chemically characterized to determine the percentage of zinc in them, as well as to evaluate the existence of certain chemical elements that could impair the leaching process.

For the chemical characterization, the first of the tests carried out was elemental analysis, detecting the percentage of carbon, hydrogen, nitrogen, and sulfur in the sample. This test was performed with LECO's TruSpec Micro (TruSpec Micro, LECO, St. Joseph, MI, USA).

Subsequently, the loss-on-ignition test of the electric arc furnace dusts was carried out at a temperature of 950 ± 5 °C. Therefore, this test quantifies the existence of organic matter, the existence of carbonates, the existence of volatile elements, and even the transformation of certain chemical compounds.

Finally, to identify and quantify the chemical elements with the highest atomic weight, the X-ray fluorescence test was carried out. This test was performed with the ADVANT'XP+ commercial equipment (ADVANT'XP+, Thermo Fisher, Waltham, MA, USA).

2.2.2. Leaching of Electric Arc Furnace Dusts in Sulfuric Acid Solutions at Ambient Temperature and Atmospheric Pressure

For the recovery of the zinc present in the electric arc furnace dusts, leaching was carried out in acidic media, at atmospheric pressure, ambient temperature (25 ± 1 °C), and in immersion with constant agitation. This leaching system was selected because it is the most suitable for the waste, since electric arc furnace dusts have a small particle size and a high percentage of zinc.

The process consists of leaching the zinc present in the sample, which is in the form of ZnO and $ZnFe_2O_4$, using different H_2SO_4 sulfuric acid solutions with different molarities. The selected molarities of sulfuric acid were 0.125, 0.25, 0.5, and 1 molar. These molarities were selected for different reasons; on the one hand, it was not desirable to use high molarities of sulfuric acid to not create a highly toxic leaching residue; on the other hand, the chemical adjustment determined that these molarities were sufficient to leach the existing zinc in the sample, and finally, it was intended to reduce the environmental impact of the process that would occur if high molarities of sulfuric acids were used.

The sample of electric arc furnace dusts was 5000 ± 5 milligrams. This sample was mixed in 100 ± 1 mL of the corresponding solution and stirred for the 5 days of the test. The pH was measured every 24 h and the electrical conductivity of the solution was determined. In addition, a sample of 5 mL was taken every 24 h for analysis by inductively coupled plasma mass spectrometry (7900, Agilent, Santa Clara, CA, USA). In this way, the variation of the zinc concentration in the leachate could be obtained, as well as the modification of the pH and conductivity of the solution.

The results of the zinc concentration in the leachate made it possible to determine the sulfuric acid solution that produced the greatest recovery of the zinc in the sample, as well as the time in which this recovery occurred. Therefore, the optimum dilution and time to recover the zinc in the electric arc furnace dusts was determined.

2.2.3. Chemical Characterization of the Leaching Residue

Once the optimum sulfuric acid solution was selected that produced the highest zinc recovery in the optimum time, large quantities of electric arc furnace dust were leached under the detailed conditions of solid/liquid, temperature, times, etc.

The leaching residue were filtered with filter paper, obtaining a solid waste that was subsequently dried at a temperature of 105 ± 5 °C for 24 h. This leaching residue was chemically characterized to determine its hazardousness, as well as the chemical elements that could be detrimental to the ceramic conformed with them.

The chemical characterization of the waste was similar to that carried out for electric arc furnace dusts. Consequently, the elemental analysis test was carried out to determine the percentage of carbon, nitrogen, hydrogen, and sulfur in the sample with LECO's TruSpec Micro commercial equipment (TruSpec Micro, LECO, St. Joseph, MI, USA). Subsequently, the loss on ignition test was performed at 950 ± 5 °C, and the X-ray fluorescence test, to quantify the elements of higher atomic weight in the leaching residue, with the ADVANT'XP+ commercial equipment (ADVANT'XP+, Thermo Fisher, Waltham, MA, USA).

2.2.4. Conformed of Ceramic Materials for Bricks with the Leaching Residue of Electric Arc Furnace Dusts

Once the leaching residue was analyzed, different families of ceramics with different percentages of clay and waste were conformed. The initial family was composed only of clay, therefore, the ceramic formed was a conventional material. This family was made to compare the physical and mechanical properties of the ceramics conformed with the leaching residue. Subsequent families were formed with increasing percentages of 10% leaching residue. Table 1 shows the various families of ceramics evaluated, as well as the percentage of clay and leaching residue for each.

Table 1. Families of ceramic samples conformed with clay and leaching residue.

Ceramic Family	% of Clay	% of Leaching Residue
RL0	100	0
RL1	90	10
RL2	80	20
RL3	70	30
RL4	60	40
RL5	50	50

For the production of the detailed ceramic samples, the dry mass of clay and leaching residue was mixed according to the family. Once the mixture was homogenized, 10% water was added to facilitate the mixing of the mixture and allow for better compaction. The detailed mixture was poured into a metal matrix of dimensions 60 × 30 mm, subsequently applying a compression pressure of 25 MPa. The use of this percentage of water and the detailed pressure was selected because of the similarity of results between the material conformed in the laboratory and that obtained in industry. Finally, the sample was extracted and dried at a temperature of 105 ± 5 °C for 24 h. A total of 6 samples were made for each family to obtain statistically reliable results.

A sintering process is essential for the shaping of ceramics. In this research, the sintering process consisted of raising the temperature of the dried specimens to 950 ± 5 °C, using a temperature ramp of 4 °C/min. The samples of all the families conformed according to the detailed procedure were subjected to the physical and mechanical tests detailed below to evaluate their quality.

The first physical tests carried out were the determination of the linear shrinkage and the mass loss that occurs in the ceramic before and after the sintering process. Subsequently, capillary water absorption (UNE-EN 772-11 standard), cold water absorption (UNE-EN 772-21 standard), and determination of open porosity and bulk density (UNE-EN 772-4) tests were carried out.

Once the physical properties were determined, the mechanical strength of the different families of samples was evaluated. The compressive strength test was carried out for each of the specimens of each family according to the UNE-EN 772-1 standard. This test is essential, since the element for which the ceramic material is intended is for bricks and must have a compressive strength greater than 10 MPa according to the standard.

3. Results and Discussions

3.1. Chemical Characterization of Electric Arc Furnace Dusts

The first of the tests carried out for the chemical characterization of the electric furnace steel dust was the elemental analysis test. The elemental analysis test shows that the percentage of nitrogen (0.027 $\pm$ 0.001) and sulfur (0.101 $\pm$ 0.003) in the sample is quite low. Consequently, it is corroborated that the chemical composition of the electric furnace dusts is mainly composed of oxides, and that there are no more volatile elements such as nitrogen that could complicate the leaching process. At the same time, the low percentage of carbon (2.053 $\pm$ 0.048) and hydrogen (0.270 $\pm$ 0.009) shows that the dusts have practically no percentage of organic matter. On the other hand, the carbon in the sample can be derived directly from the carbonation of the oxides, as well as the percentage of hydrogen from the hydration of certain compounds.

The loss on ignition test of the electric arc furnace dusts showed a value of 14.87 $\pm$ 0.348%. This variation in mass before and after subjecting the sample to the temperature of 950 $\pm$ 5 °C may be due to the existing organic matter, carbonates, hydration of certain compounds, and even the transformation of certain chemical compounds. In this case, and on the basis of the results obtained in the other chemical tests, it could be affirmed that this mass variation corresponds to the carbon in the carbonate compounds and other volatile elements, as well as the transformation of certain chemical compounds.

The X-ray fluorescence test of the electric arc furnace dusts showed the results detailed in Table 2.

Table 2. X-ray fluorescence of electric arc furnace dusts.

Element	Wt %	Est. Error
Fe	23.12	0.16
Zn	18.83	0.17
Ca	4.78	0.09
Si	2.06	0.05
Cl	3.93	0.10
Na	2.52	0.15
Mg	1.66	0.05
K	1.42	0.05
Mn	1.18	0.05
Al	0.752	0.03
Pb	0.805	0.040
Sx	0.329	0.016
Cr	0.186	0.0093
Cu	0.194	0.0097
Px	0.0766	0.0038
Ti	0.0656	0.0033
Cd	0.0646	0.0032
Sn	0.0512	0.0061
Ni	0.0267	0.0014
W	0.0259	0.0078
Pt	0.0204	0.0059
Pd	0.0181	0.0026
Co	0.0143	0.0021
Zr	0.0135	0.0023
Ru	0.0100	0.0014
Ag	0.0104	0.0029

As can be seen, the chemical element with the highest percentage in the sample is iron, which is to be expected if we take into account that this waste comes from the filtering of gases from the electric arc furnace fed with ferrous scrap. In turn, zinc makes up a very high percentage of the electric furnace dust, and it is this zinc in the form of oxide that is to be recovered in this research through leachates. The other chemical elements calcium, silicon, chlorine, sodium, magnesium, potassium, and manganese are found in much smaller proportions, all of them being common in the waste produced in the iron and steel industry, as they are common elements found in the scrap used as raw material. The remaining chemical elements are found in much smaller proportions, and in principle, do not pose a problem for the leaching process.

3.2. Leaching of Electric Arc Furnace Dusts in Sulfuric Acid Solutions at Ambient Temperature and Atmospheric Pressure

The characterized electric arc furnace dust was subjected to the leaching process detailed in the methodology with 0.125, 0.25, 0.5, and 1 molar sulfuric acid solutions. This leaching process, carried out with a 1:20 solid/liquid ratio and with continuous agitation at 500 rpm, was evaluated to corroborate the extraction of zinc from the sample of electric arc furnace dusts. To evaluate the process, the pH, electrical conductivity and zinc concentration of the solution were measured every 24 h. These measurements, always carried out at a temperature of 25 °C, are detailed below.

The variation of the pH for the different leaching solutions at different times and according to the detailed process are shown in Figure 1.

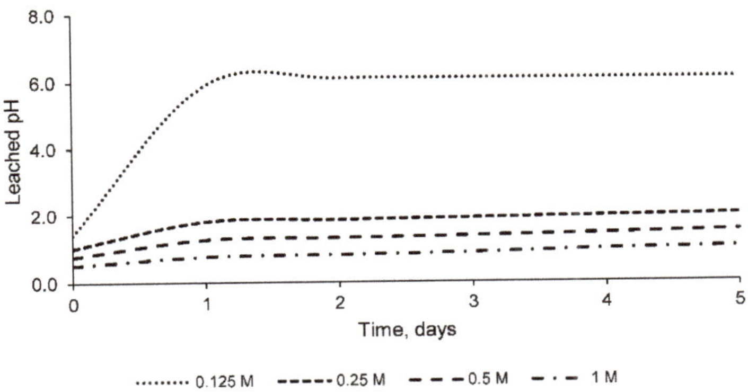

Figure 1. pH of 0.125, 0.25, 1, and 2 molar solutions of sulfuric acid at different times during leaching process with electric furnace dusts.

Figure 1 shows how the 0.125 molar solution of sulfuric acid undergoes a rapid rise in pH in the first 24 h, stabilizing at a pH of 6 over the time evaluated. This elevation of pH is obviously due to the mixing of the acidic solution with the basic electric arc furnace dust sample. Consequently, after 24 h, no further leaching of any chemical element will occur, as the solution is practically neutralized. Therefore, the leaching of the chemical elements from the electric arc furnace dusts that occurred in the 24 h is almost constant.

Something similar happens in the 0.25 molar solution, since in the first 24 h there is a considerable elevation of the pH and after this period of time the value stabilizes. However, in the 0.5 and 1 molar solutions of sulfuric acid, a different situation occurs according to the results shown in Figure 1. On the one hand, and just like the 0.125 molar and 0.25 molar solutions, the most drastic increase in pH occurs in the first 24 h. However, over the course of the days, the pH continues to rise progressively much less rapidly. Therefore, it can be estimated that there is still a process of leaching of the chemical elements from the sample of the electric furnace dusts in the following days, in contrast to the 0.125 molar solution which is saturated 24 h into the process.

In turn, the results of the electrical conductivity for the 4 sulfuric acid solutions (0.125, 0.25, 0.5, and 1 molar) are shown in Figure 2 for different times, measured at a temperature of 25 °C.

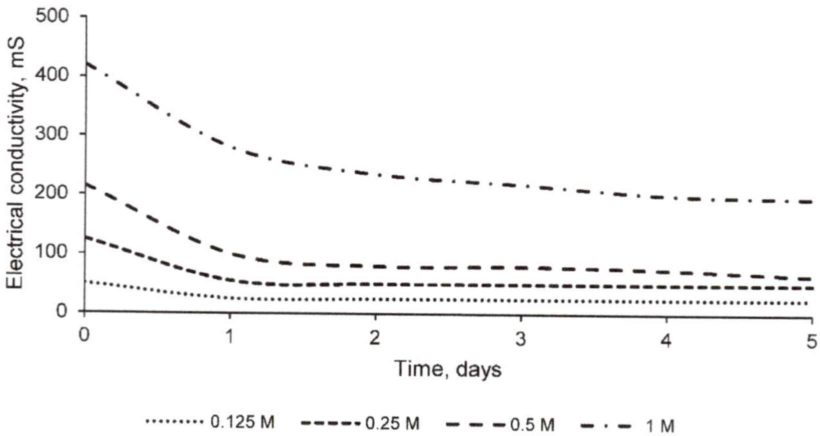

Figure 2. Electrical conductivity of 0.125, 0.25, 1, and 2 molar solutions of sulfuric acid at different times during leaching process with electric furnace dusts.

The results shown in Figure 2 of the electrical conductivity of the solution during the leaching process reflect similar conclusions to Figure 1 in relation to pH. Initially, the sulfuric acid solutions with a higher molarity have a higher electrical conductivity due to the existence of a higher percentage of free ions. The 0.125 and 0.25 molar solutions of sulfuric acid show a significant decrease in conductivity in the first 24 h, obtaining a value that is practically maintained over time. On the other hand, the 0.5 and 1 molar sulfuric acid solutions show a significant decrease in conductivity in the first 24 h but continue to experience a decrease in electrical conductivity during the test time. This shows that chemical elements continue to leach from the sample throughout the test time. However, the interesting element to obtain in the present investigation is zinc. Consequently, if a significant recovery occurred in the first hours, maintaining the time of the leaching process for longer is counterproductive, since other chemical elements are being leached that are of no interest and that also harm the subsequent process of obtaining the zinc by hydrometallurgy of the leachate.

Therefore, the determination of the of the zinc recovery rate in the different sulfuric acid solutions and at different times will establish which is the appropriate leaching time to obtain zinc, as well as the solution that provides a greater recovery of zinc from the electric arc furnace dusts. The results of this test carried out with the equipment detailed in the methodology are shown in Figure 3.

In Figure 3, the 0.125 molar sulfuric acid solution achieves a zinc recovery rate of over 30% during the first 24 h, with this ratio stabilizing over time. This fact reflects that the aforementioned solution is not capable of leaching a higher percentage of zinc, since it is almost neutralized in the first 24 h, as shown by the pH and electrical conductivity values. On the other hand, the 0.25 molar sulfuric acid solution experiences a rapid increase in zinc recovery rate in the first 24 h and later a much slower increase thereafter, since, as observed in the pH test, this solution after 24 h it remains acidic and still has the potential for zinc recovery. However, for the recovery of zinc with this solution in a viable percentage, a long time would be needed, which would make the process economically unfeasible.

The 0.5 and 1 molar solutions allow a complete recovery of the zinc in the electric arc furnace dusts. The 0.5 molar solution leaches the zinc from the sample after approximately 72 h, so this would be the test time for the recovery of the element. In turn, the 1 molar sulfuric acid solution allows complete leaching of the zinc in a time between 24 and 48 h, approximately 36 h. Consequently, and due to the fact that this process of constant agitation of the sample immersed in the solution has a higher economic cost than other hydrometallurgical processes and, therefore, it is used in samples with a high percentage of zinc and in reduced times of days, the optimum solution is the 1 molar solution of sulfuric acid with a leaching time of 36 h. Longer times would mean an absurd higher economic cost, as other chemical elements of no economic interest would be leached and could even harm the subsequent hydrometallurgical process.

To quantify the concentration of zinc, as well as the chemical elements potentially unfavorable for subsequent hydrometallurgical techniques, Table 3 is presented. Table 3 shows the concentration in the leachate of the chemical elements mentioned for the aforementioned solution and for the determined test time.

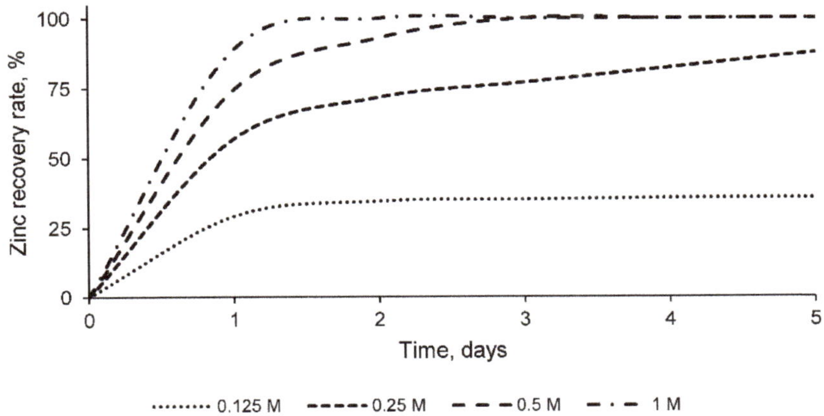

Figure 3. Zinc recovery rate of 0.125, 0.25, 0.5, and 1 molar solutions of sulfuric acid at different times during leaching process with electric furnace dusts.

Table 3. Concentration in leachate of zinc and chemical elements potentially detrimental to development of hydrometallurgical techniques in a 1 molar sulfuric acid solution and a leaching time of 36 h.

Chemical Element	Concentration in Leachate, mg/g
Fe	96.4 ± 1.1
Zn	189.7 ± 3.0
Ca	13.3 ± 0.2
Al	3.7 ± 0.0
Mn	7.9 ± 0.1
Mg	13.1 ± 0.2

As can be seen in Table 3, the concentration of zinc in the leachate, for the selected sulfuric acid solution and the determined test time, coincides with a complete recovery rate. At the same time, the concentration of iron in the leachate is not total, so this fact corroborates that the selected leaching time is adequate, as it recovers a very high percentage of zinc and avoids the leaching of other harmful elements such as iron. The rest of the chemical elements, which can be detrimental to the correct performance of subsequent zinc extraction techniques, are found in very low concentrations.

3.3. Chemical Characterization of the Leaching Residue

The solution of 1 molar sulfuric acid was selected for the leaching of the zinc existing in the electric arc furnace dusts during a test time of 36 h. Therefore, the electric arc furnace dust was leached with this solution and with the detailed time, the solution was subsequently filtered and a leaching residue was obtained. This leaching residue contains, as mentioned above, a series of chemical elements that are very negative for the environment and must be evaluated.

The first of the chemical tests carried out was the elemental analysis test, performed on the leaching residue, which was filtered and then dried at a temperature of 105 ± 5 °C for 24 h. The results of this test are shown in Table 4.

Table 4. Results of elemental analysis of leaching residue.

Sample	Nitrogen, %	Carbon, %	Hydrogen, %	Sulphur, %
Leaching residue	0.024 ± 0.001	0.053 ± 0.001	5.451 ± 0.177	3.745 ± 0.088

The elemental analysis test of the leach residue, shown in Table 4, shows that the percentage of nitrogen is similar to that obtained in the elemental analysis test of the electric arc furnace dusts. At the same time, the percentage of carbon was considerably reduced in the leaching residue compared to that of the electric arc furnace dusts, existing a much higher percentage of hydrogen and sulfur in the leaching residue. This higher percentage of hydrogen and sulfur derives largely from the 1 molar sulfuric acid solution with which the sample was in contact.

In turn, the loss on ignition test of the leaching residue determined a value of 7.57 ± 0.27%. This percentage of mass variation before and after the sample was subjected to the temperature of 950 ± 5 °C corresponds with high probability to the loss of volatile elements and to the transformation of certain chemical compounds. However, the X-ray fluorescence test provides the most information on the chemical elements of higher atomic weight in the leaching residue. The results of this test for the leaching residue are shown in Table 5.

Table 5. X-ray fluorescence of leaching residue.

Element	Wt %	Est. Error
Fe	28.85	0.17
Ca	8.26	0.11
Sx	4.53	0.06
Si	2.12	0.05
Pb	2.01	0.07
Al	1.01	0.04
Mg	1.12	0.04
Mn	1.31	0.05
Zn	0.84	0.02
Cr	0.267	0.013
Cu	0.148	0.0074
K	0.122	0.0061
Ba	0.119	0.040
Ti	0.0759	0.0038
Cl	0.123	0.0062
Sn	0.0795	0.0067
Ni	0.0367	0.0018
Px	0.0195	0.0025
Zr	0.0262	0.0031
Co	0.0206	0.0025
Sr	0.0189	0.0024
Ru	0.0091	0.0019
Bi	0.0107	0.0040
Mo	0.0068	0.0015

The X-ray fluorescence test shows that the percentage of zinc decreased considerably in the sample, thus confirming that a correct zinc leaching process took place for subsequent hydrometallurgical processes. The leaching residue shows a high percentage of iron, which was initially present in the electric furnace dusts and which was not fully leached out. This is interesting, as iron leaching is not as economically interesting as zinc leaching and also impairs the concentration process. The same applies to the chemical elements calcium, silicon, lead, aluminum, magnesium, and manganese, which were present in the initial sample of electric arc furnace dusts and are still present in the leaching residue, and were therefore not fully leached out in the detailed leaching process with the 1 molar solution. The existing the percentage of sulfur derived from the sulfuric acid leaching process which was not present in such a high proportion in the initial sample of electric arc furnace dust. Consequently, the leaching residue after leaching of the zinc in the sample presents certain potentially polluting chemical elements that must be treated or incorporated into a new material to avoid its deposition in landfill and possible environmental problems. The potentially toxic elements determined by the EPA for ceramic materials are Cr, Pb, As, Cd, and Ba, being these elements the ones that should be evaluated with leaching tests (TCLP) in the material that incorporates this leaching residue.

3.4. Conformed of Ceramic Materials for Bricks with the Leaching Residue of Electric Arc Furnace Dusts

The electric arc furnace dusts were leached in large quantities with a 1 molar solution of sulfuric acid for 36 h and to be subsequently filtered to obtain a sufficient amount of leaching residue for ceramics conforming. The conforming ceramics families are detailed in Table 1, and a total of 6 samples were made for each group according to the procedure detailed in the methodology. These samples, with up to 50% of the leaching residue, were physically and mechanically evaluated through different tests.

The first of the tests performed was the linear shrinkage test, which determined the variation in dimensions that occurred in the samples before and after the sintering process. The results of this test for the different families of ceramics conformed with the leaching residue are shown in Figure 4.

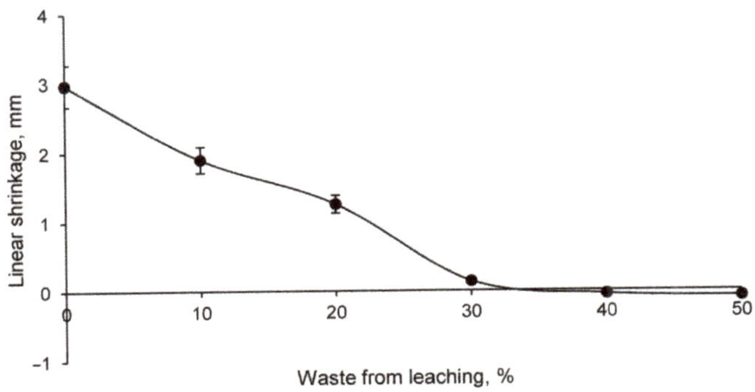

Figure 4. Linear shrinkage of different families of conformed ceramics with clay and leaching residue.

The linear shrinkage of the families of ceramics conformed with the leaching residue decreases with increasing percentage of the waste. So much so that the samples with 40% and 50% leaching residue show an expansion after the leaching process. These results are essential for the industrialization of the ceramic with the leaching residue, since the conformed ceramic must have certain dimensions. To be marketed, so that the dimensions of the material before sintering must be fixed. In addition, a linear expansion reflects an increase in the size of the material and, consequently, the formation of a more open structure, conditioning porosity, density, and even strength.

On the other hand, the weight loss experienced by the conformed ceramic material with various percentages of leaching residue before and after the leaching process is shown in Figure 5.

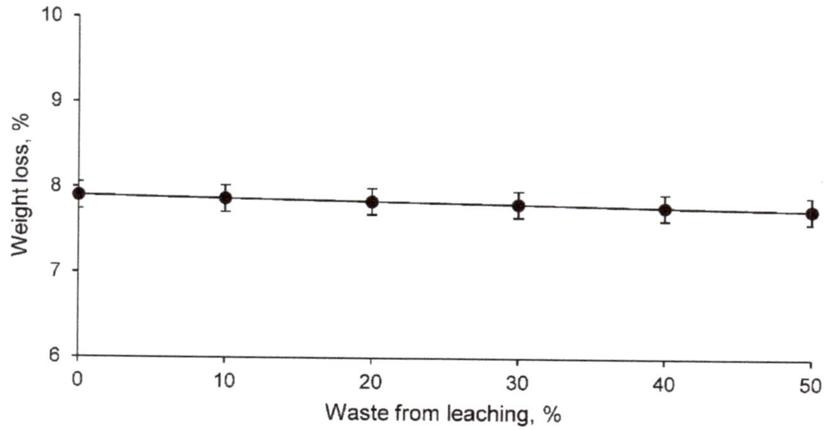

Figure 5. Weight loss of different families of conformed ceramics with clay and leaching residue.

The weight loss in the different families is linear, which is to be expected if one takes into account that the percentage of waste increases in increments of 10%. The result for the clay-only family coincides approximately with the value of loss on ignition of the clay, so that the addition of percentages of leaching residue makes this value vary proportionally based on the loss by calcination of the waste. The weight loss of ceramic families with higher percentages of leaching residue is higher. Consequently, if there is a higher weight loss and a linear expansion, a more open ceramic structure is obviously being created.

This fact can be verified with the capillary water absorption test of the different ceramic families shown in Figure 6.

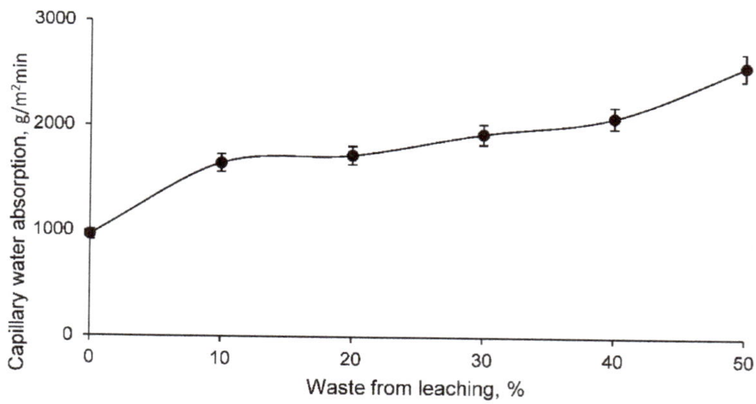

Figure 6. Capillary water absorption of different families of clay-conformed ceramics and leaching residue.

As expected from the results of linear shrinkage and weight loss, the capillary water absorption of the clay-conformed ceramics families and the leaching residue is higher the higher the percentage of waste. Consequently, there is a more open structure in the ceramics with leaching residue, in which there is a series of interconnected pores capable of absorbing a higher percentage of water by capillary action.

The cold water absorption test of the conformed ceramics shows the results shown in Figure 7.

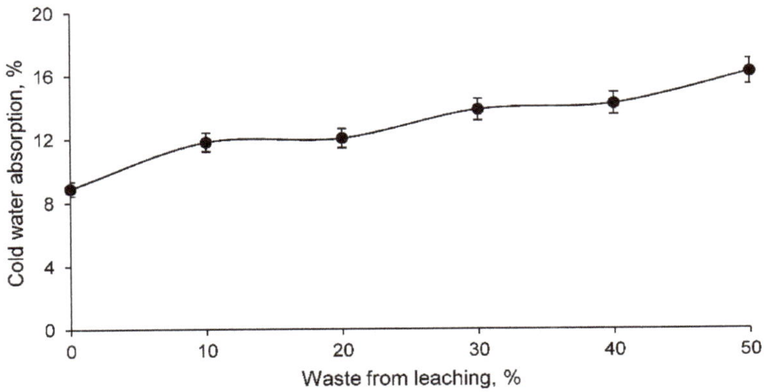

Figure 7. Cold water absorption of different families of clay-conformed ceramics and leaching residue.

The cold water absorption of the conformed ceramic families increases as the percentage of leaching residue in the ceramic increases. This fact is very interesting and must be controlled for ceramic elements that are outdoors, since rainwater can cause it to be absorbed by the ceramic and increase its weight, uselessly overloading the structure that supports these materials. However, a higher water absorption also presents some special characteristics for ceramics that are very interesting for construction.

In turn, the results of the open porosity test for all families of clay-conformed ceramics with leaching residue are shown in Figure 8.

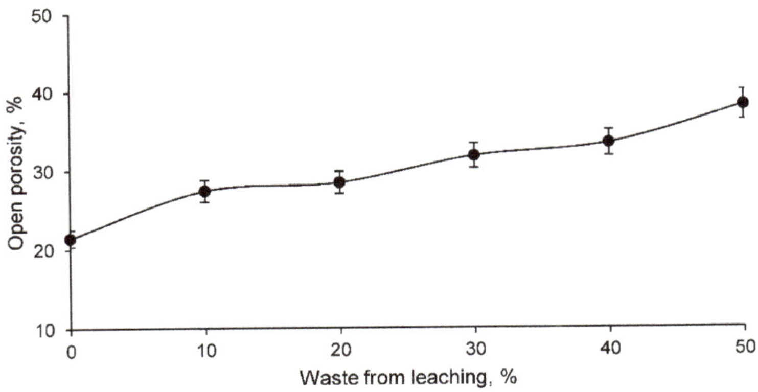

Figure 8. Open porosity of different clay-conformed ceramics families and leaching residue.

The open porosity of the clay-conformed ceramics with leach residue increases as the percentage of leaching residue increases. This was to be expected based on the test results detailed above and obviously conditions the strength of the material. However, higher porosity can improve certain properties of traditional ceramics such as thermal or acoustic insulation. Therefore, there is a wide variety of physical test results depending on the percentage of leaching residue in the ceramic, allowing to obtain ceramic materials for different applications.

Finally, the last physical property that characterizes the family of clay-conformed ceramics with leaching residue, bulk density, was calculated. The results of this test are shown in Figure 9.

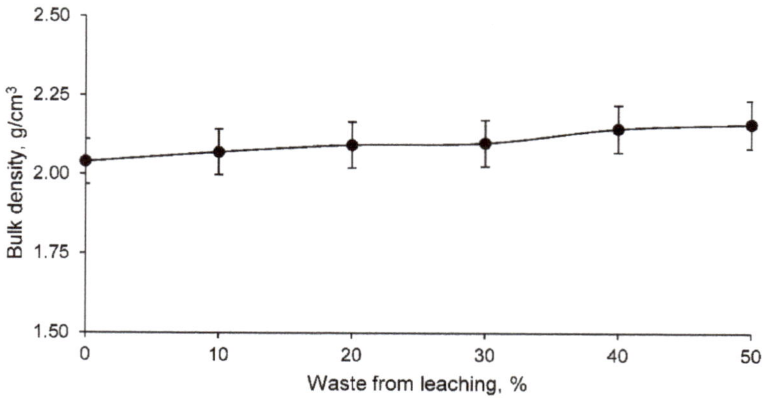

Figure 9. Bulk density of different families of clay-conformed ceramics and leaching residue.

The results shown in Figure 9 on bulk density show that the families of ceramics conformed with a higher percentage of leaching residue have a higher bulk density. This fact, far from being contradictory to previous tests in which a ceramic with higher porosity was observed, reflects the higher density of the leaching residue, which is to be expected if one takes into account the X-ray fluorescence test of this waste in which heavy metals appear in high proportion. Consequently, the ceramic conformed with a higher leaching residue, even presenting a higher porosity, has a higher apparent density due to the chemical composition of the waste.

Finally, to obtain the mechanical strength of the ceramics intended for construction bricks, Figure 10 shows the compressive strength of the various families of clay-conformed ceramics and the leaching residue.

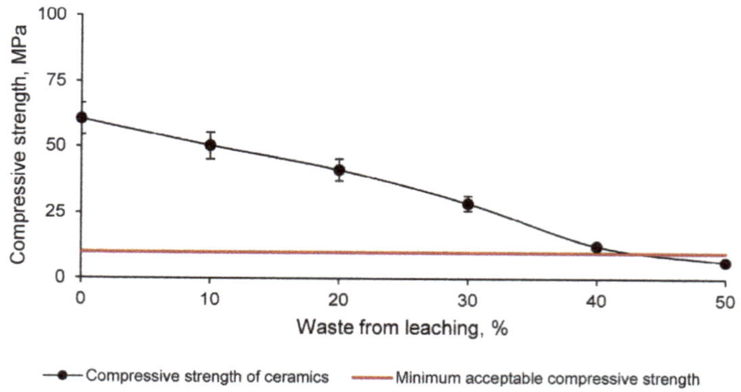

Figure 10. Compressive strength of different families of clay-conformed ceramics and leaching residue.

The compressive strength of ceramic materials intended for bricks is limited to 10 MPa according to European standards. Therefore, the family of ceramics conformed with 50% leaching residue and 50% clay is unusable. The other families of ceramics present compressive strengths higher than those established by the standards, taking into account that this strength decreases the higher the percentage of leaching residue in the ceramic. Consequently, the families conformed with clay and percentages of leaching residue of 10%, 20%, 30%, and 40% present physical and mechanical properties, according to the tests carried out, suitable for use in ceramic materials intended for the manufacture of bricks for construction.

4. Conclusions

The results of the tests mentioned in the methodology of this research allow us to obtain a series of partial conclusions that derive in the confirmation of the final objective. The final objective of this article is to extract by leaching in acid media, atmospheric pressure, and ambient temperature the zinc existing in the electric arc furnace dusts, as well as to reuse the leaching residue from the process as a raw material for ceramics. Consequently, the partial conclusions obtained are as follows:

- Electric arc furnace dusts contain a significant percentage of zinc, around 18%. This zinc seems to exist in the form of oxides, due to the formation process of this waste. There are also other elements such as iron (in high proportion), calcium, silicon, chlorine, sodium, magnesium, potassium, and manganese;
- The 0.5 and 1 molar sulfuric acid solutions leach the total percentage of zinc existing in the sample as determined by X-ray fluorescence analysis (188.3 mg/g). This leaching process was carried out with a solid/liquid ratio of 1:20, constant agitation, ambient temperature, and atmospheric pressure;
- The 1 molar solution of sulfuric acid leaches the zinc from the electric arc furnace dusts in its entirety in 36 h, according to the detailed procedure. Therefore, this solution was selected as optimal and a leaching time of 36 h was set;
- The waste produced in the leaching process with the selected dilution and time was chemically characterized, presenting a higher percentage of sulfur and hydrogen than the electric arc furnace dust. At the same time, the leaching residue contains other chemical elements such as iron (in high proportion), calcium, silicon, lead, aluminum, magnesium, and manganese. In addition, the percentage of zinc in the leaching residue was very low, corroborating the effectiveness of the leaching process;
- Ceramics conformed with clay and leaching residue showed a more open structure the higher the percentage of waste. Consequently, ceramics with leaching residue show lower linear shrinkage, higher capillary water absorption, higher cold water absorption, and higher porosity compared to that of a traditional ceramic consisting only of clay. However, the bulk density of ceramics conformed with the leaching residue increases as the percentage of waste increases, mainly due to the existence of heavy metals;
- The strength of ceramics conformed with clay and leaching residue from electric arc furnace dusts is lower the higher the percentage of waste, with unacceptable strength values for samples conformed with 50% clay and 50% leaching residue. Consequently, and according to the physical and mechanical tests carried out, the families of ceramics conformed with the leaching residue that are suitable for use are those incorporating 10%, 20%, 30%, and 40% of waste.

Based on the partial conclusions obtained, it is possible to leach the zinc existing in the electric arc furnace dusts in its entirety with sulfuric acid solutions, constant agitation, ambient temperature, and atmospheric pressure, mainly with the 1 molar solution of sulfuric acid and 36 h of process. Furthermore, with the waste from the leaching process, it is possible to manufacture ceramic materials for use as bricks with percentages of up to 40% of the waste, complying with European regulations in this respect for the tests carried out. This research is therefore an example of the potential that waste still has for the valorization of economically viable elements and their reuse as raw materials for new materials. Consequently, a new environmental hydrometallurgy is developed in which there is no waste and in which environmental pollution caused by the deposition of hazardous waste in landfills is avoided along with extracting new raw materials for the manufacture of ceramics.

Author Contributions: Conceptualization, F.A.C.-I., E.R.M.-L., J.M.T.-S. and J.S.-M.; methodology, F.A.C.-I., E.R.M.-L., J.M.T.-S. and J.S.-M.; software, J.M.T.-S. and J.S.-M.; validation, F.A.C.-I. and E.R.M.-L.; formal analysis, F.A.C.-I. and E.R.M.-L.; investigation, J.M.T.-S. and J.S.-M.; resources, F.A.C.-I.; data curation, E.R.M.-L.; writing—original draft preparation, J.S.-M.; writing—review and editing, J.M.T.-S.; visualization, J.M.T.-S.; supervision, F.A.C.-I.; project administration, J.S.-M.; funding acquisition, F.A.C.-I. All authors have read and agreed to the published version of the manuscript.

Funding: This research received no external funding.

Institutional Review Board Statement: Not applicable.

Informed Consent Statement: Not applicable.

Data Availability Statement: Data are contained within the article.

Acknowledgments: In Technical and human support provided by CICT of Universidad de Jaén (UJA, MINECO, Junta de Andalucía, FEDER) is gratefully acknowledged.

Conflicts of Interest: The authors declare no conflict of interest.

References

1. Graedel, T.E.; Harper, E.M.; Nassar, N.T.; Reck, B.K. On the materials basis of modern society. *Proc. Natl. Acad. Sci. USA* **2015**, *112*, 6295–6300. [CrossRef]
2. Hagelüken, C.; Lee-Shin, J.U.; Carpentier, A.; Heron, C. The EU Circular Economy and Its Relevance to Metal Recycling. *Recycling* **2016**, *1*, 242–253. [CrossRef]
3. Broadbent, C. Steel's recyclability: Demonstrating the benefits of recycling steel to achieve a circular economy. *Int. J. Life Cycle Assess.* **2016**, *21*, 1658–1665. [CrossRef]
4. Madias, J. Electric Furnace Steelmaking. *Treatise Process Metall.* **2014**, *3*, 271–300.
5. Naito, M.; Takeda, K.; Matsui, Y. Ironmaking Technology for the Last 100 Years: Deployment to Advanced Technologies from Introduction of Technological Know-how, and Evolution to Next-generation Process. *ISIJ Int.* **2015**, *55*, 7–35. [CrossRef]
6. Jaimes, W.; Maroufi, S. Sustainability in steelmaking. *Curr. Opin. Green Sustain. Chem.* **2020**, *24*, 42–47. [CrossRef]
7. Das, B.; Prakash, S.; Reddy, P.S.R.; Misra, V.N. An overview of utilization of slag and sludge from steel industries. *Resour. Conserv. Recycl.* **2007**, *50*, 40–57. [CrossRef]
8. De Buzin, P.J.W.K.; Heck, N.C.; Vilela, A.C.F. EAF dust: An overview on the influences of physical, chemical and mineral features in its recycling and waste incorporation routes. *J. Mater. Res. Technol.* **2017**, *6*, 194–202. [CrossRef]
9. Guézennec, A.G.; Huber, J.C.; Patisson, F.; Sessiecq, P.; Birat, J.P.; Ablitzer, D. Dust formation in Electric Arc Furnace: Birth of the particles. *Powder Technol.* **2005**, *157*, 2–11. [CrossRef]
10. Gupta, S.; French, D.; Sakurovs, R.; Grigore, M.; Sun, H.; Cham, T.; Hilding, T.; Hallin, M.; Lindblom, B.; Sahajwalla, V. Minerals and iron-making reactions in blast furnaces. *Prog. Energy Combust. Sci.* **2008**, *34*, 155–197. [CrossRef]
11. Agency, E.P. *European Waste Catalogue and Hazardous Waste List*; Environmental Protection Agency Ireland: Wexford, Ireland, 2002; ISBN 1840950838.
12. Xanthopoulos, P.; Agatzini-Leonardou, S.; Oustadakis, P.; Tsakiridis, P.E. Zinc recovery from purified electric arc furnace dust leach liquors by chemical precipitation. *J. Environ. Chem. Eng.* **2017**, *5*, 3550–3559. [CrossRef]
13. Martins, F.M.; Neto, J.M.; da Cunha, C.J. Mineral phases of weathered and recent electric arc furnace dust. *J. Hazard. Mater.* **2008**, *154*, 417–425. [CrossRef] [PubMed]
14. Oustadakis, P.; Tsakiridis, P.E.; Katsiapi, A.; Agatzini-Leonardou, S. Hydrometallurgical process for zinc recovery from electric arc furnace dust (EAFD). Part I: Characterization and leaching by diluted sulphuric acid. *J. Hazard. Mater.* **2010**, *179*, 1–7. [CrossRef]
15. Guo, X.; Zhong, J.; Song, Y.; Tian, Q. Substance flow analysis of zinc in China. *Resour. Conserv. Recycl.* **2010**, *54*, 171–177. [CrossRef]
16. Yu, G.; Peng, N.; Zhou, L.; Liang, Y.J.; Zhou, X.Y.; Peng, B.; Chai, L.Y.; Yang, Z.H. Selective reduction process of zinc ferrite and its application in treatment of zinc leaching residues. *Trans. Nonferrous Met. Soc. China* **2015**, *25*, 2744–2752. [CrossRef]
17. Su, Y.M.; Huang, W.C.; Liu, Y.C.; Chang, C.K.; Kuo, Y.L. Utilization of electric arc furnace dust as regenerable sorbents for the removal of hydrogen sulfide. *Ceram. Int.* **2017**, *43*, S694–S699. [CrossRef]
18. Lin, X.; Peng, Z.; Yan, J.; Li, Z.; Hwang, J.Y.; Zhang, Y.; Li, G.; Jiang, T. Pyrometallurgical recycling of electric arc furnace dust. *J. Clean. Prod.* **2017**, *149*, 1079–1100. [CrossRef]
19. Wang, J.; Zhang, Y.; Cui, K.; Fu, T.; Gao, J.; Hussain, S.; AlGarni, T.S. Pyrometallurgical recovery of zinc and valuable metals from electric arc furnace dust—A review. *J. Clean. Prod.* **2021**, *298*, 126788. [CrossRef]
20. Youcai, Z.; Stanforth, R. Extraction of zinc from zinc ferrites by fusion with caustic soda. *Miner. Eng.* **2000**, *13*, 1417–1421. [CrossRef]
21. Langová, Š.; Leško, J.; Matýsek, D. Selective leaching of zinc from zinc ferrite with hydrochloric acid. *Hydrometallurgy* **2009**, *95*, 179–182. [CrossRef]

22. Havlík, T.; Souza, B.V.E.; Bernardes, A.M.; Schneider, I.A.H.; Miškufová, A. Hydrometallurgical processing of carbon steel EAF dust. *J. Hazard. Mater.* **2006**, *135*, 311–318. [CrossRef] [PubMed]
23. Zhang, L.; Liu, B.; Du, J.; Liu, C.; Wang, S. CO_2 emission linkage analysis in global construction sectors: Alarming trends from 1995 to 2009 and possible repercussions. *J. Clean. Prod.* **2019**, *221*, 863–877. [CrossRef]
24. Almeida, M.I.; Dias, A.C.; Demertzi, M.; Arroja, L. Contribution to the development of product category rules for ceramic bricks. *J. Clean. Prod.* **2015**, *92*, 206–215. [CrossRef]
25. Suárez-Macías, J.; Terrones-Saeta, J.M.; Iglesias-Godino, F.J.; Corpas-Iglesias, F.A. Retention of Contaminants Elements from Tailings from Lead Mine Washing Plants in Ceramics for Bricks. *Minerals* **2020**, *10*, 576. [CrossRef]
26. Terrones-Saeta, J.M.; Suárez-Macías, J.; Iglesias-Godino, F.J.; Corpas-Iglesias, F.A. Study of the incorporation of biomass bottom ashes in ceramic materials for the manufacture of bricks and evaluation of their leachates. *Materials* **2020**, *13*, 2099. [CrossRef] [PubMed]

metals

MDPI

Article

Effect of Temperature on Biobeneficiation of Bulk Copper-Nickel Concentrate with Thermoacidophilic Microbial Communities

Anna Panyushkina [1], Natalya Fomchenko [1], Vladislav Babenko [2] and Maxim Muravyov [1,*]

1 Winogradsky Institute of Microbiology, Research Centre "Fundamentals of Biotechnology", Russian Academy of Sciences, Leninsky Ave., 33, Bld. 2, 119071 Moscow, Russia; zhuravleva-inmi@mail.ru (A.P.); natalya.fomchenko@gmail.com (N.F.)

2 Federal Research and Clinical Centre of Physical-Chemical Medicine, Federal Medical Biological Agency, 119435 Moscow, Russia; daniorerio34@gmail.com

* Correspondence: maxmuravyov@inmi.ru; Tel.: +7-495-135-6596

Abstract: Bioleaching of the bulk copper–nickel sulfide concentrate was proposed as a method to remove nickel from it and to obtain a concentrate containing copper as chalcopyrite. This approach is based on the different refractoriness of sulfide minerals in ferric sulfate solutions and oxidation by acidophilic microorganisms. The bulk concentrate contained 10.8% copper in the form of chalcopyrite ($CuFeS_2$) and 7.2% nickel that occurred in pentlandite ($(Ni,Fe)_9S_8$) and violarite ($FeNi_2S_4$). Three microbial communities grown at 35, 40, and 50 °C were used for bioleaching. The microbial community at 40 °C was the most diverse in the genus and species composition. At all temperatures of the process, the key roles in bioleaching belonged to mixotrophic and heterotrophic acidophiles. The highest levels of nickel leaching of 97.2 and 96.3% were observed in the case of communities growing at 40 and 50 °C, respectively. At the same time, the bioleach residue, which could be characterized as a marketable high-grade copper (chalcopyrite) concentrate, was obtained only at 40 °C. This solid contained 15.6% copper and 0.54% nickel. Thus, the biobeneficiation of bulk sulfide concentrates can be a promising field of biohydrometallurgy.

Keywords: bulk concentrate; biobeneficiation; selective bioleaching; biohydrometallurgy; nickel dissolution; acidophilic microbial communities

Citation: Panyushkina, A.; Fomchenko, N.; Babenko, V.; Muravyov, M. Effect of Temperature on Biobeneficiation of Bulk Copper-Nickel Concentrate with Thermoacidophilic Microbial Communities. *Metals* **2021**, *11*, 1969. https://doi.org/10.3390/met11121969

Academic Editor: Stefano Ubaldini

Received: 30 October 2021
Accepted: 3 December 2021
Published: 7 December 2021

1. Introduction

The beneficiation of bulk concentrates of nonferrous metals, such as copper–nickel and copper–zinc sulfide concentrates, is aimed at obtaining high-grade selective concentrates, which are of current importance for the metallurgical industry [1]. However, the selective separation of bulk concentrates is often difficult due to the textural and structural properties of raw materials. As a result of the difficulties of selective metal concentration, relatively simply produced bulk concentrates can circulate in flotation circuits for a long time [1,2].

To achieve selective metal leaching, environmentally sound cost-effective and easy-to-use technologies are required. These technologies include biohydrometallurgical approaches (particularly, bioleaching/biooxidation) that are applied at industrial mining and metallurgical enterprises worldwide for the recovery of metals from sulfidic raw materials [3–6]. These bio-approaches are based on the activity of communities of acidophilic chemolithotrophic microorganisms that oxidize ferrous iron, elemental sulfur (S^0), reduced inorganic sulfur compounds (RISCs), and sulfide minerals [7]. The significance of the application of biohydrometallurgy is associated with relatively low environmental hazards in comparison with conventional pyrometallurgical methods [8,9]. Biohydrometallurgy allows the processing of low-grade raw materials, products, and metallurgical wastes, with reduced emissions and discharges of toxic substances to air [6,10].

Low-grade copper ores (copper is mainly in secondary sulfides) and gold-bearing (pyrite-arsenopyrite and pyrite) ores are the best-studied raw materials involved in bioprocessing [11,12]. Chalcopyrite is one of the most refractory minerals in ferric sulfate solutions, which determines the low efficiency/intensity of biohydrometallurgical approaches used for chalcopyrite processing [13]. In that regard, the idea of selective mineral leaching, associated with chalcopyrite concentration in the solids, seems promising for the processing of refractory bulk concentrates [14]. The possibility of the selective removal of zinc from the copper–zinc concentrate during leaching with a ferric sulfate solution obtained with an acidophilic chemolithotrophic microbial community, as well as during bioleaching, has been previously reported [15–17]. The varying composition of minerals in the copper–zinc concentrate has shown that a higher chalcopyrite/sphalerite ratio improves the rate of the oxidation of sphalerite [18].

Polymetallic copper–nickel sulfide concentrates are also promising raw materials for biohydrometallurgy. They are characterized by the relatively simple nickel leaching from pentlandite and violarite, while chalcopyrite is highly stable in ferric solutions [19]. The following simplified equations [20,21] summarize the oxidation of chalcopyrite ($CuFeS_2$), pentlandite ($(Ni,Fe)_9S_8$), and violarite ($FeNi_2S_4$) during bioleaching:

$$CuFeS_2 + 4Fe^{3+} = Cu^{2+} + 2S^0 + 5Fe^{2+}, \quad (1)$$

$$2(Ni,Fe)_9S_8 + 36Fe^{3+} = 9Ni^{2+} + 16S^0 + 45Fe^{2+}, \quad (2)$$

$$FeNi_2S_4 + 18Fe^{3+} + 8H_2O = 2Ni^{2+} + 2SO_4{}^{2-} + 2S^0 + 19Fe^{2+} + 16H^+. \quad (3)$$

Microorganisms oxidize the reaction products:

$$2S^0 + 2H_2O + 3O_2 = 2H_2SO_4, \quad (4)$$

$$4Fe^{2+} + O_2 + 4H^+ = 4Fe^{3+} + 2H_2O. \quad (5)$$

Previous studies have indicated that sulfide nickel-containing concentrates could be leached using mesophilic bacteria [22]. Batch bioleaching of the nickel ore concentrate containing pentlandite as a major nickel-bearing mineral at 30 °C has resulted in the copper, cobalt, and nickel extraction of 22.4, 67.5, and 76%, respectively [22].

The study of the bioleaching of the violarite–pentlandite–chalcopyrite concentrate at 40 °C allowed the removal of nickel and the obtainment of a high-grade copper concentrate. Violarite has been shown to dissolve more rapidly than pentlandite during bioleaching of the copper–nickel concentrate, with higher rates of nickel sulfide dissolution at 40 °C as compared to 30 °C [23]. At 40 °C, the community of microorganisms was dominated by bacteria of the species *Leptospirillum* (*L.*) *ferriphilum* and *Acidithiobacillus (At.) caldus*, as well as *Sulfobacillus* (*S.*) spp. and archaea of the genera *Cuniculiplasma* and *Ferroplasma (Fr.)* as minor components. After 22 days of the bioleaching, the efficiency of nickel leaching reached 67.6% (30 °C) and 86.7% (40 °C), while the copper content in the bioleach residue increased by 2.7 and 5.7%, respectively [23].

Thermophilic bioleaching of a low-grade copper–nickel concentrate at high temperatures (65, 70, 72, 73, 74, and 75 °C) has also been compared [24]. The process at 70 °C allowed the highest copper and nickel extractions of 95% and 99%, respectively. More than 98% of nickel could be extracted at both 45 and 70 °C; however, the kinetics of leaching was slower at the former temperature. While 95% Cu extraction was shown at 70 °C, the final copper extraction of only 74% could be obtained at 45 °C [24].

Nevertheless, bioleaching of nickel-containing sulfidic raw materials has been significantly less studied than bioprocesses for copper and zinc extraction. Moreover, the roles of the members of the microbial communities, as well as microbial interactions during the biooxidation of raw materials containing nickel sulfides, are not entirely clear. The determination of the microbial diversity in the communities formed during the biooxidation of sulfidic copper–nickel raw materials, as well as the possible roles of the community members, are decisive in assessing whether the biohydrometallurgical approach is promising for particular raw materials under the selected conditions. Therefore, efficient bioleaching of metals from copper–nickel concentrates by microbial communities is of high importance from both the fundamental and practical standpoints. The industrial biooxidation of sulfide raw materials is usually associated with exothermal reactions leading to increased temperatures of the medium in bioreactors. In this research, we used thermoacidophilic microbial communities for the biobeneficiation of bulk copper–nickel concentrates for the first time.

The goal of this research was to study changes in solids composition depending on the temperature during bioleaching of the copper–nickel concentrate by thermoacidophilic microbial communities. In addition, this study aimed to choose the temperature that allows obtaining a residue considered a marketable copper (chalcopyrite) concentrate.

2. Materials and Methods

2.1. Copper–Nickel Concentrate

The bulk sulfide concentrate of the copper–nickel ore (Shanuch district, Kamchatka territory, Russia) flotation was the object of this study. The concentrate was composed of chalcopyrite, pentlandite, violarite, pyrrhotite, and pyrite. Quartz, chabazite, jarosite, plagioclase, chlorite, and albite were gangue minerals. The content of the main elements is shown in Table 1. Table 2 shows the content of sulfide minerals in the concentrate. The particle size of the concentrate was d_{80} < 44 µm. The methods used for obtaining these data are provided in Section 2.5.

Table 1. Main element content (wt%) in concentrate.

Cu	Ni	Fe	S	Ca	Si
10.8	7.2	21.8	26.5	0.96	18.0

Table 2. Sulfide mineral content (wt%) in concentrate.

Chalcopyrite	Pentlandite	Violarite	Pyrrhotite	Pyrite
31	12	8	4	6

2.2. Microbial Communities and Cultivation Conditions

For bioleaching experiments at 35, 40, and 50 °C, three acidophilic microbial consortia were formed from the stirred-tank biotreatment of the waste containing zinc [25], as well as copper–zinc and copper–nickel concentrates [23]. The original community contained chemolithoautotrophic and mixotrophic strains of iron and sulfur oxidizers assigned to various species: *At. caldus*, *At. thiooxidans*, *At. ferrooxidans*, *Fr. acidarmanus*, *Fr. acidiphilum*, *L. ferriphilum*, *Acidiplasma* (*Ad.*) sp., *S. thermosulfidooxidans* and *S. thermotolerans*, as well as *Acidibacillus* (*Ac.*) sp. and *Ferrimicrobium* (*Fm.*) sp. The community was also composed of acidophilic heterotrophs: *Alicyclobacillus* (*Al.*) spp. and *Acidiphilium (Ap.) multivorum* [26], as well as the archaeon *Cuniculiplasma* sp. [25].

This original community was subsequently grown at 35, 40, and 50 °C to obtain inocula for bioreactor experiments. Flasks (culture medium, 900 mL; inoculum, 100 mL) were incubated for two days at 30, 40, and 50 °C (Redline RI53 thermostats, Binder, Germany). The culture medium contained the following (g/L): $(NH_4)_2SO_4$, 0.75; KCl, 0.025; $K_2HPO_4 \cdot 3H_2O$, 0.125; $MgSO_4 \cdot 7H_2O$, 0.125; $Ca(NO_3)_2 \cdot 4H_2O$, 0.0025; $FeSO_4 \cdot 7H_2O$, 30.0; S^0, 10.0; yeast extract, 0.1. The initial pH was adjusted to 1.6 with 5 M H_2SO_4. The experiments were carried out under the conditions of intensive aeration. The air was supplied at a flow rate of 1 v/(v·min). The obtained solutions (pH 1.3) contained microbial communities of acidophiles (0.1×10^9 cells/mL at 50 °C and 0.8×10^9 cells/mL at 35 and 40 °C), Fe^{3+} (6.0, 5.9, and 5.9 g/L at 35, 40, and 50 °C, respectively) and the unoxidized part of S^0. These solutions were separated from the sulfur precipitates by filtration on paper filters and used for bioleaching processes.

2.3. Analysis of Microbial Community

The structure of the original microbial consortia (inocula) grown at 35, 40, and 50 °C, as well as microbial communities formed at the end of the process, were assessed by cell morphology using phase-contrast microscopy and metabarcoding using the 16S rRNA gene hypervariable region (V3–V4). To obtain biomass for metabarcoding analysis, a suspension was centrifuged (100× *g*, 2 min) to precipitate solids, and cell pellets were further collected by centrifugation (5000× *g*, 15 min) of the obtained supernatant. The biomass was sequentially washed by centrifugation (10,000× *g*, 5 min) with the iron-free 9K medium First, the iron-free 9K medium with pH 1.5 was used to wash the biomass by centrifugation (10,000× *g*, 5 min). In the next step, the biomass was washed using the same medium but with a neutral pH [27]. Homogenization of the samples was carried out in a MagNA Lyser (Roche, Basel, Switzerland) device. DNA was extracted according to the standard protocol for MagNA Pure Bacteria Lysis Buffer (Roche, Basel, Switzerland). The 16S library preparation and sequencing were carried out according to Illumina protocol. Briefly, amplification of the extracted DNA with standard 16S rRNA gene primers, complementary to the V3–V4 region and containing 5′-Illumina adapter sequences, was performed. After that, individual amplicons were PCR-indexed and pooled. Sequencing of DNA libraries was carried out with a MiSeq instrument (Illumina, San Diego, CA, USA) and the Miseq reagent kit v3 (Illumina, San Diego, CA, USA). After demultiplexing was performed [28], reads containing the 16S rRNA primer sequences were filtered out with the CLC Genomics Workbench version 10.0 (Qiagen, Redwood City, CA, USA). To choose operational taxonomic units (OTU) and to carry out the taxon-based assignment, the QIIME open-source software pipeline [29] and the Silva132 database [30] were used. All parameters were set to default values. The analysis included ≈12,000–30,000 fragments (an average length, 449 nucleotides).

To identify microorganisms at the species level, analysis of the similarity of nucleotide sequences of the 16S rRNA gene fragments was carried out using the Nucleotide Basic Local Alignment Search Tool (BLAST; Bethesda, MD, USA, https://blast.ncbi.nlm.nih.gov/Blast.cgi, accessed on 25 October 2021).

2.4. Bioleaching Procedure

Batch bioleaching of the concentrate was carried out in a 2.0 L bioreactor (working volume, 1.0 L). Before the beginning of the bioleaching, the bioreactor containing 10 g of the solid sample was supplemented with 1.0 L of inoculum containing $\approx 0.1 \times 10^9$ cells/mL at 50 °C and 0.8×10^9 cells/mL at 35 and 40 °C (the pulp density, 1%(*w*/*v*)). Thus, the inoculum volume was 100%(*v*/*v*). The initial pH value was 1.1–1.2. The pH of the pulp was monitored, and concentrated sulfuric acid was added when necessary to maintain the pH values. The experiments were carried out at a stirring speed of 500 rpm using an overhead 8-blade Rushton turbine impeller, with aeration (4 v/(v·min)). Once a day, distilled water was added to the bioreactor to compensate for evaporation losses. At the end of bioleaching, the pulp from the bioreactor was centrifuged at 2000× *g* rpm for 5 min, and the residue was washed three times with distilled water and dried to constant weight.

2.5. Analytical Methods

The relative number of cells in 1 mL was determined by ten-fold dilutions and subsequent direct counting in the Goryaev chamber using a Mikmed-2 microscope equipped with a phase contrast device (LOMO, Moscow, Russia).

The values of the pH and redox potential were measured with a pH-150MI pH meter–millivoltmeter (Izmeritelnaya tekhnika, Moscow, Russia). The concentration of the Fe^{3+} ions in the liquid phase was determined by complexometric titration [31]. The same method was used to determine the total concentration of iron after the reaction with ammonium peroxydisulfate. The concentration of ferrous iron ions was calculated as the difference between the total and ferric iron. Flame atomic absorption spectroscopy (FAAS) (Perkin

Elmer 3100, PerkinElmer, Waltham, MA, USA) was used to determine the concentrations of copper and nickel.

Solids were analyzed by a wet chemical method, using 1:3 concentrated nitric acid/concentrated hydrochloric acid at the boiling temperature of the mixture. This analysis was followed by FAAS for metals and gravimetric analysis with $BaCl_2$ for sulfur [32]. The mineralogical composition of solids was determined with X-ray diffraction microscopy (DRON-2 diffractometer, Burevestnik, St. Petersburg, Russia). The Rietveld refinement of the XRD patterns was used to evaluate the relative mineral phase amounts [33]. For image acquisition, a reflected light optical microscopy (Olympus BX51 microscope, Olympus, Tokyo, Japan) was used. The microscope was set to operate using polarized reflected light. The samples of solids were mounted in polished sections embedded in epoxy resin.

According to the following equation, mass loss (wt %) was calculated:

$$\gamma = 100 - \frac{m_r}{m_{in}} \cdot 100, \tag{6}$$

where m_r—the weight of bioleach residue (g); m_{in}—the weight of the original sample (g).

Metal recovery (wt %) was calculated from the following equation:

$$\varphi = \frac{C \cdot V}{10 \cdot \beta_{in} \cdot m_{in}}, \tag{7}$$

where C is the metal concentration in pregnant leach solution (mg/L); V is a solution volume (L); β_{in} is the metal content in the original sample (wt%).

All experiments were carried out in duplicate. Statistical processing was performed using Microsoft Excel 2013. The standard deviation (SD) of the arithmetic mean was calculated, and the significance of the results was assessed using Student's t-test at the significance level $p \leq 0.1$.

3. Results and Discussion

3.1. Pregnant Leach Solution Characteristics

The physicochemical parameters of the liquid phase of the pulp were measured during the concentrate bioleaching. These were the pH value, the ambient redox potential, the concentration of ferrous and ferric iron, as well as the copper and nickel concentrations. Based on these values, the extraction of metals from solids was calculated. Figure 1 shows changes in the pH and redox potential values in different temperature modes. An abrupt increase in the pH value on the first day of the bioleaching required the addition of 1 mL of concentrated sulfuric acid for the processes at 35 and 50 °C and 2 mL at 40 °C. The dissolution of gangue minerals could lead to an increase in pH according to the reaction:

$$Fe_{1-x}S + 2(1-x)H^+ = (1-x)Fe^{2+} + (1-x)H_2S + xS^0. \tag{8}$$

In all cases, after sulfuric acid was added on the first day of bioleaching, the pH value remained relatively constant in the range of 1.1–1.2. This was due to the simultaneous microbial oxidation of ferrous iron increasing pH, and elemental sulfur oxidation producing sulfuric acid. The redox potential value decreased sharply on the first day of the process due to the consumption of ferric iron for the oxidation of sulfide minerals according to Equations (1)–(3), as well as the following reaction:

$$Fe_{1-x}S + (2-2x)Fe^{3+} = (3-3x)Fe^{2+} + S^0. \tag{9}$$

After that, it increased relatively rapidly, within 1–7 days, due to the microbial oxidation of ferrous iron (Equation (5)), and stabilized at the level of 860–911, 930–943, and 845 mV (vs. standard hydrogen electrode (SHE)) at 35, 40, and 50 °C, respectively. A lower value of the redox potential at 50 °C was due to the low concentration of ferric iron in the liquid phase of the pulp, as a result of its more pronounced precipitation in the form of jarosite (see Table 3). An increase in the temperature during the bioleaching of sulfidic raw materials is known to lead to a greater jarosite formation [34,35]. Changes in the ferrous and ferric iron concentrations in the liquid phase of the pulp at different temperatures (Figure 2) confirmed this conclusion. At all temperatures studied, the microbial communities oxidized ferrous iron completely and relatively rapidly. Beginning from the fourth day of bioleaching, the ferric iron concentration remained almost constant due to the low pulp density and relatively low iron content (21.8 wt %) in the concentrate.

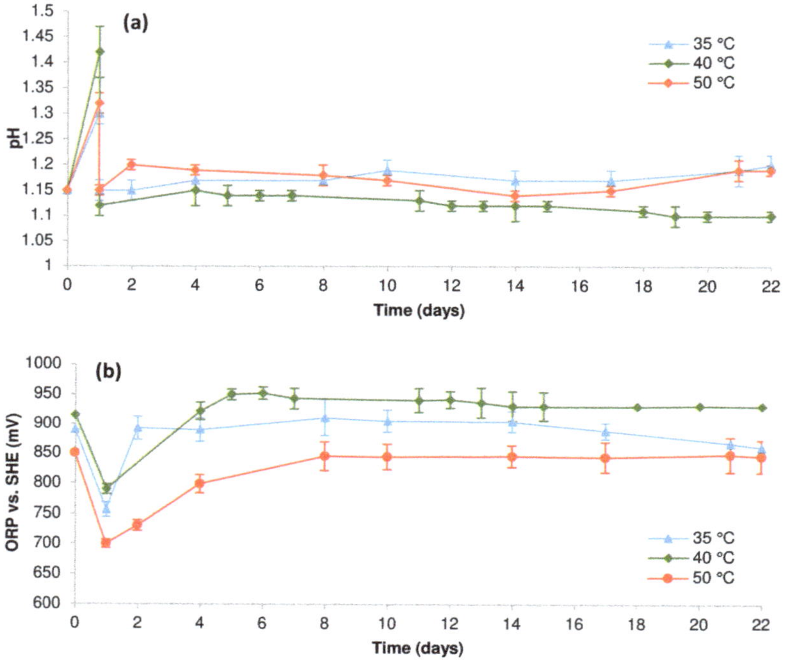

Figure 1. Values of (**a**) pH and (**b**) redox potential during stirred-tank bioleaching (pulp density, 1 wt%).

Table 3. Content of main minerals in the bioleach residues (pulp density, 1 wt%; duration, 22 days).

Temperature at Bioleaching (°C)	Content (wt%)						
	Chalcopyrite	Pentlandite	Violarite	Pyrite	Pyrrhotite	Jarosite	Elemental Sulfur
35	35	2	0	2	0	25	1
40	45	traces	0	1	0	17	1
50	30	traces	0	traces	0	40	1

Figure 3 shows the dissolution of nickel and copper during bioleaching of the concentrate at different temperatures. The lowest dissolution of nickel was recorded at 35 °C. At 40 and 50 °C, the final nickel dissolution was close and reached 97.2 and 96.3 wt%, respectively. The highest dissolution of copper was observed at 50 °C. A greater dissolution of chalcopyrite (as well as other sulfide minerals) by thermotolerant and moderately thermophilic microorganisms in comparison with mesophiles is well known. Among other factors, it is caused by an increase in the rate of chemical reactions with increasing temperatures [36–38]. Moreover, an increase in the solubility of chalcopyrite in ferric sulfate solutions has been repeatedly shown at relatively low values of the redox potential [39–41]. Figure 2 indicates that the value of the redox potential was, on average, 100 mV less during bioleaching of the concentrate at 50 °C than at 40 °C. Choosing a more suitable temperature of these two and taking into account that the goal of bioleaching was to achieve the greatest removal of nickel and to obtain a high-grade copper concentrate as solids, a community growing at 40 °C should be selected.

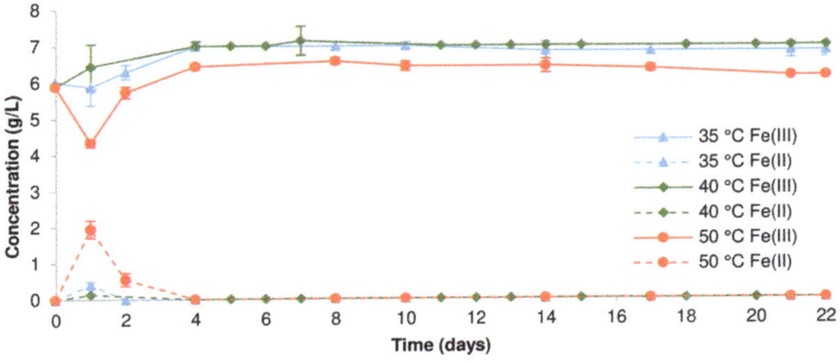

Figure 2. Ferric and ferrous iron concentration in the aqueous phase during stirred-tank bioleaching of the concentrate (pulp density, 1 wt%).

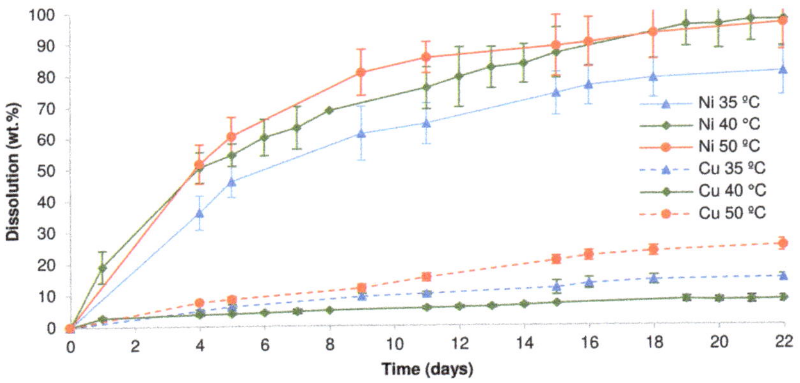

Figure 3. Rate of copper and nickel dissolution during stirred-tank bioleaching (pulp density, 1 wt%).

3.2. Solids Composition

The solids obtained after 22 days of bioleaching of the copper–nickel concentrate were subjected to chemical and mineralogical analyses to determine their compositions. The results of powder X-ray diffraction and light microscopy of polished sections of solids in comparison with the original concentrate are shown in Figures 4 and 5, respectively. The calculations of the mineral contents are presented in Table 3. The content of the main elements in bioleach residues is shown in Table 4. Analysis of the original concentrate composition indicated the presence of the following sulfide minerals: pentlandite, violarite, pyrrhotite, pyrite, and chalcopyrite. According to the data presented, pyrrhotite and violarite were almost completely removed from the concentrate during the bioleaching at all three temperatures. At 35 °C, a small amount of pentlandite was found in the solids, while it was almost completely dissolved at other temperatures. Pyrite was present in all three bioleach residues. However, its content was the highest in the solid obtained at 35 °C. These patterns agree with the refractoriness of sulfide minerals in ferric sulfate solutions due to the rest potential of the sulfide minerals [42]. Mason and Rice [43] have previously shown that the solubility of sulfide minerals during the bioleaching of nickel concentrate may be ranked in the following order: pyrrhotite > violarite > pyrite > pentlandite > chalcopyrite.

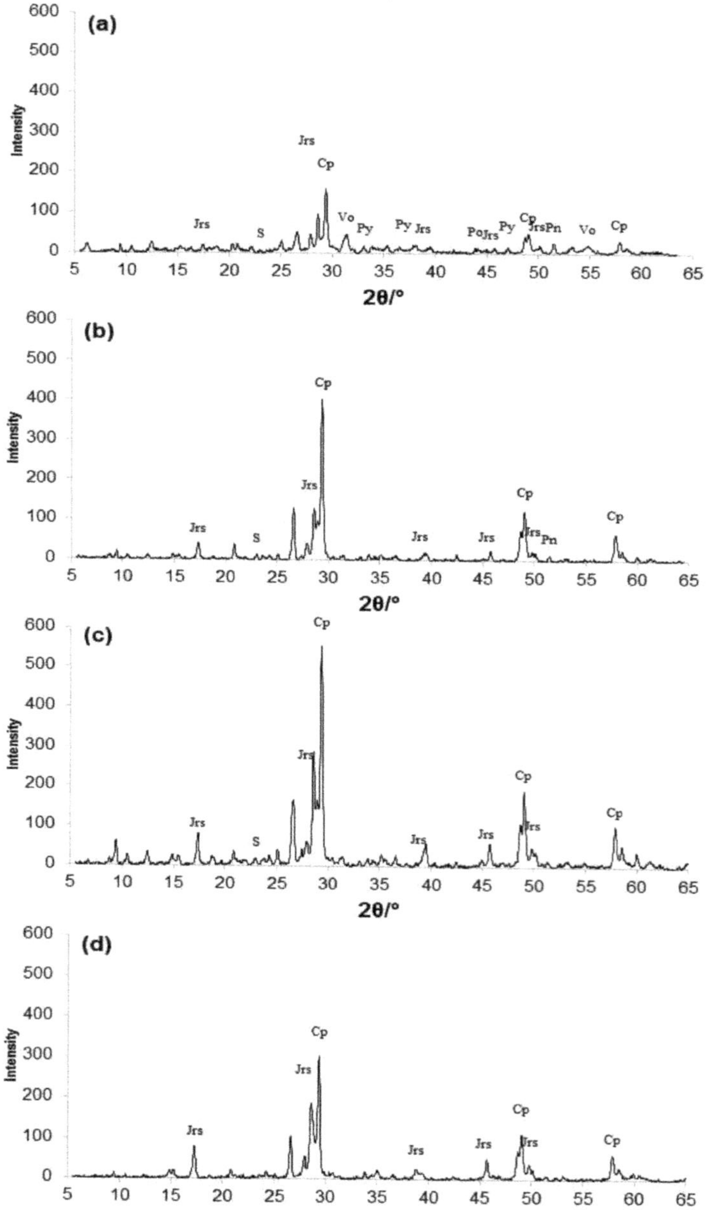

Figure 4. XRD patterns (**a**) of the original concentrate and bioleach residues at (**b**) 35, (**c**) 40, and (**d**) 50 °C (Pn, pentlandite; Vo, violarite; Cp, chalcopyrite; Py, pyrite; Po, pyrrhotite; S, elemental sulfur; Jrs, jarosite).

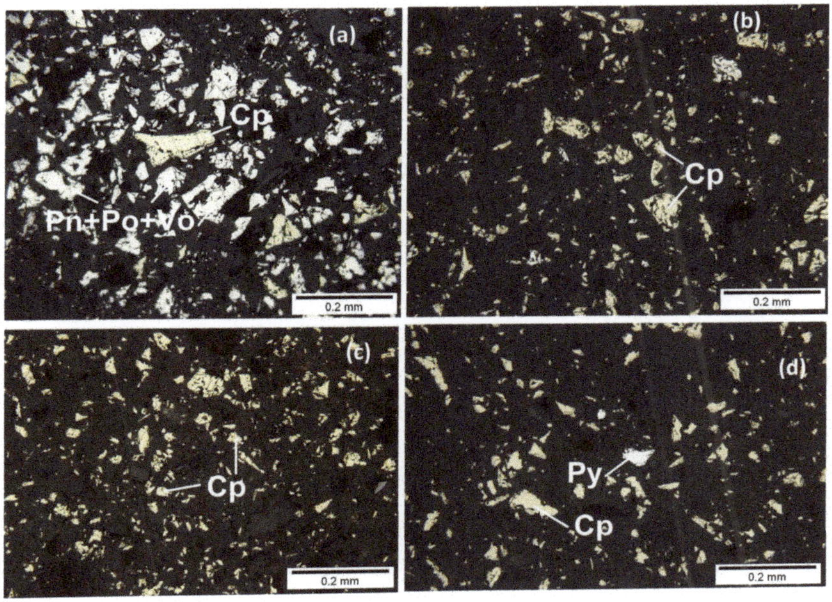

Figure 5. Optical micrographs of (**a**) the original concentrate and solids after their bioleaching at (**b**) 35, (**c**) 40, and (**d**) 50 °C, respectively (Pn, pentlandite; Vo, violarite; Cp, chalcopyrite; Py, pyrite; Po, pyrrhotite).

Table 4. Mass loss of concentrate during bioleaching and content of main elements in bioleach residues (pulp density, 1 wt %; duration, 22 days).

Temperature at Bioleaching (°C)	Mass Loss (wt %)	Content (wt %)			
		Cu	Ni	Fe	S
35	27	12.3	1.54	20.2	18.1
40	40	15.6	0.54	21.4	19.8
50	27	10.3	0.18	20.6	15.7

In general, bioleach residues obtained with thermotolerant and moderately thermophilic microbial cultures at 40 and 50 °C contained mainly chalcopyrite, jarosite, quartz, and chabazite. In fact, these residues were copper concentrates. At the same time, according to the Russian Federation standard (GOST R 52998–2008), only the solid obtained at 40 °C was a high-grade copper concentrate, which contained 15.6 wt% copper.

3.3. *Microbial Community Structure*

The original microbial consortium was composed of acidophilic bacteria and archaea that were different in their phylogenetic position (see Section 2.2 and Table 5) and physiological characteristics. This community included sulfur- and iron-oxidizing microorganisms as well as heterotrophic bacteria and archaea, which showed active growth within the temperature range of 35–50 °C. After subsequent culturing in the medium containing ferrous iron and elemental sulfur, certain changes in the community structure were recorded (Table 5). During 22 days of bioleaching, the cell populations were monitored using microscopy and metabarcoding analysis at the beginning and the end of the process. Although closely related microorganisms are usually not differentiated only based on the differences in the V3–V4 region, some bacterial and archaeal species were predicted due to the high variability of the V3–V4 regions within the corresponding genera [25].

Table 5. Structure of the microbial communities at the beginning and the end of the bioleaching of the concentrate at different temperatures.

Species	Original consortium 1	Inoculum 2, 35 °C	Final 3, 35 °C	Inoculum, 40 °C	Final, 40 °C	Inoculum, 50 °C	Final, 50 °C
At. ferrooxidans	+ 4	N/A	N/A	N/A	N/A	− 5	−
At. thiooxidans	+	N/A	N/A	N/A	N/A	−	−
At. caldus	+	N/A	N/A	N/A	N/A	−	−
Acidithiobacillus spp.	+	+	+	+	+	−	−
S. thermotolerans	+	+	+	+	+	+	+
S. thermosulfidooxidans	+	−	−	+	+	+	+
Al. tolerans	+	+	+	+	+	+	+
Ac. ferrooxidans	+	−	−	+	+	−	−
Fr. acidiphilum	+	+	+	+	−	−	−
Fr. acidarmanus	+	−	−	−	−	−	−
Acidiplasma sp.	+	−	−	−	−	+	+
Cuniculiplasma sp.	+	−	−	+	−	−	−
L. ferriphilum	+	+	−	+	+	−	−
Ap. multivorum	+	−	−	−	−	−	−
Fm. acidiphilium	+	−	−	+	−	−	−

[1] Original microbial consortium structure used to form subsequent original microbial communities at 35, 40, and 50 °C. [2] Original community structure in inocula grown at 35, 40, or 50 °C. [3] Structure of the communities at the end of the process (day 22). [4] +, present. [5] −, absent. Abbreviations: N/A, not available; *At. Acidithiobacillus*; *S. Sulfobacillus*; *Al. Alicyclobacillus*; *Ac. Acidibacillus*; *Ap. Acidiphilium*; *Fr. Ferroplasma*; *Fm. Ferrimicrobium*.

While the initial cell numbers at 35 and 40 °C were similar (0.8×10^9 cells/mL), the original microbial community was more diverse in genera at 40 C. At 50 °C, the cell number was approximately one order of magnitude lower (0.1×10^9 cells/mL), and the microbial diversity was the lowest among all three variants (Table 5). In the course of copper–nickel concentrate bioleaching at 35 and 40 °C, the total cell numbers increased up to 2.5×10^9 cells/mL during the first seven days of bioleaching, and this pattern did not change until the end of the process. At 50 °C, the cell abundance increased until the last day of the process, reaching 3.0×10^8 cells/mL after seven days of cultivation and 6.0×10^8 cells/mL on day 22. Such variations in microbial growth patterns are mainly due to the different microbial structures of the communities at these temperatures. At temperatures over 45 °C, the most abundant sulfur-oxidizing and iron-oxidizing members, such as *Acidithiobacillus* spp. that reach the numbers of $>10^9$ cells/mL, are usually eliminated due to their temperature growth limit [7]. At the end of the process, communities formed at 35 and 40 °C included *Acidithiobacillus* spp., *Sulfobacillus* spp., and *Alicyclobacillus* spp.

It is noteworthy that metabarcoding analysis indicated the presence of *Sulfobacillus* spp. and *Alicyclobacillus* spp. at all temperatures studied (Table 5). Strains of different *Sulfobacillus* species are mixotrophic bacteria capable of both sulfur and iron oxidation in the presence of small amounts of organic substances in the temperature range of 20–60 °C, with an optimum of >35–55 °C [7,44]. Except for 35 °C, the genus *Sulfobacillus* in the microbial communities was represented by the species *S. thermosulfidooxidans*. Nucleotide sequences of the 16S rRNA gene fragment showed 100% similarity with those of *S. thermosulfidooxidans* strains and only 99.1 and 98.7% similarity to other closest species: *S. sibiricus* and *S. benefaciens*, respectively. In contrast, another species, *S. thermotolerans* (100% similarity to the type strain *S. thermotolerans* Kr1^T, as well as other *S. thermotolerans* strains, and only 97.5% similarity to the second closest species *S. thermosulfidooxidans*), was identified at all temperatures: 35, 40, and 50 °C. Previously, this species has been shown to play a crucial role in biotechnological processes and to possess high metabolic and resistance capacities [25,44].

As mentioned above, in addition to *Sulfobacillus* spp., bacteria assigned to the genus *Alicyclobacillus* were also detected at all temperatures. Most *Alicyclobacillus* species are known to possess organoheterotrophic metabolism. However, *Al. tolerans* can grow heterotrophically or mixotrophically, oxidizing elemental sulfur, ferrous iron, and/or sulfide minerals in the presence of organic substances in the medium, with the optimum growth temperature of 37–42 °C [45]. Most probably, it was this species that was present in the microbial communities in these bioleaching experiments. This conclusion was based on 100% similarity to nucleotide sequences of the 16S rRNA gene fragments of *Al. tolerans* strains, including the type strain *Al. tolerans* K1^T [45] and another close strain *Al. tolerans* PCG-3 [46], and only 93.8% similarity to the second closest species *Al. disulfidooxidans*. Thus, in all communities, the genus *Alicyclobacillus* was represented by the species *Al. tolerans* characterized by mixotrophic/heterotrophic metabolism.

Iron-oxidizing bacteria *Leptospirillum* spp. (most probably, *L. ferriphilum*, according to the results of metabarcoding analysis) were identified in the original communities grown at 35 and 40 °C. At 40 °C, *Leptospirillum* cells reached ≈50% of the total microbial population at the beginning of cultivation, substantially decreasing in number in the course of the process due to the prevailing oxidation of sulfur by the community in the middle and at the end of the bioleaching. Moreover, at this temperature, the original microbial community and the community formed at the end of the bioleaching process showed the highest diversity and was also represented by facultatively chemolithoheterotrophic bacterium *Acidibacillus* sp. The latter was most probably related to the species *Ac. ferrooxidans* based on the nucleotide sequences of the 16S rRNA gene fragment, similarly to the data on the bioleaching of another copper–nickel concentrate at 40 °C [23]. *Acidibacillus* spp. belong to the phylum *Firmicutes*, which is well separated from *Sulfobacillus* and *Alicyclobacillus* spp. and capable of sulfur and/or ferrous iron oxidation [47]. At 40 °C, a heterotrophic iron-oxidizing extremely acidophilic actinobacterium assigned to the genus *Ferrimicrobium*, *Fm. acidiphilium* [48], was also identified. *Fm. acidiphilium* is the only known species belonging to this genus. In addition to heterotrophic/mixotrophic *Alicyclobacillus* strains, another heterotrophic member of the community, archaeon *Cuniculiplasma* sp. [49], was also detected in the original community formed at 40 °C. At 50 °C, an iron-oxidizing archaeon *Acidiplasma* sp., which is known to oxidize ferrous iron at increased temperatures [50], was functioning in the community during bioleaching of the concentrate.

Thus, the microbial community at 40 °C (both the original one and the community formed at the end of the bioleaching process) was the most diverse in the genus and species composition. At the same time, the crucial roles in bioleaching belonged to mixotrophic and heterotrophic acidophiles at all temperatures of the bioleaching process. This result underlines the pivotal role of mixotrophs and the underestimated contribution of heterotrophic microorganisms that facilitate the growth of chemolithotrophic community members and oxidation of the concentrate components by them.

4. Conclusions

In this study, bioprocessing of the bulk copper–nickel concentrate with different thermoacidophilic microbial communities was compared at three different temperatures: 35, 40, and 50 °C. According to the results, the optimum temperature for copper–nickel concentrate bioleaching proved to be 40 °C, at which a high-grade copper concentrate was obtained and nickel was almost completely leached. At increased temperature (50 °C), the oxidation level of chalcopyrite was higher and yielded a decreased content of copper in the final solid. At a lower temperature (35 °C), relatively high content of nickel remained in the bioleach residue. Analysis of the microbial communities involved in the process indicated that the crucial roles in bioleaching belonged to mixotrophic and heterotrophic acidophiles. This result underlined an underestimated role of heterotrophic microorganisms that indirectly contributed to the efficiency of the bioprocess, facilitating the growth and oxidation of the concentrate components by other chemolithotrophic community members. Insights into the patterns of the biooxidation of nickel sulfidic minerals with acidophilic microbial communities in different modes of the process are the key issues to intensify the biohydrometallurgical treatment of nickel-containing raw materials. Such studies can increase the attractiveness of biotechnologies to industrialists and improve their commercial applications.

Author Contributions: Conceptualization, M.M. and A.P.; methodology, M.M, A.P. and V.B.; validation, M.M. and A.P.; formal analysis, N.F.; investigation, M.M., A.P., N.F. and V.B.; data curation, M.M. and A.P.; writing—original draft preparation, M.M. and A.P.; writing—review and editing, M.M. and A.P.; visualization, M.M. and A.P.; supervision, A.P.; project administration, M.M.; funding acquisition, M.M. All authors have read and agreed to the published version of the manuscript.

Funding: This research was funded by Russian Science Foundation, grant number 21-14-00077.

Institutional Review Board Statement: Not applicable.

Informed Consent Statement: Not applicable.

Data Availability Statement: Not applicable.

Conflicts of Interest: The authors declare no conflict of interest. The funder had no role in the design of the study; in the collection, analyses, or interpretation of data; in the writing of the manuscript, or in the decision to publish the results.

References

1. Wills, B.A.; Finch, J.A. *Wills' Mineral Processing Technology: An Introduction to the Practical Aspects of Ore Treatment and Mineral Recovery*, 8th ed.; Butterworth-Heinemann: Oxford, UK, 2015.
2. Frias, C.; Sánchez, F.; Van Staden, P.; Milanovic, D.; Kolehmainen, E. The IntMet project provides innovative hydro and biohydro-Technologies to deal efficiently with polymetallic and complex suphide ores. In Proceedings of the IMPC 2018–29th International Mineral Processing Congress, Moscow, Russia, 17–21 September 2018; Canadian Institute of Mining, Metallurgy and Petroleum: Montreal, QC, Canada, 2019; pp. 2690–2699.
3. Kaksonen, A.H.; Deng, X.; Bohu, T.; Zea, L.; Khaleque, H.N.; Gumulya, Y.; Boxall, N.J.; Morris, C.; Cheng, K.Y. Prospective directions for biohydrometallurgy. *Hydrometallurgy* **2020**, *195*, 105376. [CrossRef]
4. Yin, S.-H.; Chen, W.; Fan, X.-L.; Liu, J.-M.; Wu, L.-B. Review and prospects of bioleaching in the Chinese mining industry. *Int. J. Miner. Metall. Mater.* **2021**, *28*, 1397–1412. [CrossRef]
5. Kaksonen, A.H.; Lakaniemi, A.-M.; Tuovinen, O.H. Acid and ferric sulfate bioleaching of uranium ores: A review. *J. Clean. Prod.* **2020**, *264*, 121586. [CrossRef]
6. Li, J.; Yang, H.; Tong, L.; Sand, W. Some aspects of industrial heap bioleaching technology: From basics to practice. *Miner. Process. Extr. Metall. Rev.* **2021**, 1–9. [CrossRef]
7. Kondrat'eva, T.F.; Pivovarova, T.A.; Tsaplina, I.A.; Fomchenko, N.V.; Zhuravleva, A.E.; Murav'ev, M.I.; Melamud, V.S.; Bulayev, A.G. Diversity of the communities of acidophilic chemolithotrophic microorganisms in natural and technogenic ecosystems. *Microbiology* **2012**, *81*, 1–24. [CrossRef]
8. Pathak, A.; Morrison, L.; Healy, M.G. Catalytic potential of selected metal ions for bioleaching, and potential techno-economic and environmental issues: A critical review. *Biores. Technol.* **2017**, *229*, 211–221. [CrossRef]
9. Potysz, A.; Pędziwiatr, A.; Hedwig, S.; Lenz, M. Bioleaching and toxicity of metallurgical wastes. *J. Environ. Chem. Eng.* **2020**, *8*, 104450. [CrossRef]
10. Fomchenko, N.V.; Muravyov, M.I. Two-step biohydrometallurgical technology of copper-zinc concentrate processing as an opportunity to reduce negative impacts on the environment. *J. Environ. Manag.* **2018**, *226*, 270–277. [CrossRef]
11. Mahmoud, A.; Cezac, P.; Hoadley, A.F.A.; Contamine, F.; D'Hugues, P. A review of sulfide minerals microbially assisted leaching in stirred tank reactors. *Int. Biodeterior. Biodegradation* **2017**, *119*, 118–146. [CrossRef]
12. Kaksonen, A.H.; Mudunuru, B.M.; Hackl, R. The role of microorganisms in gold processing and recovery: A review. *Hydrometallurgy* **2014**, *142*, 70–83. [CrossRef]
13. Zhao, H.; Zhang, Y.; Zhang, X.; Qian, L.; Sun, M.; Yang, Y.; Zhang, Y.; Wang, J.; Kim, H.; Qiu, G. The dissolution and passivation mechanism of chalcopyrite in bioleaching: An overview. *Miner. Eng.* **2019**, *136*, 140–154. [CrossRef]
14. Harvey, T.J.; Yen, W.T.; Paterson, J.G. Selective zinc extraction from complex copper/zinc sulphide concentrates by pressure oxidation. *Miner. Eng.* **1992**, *5*, 975–992. [CrossRef]
15. Fomchenko, N.V.; Muravyov, M.I. Selective leaching of zinc from copper-zinc concentrate. *Appl. Biochem. Microbiol.* **2017**, *53*, 73–77. [CrossRef]
16. Fomchenko, N.; Uvarova, T.; Muravyov, M. Effect of mineral composition of sulfidic polymetallic concentrates on non-ferrous metals bioleaching. *Miner. Eng.* **2019**, *138*, 1–6. [CrossRef]
17. Muravyov, M.I.; Panyushkina, A.E.; Melamud, V.S.; Bulaev, A.G.; Fomchenko, N.V. Ferric leaching of bulk sulfidic concentrates with biologically generated solution. *Appl. Biochem. Microbiol.* **2021**, *57*, 493–499. [CrossRef]
18. Fomchenko, N.V.; Muravyov, M.I. Effect of sulfide mineral content in copper-zinc concentrates on the rate of leaching of non-ferrous metals by biogenic ferric iron. *Hydrometallurgy* **2019**, *185*, 82–87. [CrossRef]
19. Arpalahti, A.; Lundström, M. The leaching behavior of minerals from a pyrrhotite-rich pentlandite ore during heap leaching. *Miner. Eng.* **2018**, *119*, 116–125. [CrossRef]
20. Watling, H.R. The bioleaching of nickel-copper sulfides. *Hydrometallurgy* **2008**, *91*, 70–88. [CrossRef]

21. Watling, H.R. The bioleaching of sulphide minerals with emphasis on copper sulphides: A review. *Hydrometallurgy* **2006**, *84*, 81–108. [CrossRef]
22. Behera, S.K.; Manjaiah, M.; Sekar, S.; Panda, S.K.; Mavumengwana, V.; Mulaba-Bafubiandi, A.F. Optimization of microbial leaching of base metals from a South African sulfidic nickel ore concentrate by *Acidithiobacillus ferrooxidans*. *Geomicrobiol. J.* **2018**, *35*, 447–459. [CrossRef]
23. Muravyov, M.; Panyushkina, A.; Bulaev, A.; Fomchenko, N. Biobeneficiation of bulk copper-zinc and copper-nickel concentrates at different temperatures. *Miner. Eng.* **2021**, *170*, 107040. [CrossRef]
24. Gericke, M.; Govender, Y. Bioleaching strategies for the treatment of nickel-copper sulphide concentrates. *Miner. Eng.* **2011**, *24*, 1106–1112. [CrossRef]
25. Muravyov, M.; Panyushkina, A. Distinct roles of acidophiles in complete oxidation of high-sulfur ferric leach product of zinc sulfide concentrate. *Microorganisms* **2020**, *8*, 386. [CrossRef]
26. Panyushkina, A.; Bulaev, A.; Belyi, A.V. Unraveling the central role of sulfur-oxidizing *Acidiphilium multivorum* lms in industrial bioprocessing of gold-bearing sulfide concentrates. *Microorganisms* **2021**, *9*, 984. [CrossRef] [PubMed]
27. Silverman, M.P.; Lundgren, D.G. Studies on the chemoautotrophic iron bacterium *Ferrobacillus ferrooxidans*. I. An improved medium and a harvesting procedure for securing high cell yields. *J. Bacteriol.* **1959**, *77*, 642–647. [CrossRef] [PubMed]
28. Fadrosh, D.W.; Ma, B.; Gajer, P.; Sengamalay, N.; Ott, S.; Brotman, R.M.; Ravel, J. An improved dual-indexing approach for multiplexed 16S rRNA gene sequencing on the Illumina MiSeq platform. *Microbiome* **2014**, *2*, 1–7. [CrossRef] [PubMed]
29. Caporaso, J.G.; Kuczynski, J.; Stombaugh, J.; Bittinger, K.; Bushman, F.D.; Costello, E.K.; Fierer, N.; Pēa, A.G.; Goodrich, J.K.; Gordon, J.I.; et al. QIIME allows analysis of high-throughput community sequencing data. *Nat. Methods* **2010**, *7*, 335–336. [CrossRef] [PubMed]
30. Quast, C.; Pruesse, E.; Yilmaz, P.; Gerken, J.; Schweer, T.; Yarza, P.; Peplies, J.; Glöckner, F.O. The SILVA ribosomal RNA gene database project: Improved data processing and web-based tools. *Nucleic Acids Res.* **2013**, *41*, D590–D596. [CrossRef] [PubMed]
31. Davis, D.G.; Jacobsen, W.R. Determination of iron and iron-aluminum mixtures by titration with EDTA. *Anal. Chem.* **1960**, *32*, 215–217. [CrossRef]
32. Filippova, N.A. *Phazovy Analiz rud i Produktov ikh Pererabotki*, 2nd ed.; Khimiya: Moscow, Russia, 1975.
33. Rietveld, H.M. Line profiles of neutron powder-diffraction peaks for structure refinement. *Acta Cryst.* **1967**, *22*, 151–152. [CrossRef]
34. Cruz, F.L.S.; Oliveira, V.A.; Guimarães, D.; Souza, A.D.; Leão, V.A. High-temperature bioleaching of nickel sulfides: Thermodynamic and kinetic implications. *Hydrometallurgy* **2010**, *105*, 103–109. [CrossRef]
35. Soe, K.M.; Ruan, R.; Jia, Y.; Tan, Q.; Wang, Z.; Shi, J.; Zhong, C.; Sun, H. Influence of jarosite precipitation on iron balance in heap bioleaching at monywa copper mine. *J. Min. Inst.* **2021**, *247*, 102–113. [CrossRef]
36. Hedrich, S.; Joulian, C.; Graupner, T.; Schippers, A.; Guézennec, A.-G. Enhanced chalcopyrite dissolution in stirred tank reactors by temperature increase during bioleaching. *Hydrometallurgy* **2018**, *179*, 125–131. [CrossRef]
37. Ranjbar, M.; Fazaelipoor, M.H.; Ranjbar Hamghavandi, M.; Schaffie, M.; Manafi, Z. Modeling study of the bio-dissolution of copper concentrate in a continuous bioreactors system. *Miner. Eng.* **2020**, *153*, 384. [CrossRef]
38. Liu, H.-C.; Xia, J.-L.; Nie, Z.-Y.; Liu, L.-Z.; Wang, L.; Ma, C.-Y.; Zheng, L.; Zhao, Y.-D.; Wen, W. Comparative study of S, Fe and Cu speciation transformation during chalcopyrite bioleaching by mixed mesophiles and mixed thermophiles. *Miner. Eng.* **2017**, *106*, 22–32. [CrossRef]
39. Tian, Z.; Li, H.; Wei, Q.; Qin, W.; Yang, C. Effects of redox potential on chalcopyrite leaching: An overview. *Miner. Eng.* **2021**, *172*, 107135. [CrossRef]
40. Tanne, C.; Schippers, A. Electrochemical investigation of microbially and galvanically leached chalcopyrite. *Hydrometallurgy* **2021**, *202*, 105603. [CrossRef]
41. Zhao, H.; Wang, J.; Yang, C.; Hu, M.; Gan, X.; Tao, L.; Qin, W.; Qiu, G. Effect of redox potential on bioleaching of chalcopyrite by moderately thermophilic bacteria: An emphasis on solution compositions. *Hydrometallurgy* **2015**, *151*, 141–150. [CrossRef]
42. Fu, K.; Ning, Y.; Chen, S.; Wang, Z. Bioleaching of different copper sulphide minerals and their physicochemical properties dependence. *Trans. Inst. Min. Metall. Sect. C Miner. Process. Extr. Metall.* **2016**, *125*, 1–4. [CrossRef]
43. Mason, L.J.; Rice, N.M. The adaptation of *Thiobacillus ferrooxidans* for the treatment of nickel-iron sulphide concentrates. *Miner. Eng.* **2002**, *15*, 795–808. [CrossRef]
44. Panyushkina, A.E.; Babenko, V.V.; Nikitina, A.S.; Selezneva, O.V.; Tsaplina, I.A.; Letarova, M.A.; Kostryukova, E.S.; Letarov, A.V. *Sulfobacillus thermotolerans*: New insights into resistance and metabolic capacities of acidophilic chemolithotrophs. *Sci. Rep.* **2019**, 9. [CrossRef] [PubMed]
45. Karavaiko, G.I.; Bogdanova, T.I.; Tourova, T.P.; Kondrat'eva, T.F.; Tsaplina, I.A.; Egorova, M.A.; Krasil'nikova, E.N.; Zakharchuk, L.M. Reclassification of "*Sulfobacillus thermosulfidooxidans* subsp. *thermotolerans*" strain K1 as *Alicyclobacillus tolerans* sp. nov. and *Sulfobacillus disulfidooxidans* Dufresne et al. 1996 as *Alicyclobacillus disulfidooxidans* comb. nov., and emended description. *Int. J. Syst. Evol. Microbiol.* **2005**, *55*, 941–947. [CrossRef] [PubMed]
46. Panyushkina, A.E.; Tsaplina, I.A.; Grigor'eva, N.V.; Kondrat'eva, T.F. Thermoacidophilic microbial community oxidizing the gold-bearing flotation concentrate of a pyrite-arsenopyrite ore. *Microbiology* **2014**, *83*, 539–549. [CrossRef]
47. Holanda, R.; Hedrich, S.; Ñancucheo, I.; Oliveira, G.; Grail, B.M.; Johnson, D.B. Isolation and characterisation of mineral-oxidising "Acidibacillus" spp. from mine sites and geothermal environments in different global locations. *Res. Microbiol.* **2016**, *167*, 613–623. [CrossRef] [PubMed]

48. Johnson, D.B.; Bacelar-Nicolau, P.; Okibe, N.; Thomas, A.; Hallberg, K.B. *Ferrimicrobium acidiphilum* gen. nov., sp. nov. and Ferrithrix thermotolerans gen. nov., sp. nov.: Heterotrophic, iron-oxidizing, extremely acidophilic actinobacteria. *Int. J. Syst. Evol. Microbiol.* **2009**, *59*, 1082–1089. [CrossRef] [PubMed]
49. Golyshina, O.V.; Lünsdorf, H.; Kublanov, I.V.; Goldenstein, N.I.; Hinrichs, K.-U.; Golyshin, P.N. The novel extremely acidophilic, cell-wall-deficient archaeon *Cuniculiplasma divulgatum* gen. nov., sp. nov. represents a new family, *Cuniculiplasmataceae* fam. nov., of the order *Thermoplasmatales*. *Int. J. Syst. Evol. Microbiol.* **2016**, *66*, 332–340. [CrossRef] [PubMed]
50. Bulaev, A.G.; Kanygina, A.V.; Manolov, A.I. Genome analysis of *Acidiplasma* sp. MBA-1, a polyextremophilic archaeon predominant in the microbial community of a bioleaching reactor. *Microbiol. Russian Fed.* **2017**, *86*, 89–95. [CrossRef]

metals

MDPI

Article

Biosorption and Bioaccumulation Capacity of *Arthrospira platensis* toward Yttrium Ions

Nikita Yushin [1,2], Inga Zinicovscaia [1,3,4,*], Liliana Cepoi [5], Tatiana Chiriac [5], Ludmila Rudi [5] and Dmitrii Grozdov [1]

1 Department of Nuclear Physics, Joint Institute for Nuclear Research, 141980 Dubna, Russia
2 Doctoral School of Biological, Geonomic, Chemical and Technological Science, State University of Moldova, MD-2009 Chisinau, Moldova
3 Department of Nuclear Physics, Horia Hulubei National Institute for R&D in Physics and Nuclear Engineering, MG-6, Magurele, 077125 Ilfov, Romania
4 Laboratory of Physical and Quantum Chemistry, Institute of Chemistry, MD-2028 Chisinau, Moldova
5 Laboratory of Phycobiotechnology, Institute of Microbiology and Biotechnology, MD-2028 Chisinau, Moldova
* Correspondence: zinikovskaia@mail.ru; Tel.: +7-4962165609

Abstract: Yttrium is an element of critical importance for industry and technology. Cyanobacteria *Arthrospira platensis* was employed for Y(III) recovery from contaminated wastewater through biosorption and bioaccumulation processes. The effect of pH of a solution, contact time, temperature, and initial Y(III) concentration on the adsorption behaviour of *Arthrospira platensis* were studied. The maximum adsorption capacity of 719.8 mg/g was attained at a pH of 3, temperature of 20 °C, and adsorption time of 3 min. The Langmuir and Freundlich isotherm models were suitable to describe the equilibrium of the biosorption, while kinetic of the process followed the pseudo-first-order model. Thermodynamic parameters showed that the biosorption process was spontaneous and exothermic in nature. In bioaccumulation experiments, *Arthrospira platensis* was able to remove up to 70% of Y(III) from the solution. Beside biomass uptake capacity, the toxic effect of Y(III) on the biomass productivity and biochemical composition was assessed. Thus, Y(III) in concentration of 10–30 mg/L led to a decrease in the content of proteins, carbohydrates, and phycobiliproteins in the biomass and had no significant negative impact on productivity and photosynthetic pigments content. Experiments performed using *Arthrospira platensis* showed that biological objects have a great potential to be applied for the recovery of rare earth elements from wastewater.

Keywords: yttrium; bioremediation; cyanobacteria; pollution

Citation: Yushin, N.; Zinicovscaia, I.; Cepoi, L.; Chiriac, T.; Rudi, L.; Grozdov, D. Biosorption and Bioaccumulation Capacity of *Arthrospira platensis* toward Yttrium Ions. *Metals* **2022**, *12*, 1465. https://doi.org/10.3390/met12091465

Academic Editor: Felix A. Lopez

Received: 15 August 2022
Accepted: 29 August 2022
Published: 31 August 2022

1. Introduction

Rare earth elements, because of their excellent physical and chemical properties, are widely used in various fields of industry, specifically in global clean energy technology development and consumer products [1,2]. Today, rare earth elements are defined as the "industrial vitamin" [2]. Yttrium, along with the other four elements Dy, Eu, Nd, and Tb amount up to ~85% of the total weight of rare earth elements in final products [2,3]. Yttrium has been widely used as a catalyst in a wide array of industries, as a deoxidizer for non-ferrous metals, in fiber optics, and in making fireproof bricks [2,4].

Today, Y(III) released from mining activity, processing, and the disposal of wastes contaminate objects of the environment [5]. For, example, it was reported that concentrations of rare earth elements downstream of the river and wastewater of mining and refining factories can range from 1 to 200 mg/L [3]. The discharge of waste containing rare earth elements, including Y(III), in the environment not only causes the loss of rare earth resources but also negatively affect humans and other living organisms [3].

Information regarding Y(III) toxicity is very controversial. Thus, Ding and co-authors [6] demonstrated that YCl_3 treatment greatly promoted neuron cells' death in rats. Among the three freshwater organisms *Daphnia magna*, *Chironomus riparius*, and *Oncorhynchus mykiss*, only *Chironomus riparius* showed toxicity at Y(III) exposure concentrations close to environmental ones [5]. In the study of Nakamura et al. [7], Y(III) in rats was mainly accumulated in the liver, spleen, and bones, but it did not provoke any visible toxic effects.

Recovery of the rare earth elements from wastewater, in which their concentrations range from 1 to 100 mg/L is a key concern for the economy and environmental protection. Various techniques have been designed to ensure the efficient recovery of rare earth elements, including chemical precipitation, filtration, solvent extraction, membrane technology, and adsorption [2]. However, conventional methods applied for industrial wastewater treatment are considered inefficient for diluted effluents, are relatively expensive, and require a high energy and chemical consumption. Some of these methods generate large volumes of waste and consequently can lead to secondary environmental pollution, while others have limited applications [8,9].

The cyanobacterium *Arthrospira platensis* (spirulina) is a well-known model object used in bioremediation studies to elaborate effective procedures of pollutant removal from aquatic environments [10–16]. Spirulina application in bioremediation studies is explained by the fact that cyanobacteria are polyextremophile organisms able to cope with high alkalinity, temperature, salt concentration, and the presence of different pollutants in culture media. Moreover, spirulina can accumulate a large amount of biomass under these conditions. Similarly, spirulina is a technologically convenient object, being easy to handle under both laboratory and industrial conditions, which significantly facilitates the implementation of the developed bioremediation technology [17].

Microbial bioremediation of metal-polluted industrial effluents is considered one of the most effective eco-friendly tools to combat pollution [8]. Biological pathways of metal removal from wastewater are low-cost, ecologically viable, and safe for the environment [9]. It is suggested that biological objects possess high uptake capacities and tolerances toward rare earth elements.

In the present study, the biosorption and bioaccumulation capacity of *Arthrospira platensis* (*A. platensis*) toward Y(III) was tested. Biomass biosorption capacity was investigated regarding the influence of pH, contact time, initial Y(III) ions concentration, and temperature. The sorption process was described using well-known kinetics, equilibrium, and thermodynamics models. The effect of the Y(III) in different concentrations on the spirulina accumulation capacity, productivity, and biochemical parameters (proteins, carbohydrates, lipids, pigments) was assessed.

2. Materials and Methods

2.1. Chemicals

All the chemical reagents of analytical grade used in the study were purchased from Sigma-Aldrich (Darmstadt, Germany).

2.2. Object of Study

The biosorption and bioaccumulation experiments were carried out with the cyanobacterium *Arthrospira platensis* (spirulina) CNMN-CB-02 from the collection of non-pathogenic microorganisms (Institute Microbiology and Biotechnology, Chisinau, Moldova). The composition of the cultivation medium and conditions of biomass growth are presented in [10]. Spirulina biomass was used as biosorbent after the full cultivation cycle was separated from the medium by filtration, dried at 105 °C, and homogenized. The conditions of biomass growth in bioaccumulation experiments are presented in Section 2.4.

2.3. Biosorption

Biosorption experiments were performed in a 50-milileter Erlenmeyer flask containing 20 mL of yttrium solution and 100 mg of biomass. Single-factor experiments were performed in order to reveal the effect of pH (2.0–6.0), yttrium initial concentration (10–100 mg/L), contact time (2–120 min), and temperature (20–50 °C). Studying the effect of parameters, except time, experiments lasted for 30 min, then the solution was filtrated and yttrium concentration was determined using ICP-OES PlasmaQuant PQ 9000 Elite spectrometer (Analytik Jena, Jena, Germany). Results are presented as average values obtained from the triplicate analysis. The removal efficiency and adsorption capacity of the spirulina were calculated according to Equations (1) and (2):

$$q = \frac{V(C_i - C_f)}{m} \quad (1)$$

$$E = \frac{C_i - C_f}{C_i} \times 100 \quad (2)$$

where V is the volume of the solution, ml; C_i and C_f are the initial and final metal concentrations in mg/L, and m is the mass of sorbent in g.

2.4. Bioaccumulation

In order to assess the bioaccumulation capacity of spirulina biomass Y(III) in concentrations of 10–30 mg/L, metal was added to the cultivation medium on the first day of biomass growth. Then, biomass was cultivated for six days at a temperature of 25–28 °C, illumination ~37 µM photons/m^2/s, and pH 8–9. At the end of the cultivation cycle, the spirulina biomass was separated from the culture liquid by filtration. The amount of biomass, expressed in g/L, was determined spectrophotometrically based on the calibration curve. The samples were standardized to the concentration of 10 mg/mL of distilled water. For the biochemical tests, the biomass was subjected to the repeated freezing–thawing procedure. The concentration of Y(III) in experimental solutions was determined by ICP-OES.

Served as a control in both types of experiments, spirulina biomass was cultivated in the medium without the addition of Y(III).

2.5. Biochemical Tests

Protein content was determined based on the Lowry method in samples of 10 mg/mL biomass. Biomass was processed with 0.1N NaOH solution for 30 min. Then, reagent (49 parts of 2% Na_2CO_3 solution in 0.1N NaOH and 1 part of 0.5% $CuSO_4$ solution in 1.0% $Na_3C_6H_5O_7$) and 0.2 mL of Folin–Ciocalteu reagent (Sigma Aldrich, Darmstadt, Germany) were added to 0.2 mL of basic extract. The absorbance at 750 nm was recorded through 30 min. The amount of protein was calculated based on the calibration curve plotted for bovine serum albumin [18].

Carbohydrates content was determined using Anthracene-9(10H)-one. The reaction mixture consisted of 2.5 mg of biomass and 2.5 mL of 0.5% Antron reagent (Sigma Aldrich, Darmstadt, Germany) solution in 66% H_2SO_4. The obtained samples were boiled on a water bath for 30 min. After cooling and incubation in the dark for 30 min, the absorbance of the samples was recorded at 620 nm. The quantitative calculation was performed based on the calibration curve plotted for glucose [12].

Quantitative determination of lipids was performed using the phosphovanillin reagent (Sigma Aldrich, Darmstadt, Germany). The lipid extract from the spirulina biomass was obtained by adding 1 mL of a mixture of chloroform and ethanol in a ratio of 9:1 (v:v) to 10 mg of biomass. The extraction was carried out by continuous stirring for 2 h at room temperature. The lipid extract underwent hydrolysis with sulfuric acid. The reaction mixture consisted of 0.1 mL of hydrolyzate and 2.9 mL of phosphovanillin reagent. After 30 min, the absorbance at 560 nm was recorded. The quantitative calculation was performed based on the calibration curve constructed for oleic acid [19].

The content of phycobiliproteins in the biomass was calculated using formulas based on molar coefficients for pigments. The hydrous extract of phycobiliproteins was obtained as a result of applying the procedure of repeated freezing–thawing of standardized biomass with distilled water. The absorbance of the extract at 620 nm was recorded for phycocyanin and at 650 nm for allophycocyanin [20].

The content of chlorophyll α and β-carotene in the biomass was determined using ethanolic extracts. The mixture of 10 mg of biomass with 1.0 mL of 96% ethyl alcohol was prepared.

The pigments extraction was performed by continuous stirring at room temperature for 12 h. Chlorophyll content was determined at an absorbance of 665 nm, and β-carotene at an absorbance of 450 nm. The quantitative calculation was performed based on the calibration curves.

The determination of the content of malonic dialdehyde (MDA) in the biomass was carried out based on the reactive substances of thiobarbituric acid (TBA). Three milliliters of 0.76% TBA solution in 20% $C_2HCl_3O_2$ were added to 10 mg of biomass. The obtained mixture was boiled on a water bath at 95 °C for 20 min, then the mixture was cooled and the absorbance was recorded at wavelengths of 532 nm and 600 nm. The content of MDA in samples was calculated using the extinction coefficient of the complex product of the MDA-TBA [21].

2.6. Antioxidant Activity

The antioxidant activity of the ethanolic and water extracts was determined using the radical cation ABTS (2,2 azinobis 3-ethylbenzothiazoline-6-sulfonic acid) [22]. The ethanolic extract from the biomass was obtained in a similar way to the extract for the determination of pigments. The water extract was obtained identically to the extracts for the determination of phycobiliproteins. ABTS radical cation was obtained by oxidation with potassium persulfate. ABTS stock solution was prepared by mixing 7 mM ABTS in deionized water with 2.45 mM $K_2S_2O_8$ in a 1:1 ratio. The ABTS oxidation process took place in the dark at room temperature for 12–16 h. The working radical ABTS solution had an absorbance value of 0.700 ± 0.020 at 734 nm. The samples were obtained by mixing 0.3 mL of biomass extract and 2.7 mL of ABTS solution. The absorbance of the samples was measured after 6 min. The % of inhibition was calculated relative to the absorbance of the ABTS reagent.

2.7. Statistical Analysis

The values in the manuscript are presented as an average of three experiments ± standard deviation. One-way analysis of variance (ANOVA) was performed using Student's *t*-test.

3. Results and Discussion

3.1. Yttrium Removal by Biosorption

The effect of pH, Y(III) concentration, contact time, and temperature on the biosorption capacity of A. platensis is shown in Figure 1.

The pH of the solution plays an important role in adsorption experiments affecting the charge of the adsorbent surface and the chemical state of the metal. The spirulina biosorption capacity increased from 55 to 78% with a rise in the pH from 2.0 to 3.0, and at a further increase in pH, a decline in the sorption was noticed (Figure 1a). At a pH of 2.0, relatively low biomass removal efficiency (55%) can be associated with the predominance of H^+ in the solution and their competition with Y(III) for binding sites. At a pH of 4.0–6.0, the removal efficiency of spirulina significantly decreased (up to 20%), which can be explained by the increase in the concentration of OH groups in the solution and Y(III) precipitation in the form of $Y(OH)_3$ [2]. According to [23], Y(III) at a pH range of 2.0–5.5 is present in the solution in the form of Y^{3+}, and at higher pH values, $Y(OH)_3$ is formed. Thus, further experiments were performed at a pH of 3.0. For a comparison maximum, Y(III) biosorption on *Serratia marcescens* was achieved at a pH of 5.5 [2]. The process of Y(III) sorption on bacteria *Microbacterium* sp., *Curtobacterium* sp., *Bacillus subtilis*, *Pseudomonas putida*, and *Bacillus pumilis* better proceeded at a pH of 4 [24].

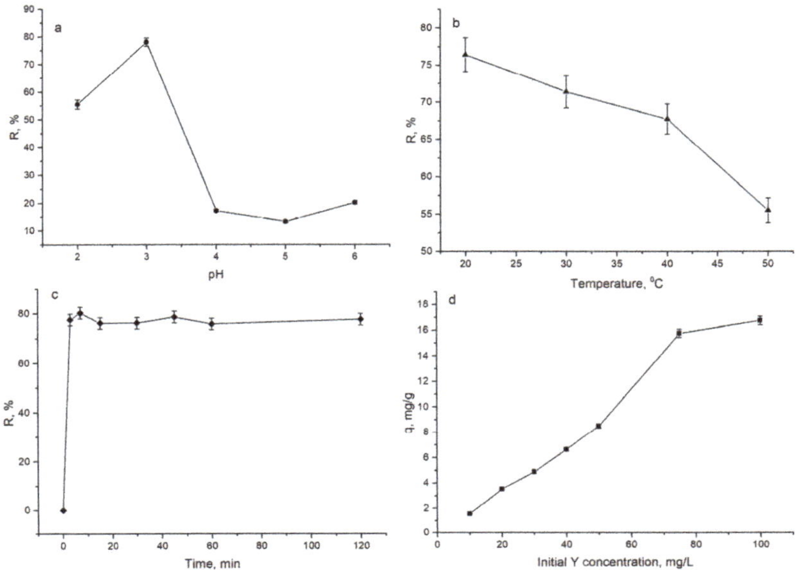

Figure 1. Influence of adsorption parameters (**a**) pH; (**b**) temperature, (**c**) time, and (**d**) Y(III) concentration on the biosorption capacity of spirulina biomass.

An increase in temperature from 20 to 50 °C resulted in a decrease in the spirulina adsorption capacity from 76 to 55 % (Figure 1b). The decrease in biomass sorption capacity occurred with the temperature point at the exothermic character of Y(III) biosorption. The result is opposite to that reported by [2], which showed that Y(III) sorption on Serratia marcescens was an endothermic process. At the same time, the results are in line with [25], which reported that a temperature increase produced a decrease in Y(III) sorption by microbial biomass.

The process of Y(III) biosorption was very rapid; the maximum metal removal was achieved in 3 min of interaction when 76% of ions were removed from the solution. An increase in the time of biosorption up to 120 min has no noticeable impact on the A. platensis adsorption efficiency. Rapid Y(III) adsorption in the first minutes of sorbent interaction with the sorbate is explained by a large number of unoccupied biosorption sites. Reduction in the rate of Y(III) sorption and equilibrium achievement is associated with the occupation of binding sites. The highest sorption rate for Y(III) was registered during the first 10 min for Microbacterium sp., Bacillus pumilis, and *Pseudomonas putida* strains, while the highest sorption rate for *Curtobacterium* sp. and *Bacillus subtilis* strains was detected after 35 min of the experiment [24].

Y(III) initial concentration in the solution was directly proportional to its sorption on the spirulina; the adsorption capacity of 1.6 mg/g recorded at a Y(III) concentration of 10 mg/L increased to 16.7 mg/g at a metal concentration of 100 mg/L. Spirulina biosorption capacity rapidly increased at a Y(III) concentration range of 10–75 mg/L, which is in agreement with [2], and then the rise was less pronounced.

Experimentally obtained data were used to describe equilibrium, kinetics, and thermodynamics of Y(III) biosorption on A. platensis biomass. Pseudo-first- and pseudo-second-order, as well as Elovich models, have been employed to assess the adsorption reaction mechanism (see Equations (3)–(5)):

$$q = q_e\left(1 - e^{-k_1 t}\right) \quad (3)$$

$$q = \frac{q_e^2 k_2 t}{1 + q_e k_2 t} \quad (4)$$

$$q_t = \frac{1}{\beta} \ln(1 + \alpha\beta t) \quad (5)$$

where q_e and q_t are the content of metal (mg/g) adsorbed at equilibrium and at t time, k_2 (g/mg·min) and k_1 (1/min) are rates constant of pseudo-first-order and pseudo-second-order, and α (g/mg·min) and β (g/mg) are the Elovich equation constants.

In addition to the coefficient of determination, the applicability of applied models was assessed using the sum of error squares (SSE, %) and nonlinear chi-square test (χ^2)-te, given by Equations (6) and (7):

$$SSE = \sum_{i=1}^{n} \left(q_{e,\,cal} - q_{e,\,exp}\right)^2 \quad (6)$$

$$\chi^2 = \sum_{i=1}^{n} \frac{\left(q_{e,\,cal} - q_{e,\,exp}\right)^2}{q_{e,\,exp}} \quad (7)$$

where *n* is the number of data points.

The graphical presentation of the data is illustrated in Figure 2a and the parameters of the applied models are listed in Table 1. According to coefficient of determination values, all three models can be considered applicable for the description of experimental data. Values of SSE and χ^2 , 0.006 and 0.003, respectively (the same values for both models were obtained), showed that the pseudo-first model and pseudo-second-order model were suitable to satisfactorily describe Y(III) biosorption. However, extremely high values of α in the Elovich model and a negative value of k_2 in the pseudo-second model exclude the possibility of their use for the description of experimental findings. Additionally, the Akaike Information Criterion (AIC) test was applied to emphasize which model best describes the experimentally obtained values. According to the AIC test, the pseudo-first-order model better describes Y(III) biosorption on spirulina biomass. The applicability of the pseudo-first-order model suggests that the process of Y(III) sorption is physical, with the formation of a monolayer on a heterogeneous surface [26–28].

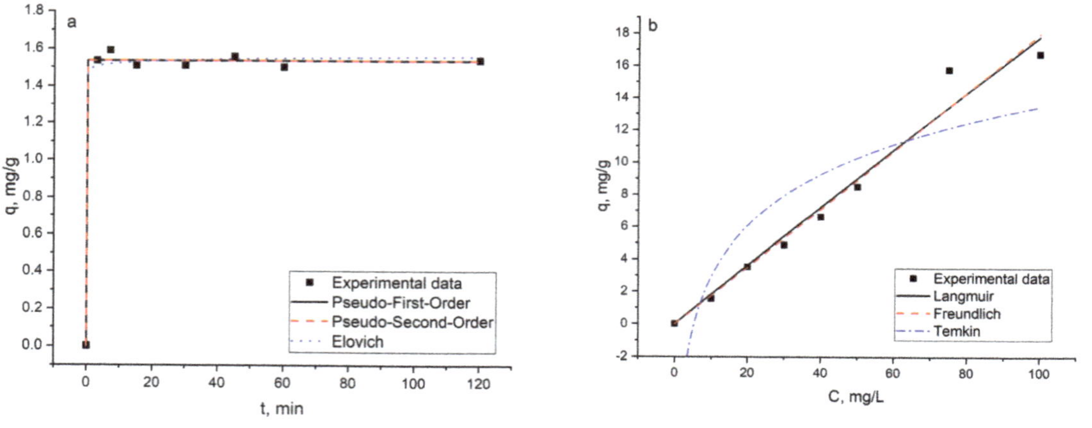

Figure 2. (**a**) Kinetics and (**b**) isotherms of Y(III) biosorption on *A. platensis* biomass.

Table 1. Parameters of the kinetics and isotherm of Y(III) sorption on *A. platensis*.

	Model	Parameters	Y(III)		Model	Parameters	Y(III)
Kinetics	Pseudo-First-Order	q_e	1.53	**Isotherms**	Langmuir	q_m	719.8
		k_1	1.79			b	0.0002
		R^2	0.99			R^2	0.97
		SSE	0.006		Freundlich	K_F	0.17
		χ^2	0.003			n	0.99
	Pseudo-Second-Order	q_e	1.5			R^2	0.97
		k_2	−2.54		Temkin	b_T	0.53
		R^2	0.99			a_T	0.2
		SSE	0.006			R^2	0.75
		χ^2	0.003				
	Elovich	α	$1.15 \cdot 10^{43}$				
		β	69.4				
		R^2	0.99				
		SSE	32286				
		χ^2	21054				

Equilibrium sorption data were modelled using Langmuir, Freundlich, and Temkin sorption isotherms (Figure 2b).

$$q_e = \frac{q_m\, bC_e}{1 + bC_e} \quad (8)$$

where C_e is Y(III) concentration at equilibrium in mg/L, q_e and q_m are amount of metal adsorbed at equilibrium and maximum adsorption capacity in mg/g, and b is Langmuir constant in L/mg.

$$q_e = K_F C^{\frac{1}{n}} \quad (9)$$

where K_F and n are Freundlich constants.

The Temkin isotherm model takes into account the effects of indirect adsorbate/adsorbate interactions on the adsorption process [13]:

$$q_e = \frac{RT}{b_T} \ln(a_T C_e) \quad (10)$$

$1/b_T$ indicates the sorption potential of the sorbent, a_T is the Temkin constant, R is the universal gas constant (8.314 J K^{-1} mol^{-1}), and T is the temperature (K).

The calculated isotherm constants are given in Table 1. Langmuir and Freundlich models well-represented the data for the Y(III) biosorption on *A. platensis* since the determination coefficients for both models were 0.97. The coefficient of determination for the Temkin model was significantly lower. The maximum adsorption capacity calculated by the Langmuir model was 719.8 mg/g, and it was considerably higher than the values obtained for the other sorbents (Table 2). Application of the Langmuir models supports the sorption on homogenous sites, facilitating chemisorption [29], while the Freundlich model is suitable for heterogeneous surfaces with a non-uniform distribution of active adsorption sites. Since both models fit well with the experimental data, it can be suggested that Y(III) biosorption by *A. platensis* follows an intermediate behaviour between mono- and multi-layer adsorption mechanisms [30]. According to the AIC test, the Langmuir model was more applicable for the description of experimental data. The Freundlich isotherm model better described the adsorption of Y(III) on magnetic nitrogen functionalized mesoporous expanded perlite [31].

Table 2. Comparison of maximum sorption capacity for Y(III) achieved in the present study with the literature data.

Sorbent	pH	q_{max}, mg/g	Reference
A. platensis	3	719.8	Present study
Serratia marcescens	5.5	123.65	[2]
Nano-thorium (IV) Oxide	6.9	10.5	[4]
Nano-zirconium (IV) Oxide	6.9	18	[4]
3-Amino-5-Hydroxypyrazole Impregnated Bleaching Clay	6.0	172.41	[32]
Magnetic Nitrogen Functionalized Mesoporous Expanded Perlite	5.5	383.2	[31]

The three thermodynamic parameters, the Gibbs free energy (ΔG°), enthalpy (ΔH°), and entropy (ΔS°) were obtained by using Equations (11) to (13) [26,27].

$$\ln K_d = \frac{\Delta S^{\circ}}{R} - \frac{\Delta H^{\circ}}{RT} \quad (11)$$

$$\Delta G^{\circ} = \Delta H^{\circ} - T\Delta S^{\circ} \quad (12)$$

$$K_d = \frac{(C_0 - C_e)V}{mC_e} \quad (13)$$

where K_d is the distribution coefficient, R is the gas constant, and T is the absolute temperature.

The thermodynamic parameters evaluated by plotting $\ln K_d$ versus 1/T are presented in Figure A1, and the thermodynamic parameters are listed in Table 3.

Table 3. Thermodynamics parameters for Y(III) sorption on *A. platensis*.

Temperature, K	ΔG°, kJ/mol	ΔH°, kJ/mol	ΔS°, J/mol·K
293	−12.1	−2.1	34.1
303	−12.5		
313	−12.8		
323	−13.2		

The values of the ΔG° ranging from −13.2 to −12.1 kJ/mol indicate the feasibility and spontaneity of the biosorption process. The negative value of ΔH° indicates the exothermic nature, and the positive value of ΔS° points to the randomness at the solid and solution interface. Since ΔH° was less than °25 kJ/mol, the sorption can be considered physical. Since the value of ΔS° was higher than −10 J/mol·K, it can be suggested that the adsorption reaction complies with a dissociative mechanism [33].

Biological sorbents, such as other types of sorbents, are characterized by the possibility of their regeneration and multiple use. Thus, using 0.2 M HCl/0.5 M $CaCl_2$, the elution of rare earth elements from new phosphorylated hydrogel (algal biomass/PEI) was complete and remarkably stable for the five cycles [34]. A number of sulphate and chloride salts were successfully used for rare earth elements desorption and the desorption efficiency of the sulphates was higher compared with chloride [35].

3.2. Yttrium Removal by Bioaccumulation

In bioaccumulation experiments, when spirulina was exposed to 10–30 mg/L of Y(III), the lowest rate of Y(III) removal of 29% was observed at the initial Y(III) concentration of 10 mg/L. At higher concentrations, the efficiency of Y(III) removal was approximately two times higher (on the level of 60–70%). Microalgae *Ulva lactuca* was able to remove 70–80% of Y(III) from the solutions containing 10–500 µg/L of metal, *Gracilaria* sp. contributed to the removal of 50–65% of Y(III), while removal efficiency of *Ulva intestinalis*, *Fucus spiralis*, *Fucus vesiculosus*, and *Osmundea pinnatifida* was not higher than 20–30% [36]. Fungus *Penicillium* sp. ZD28 accumulated up to 99% of Y(III) at all applied metal concentrations [1].

One of the main indicators of the toxicity of different chemical compounds for microorganisms is the ability to grow and multiply cells, which is expressed as the amount of biomass accumulated in a unit or per a period of time. The accumulation of cyanobacterium *A. platensis* biomass in a cycle of cultivation in a closed system was monitored both under standard conditions (control) and under the application of Y(III). According to results presented in Figure 3, at a Y(III) concentration of 10 mg/L, no difference between the control and the experimental values of biomass productivity was observed. The concentration of 20 mg/L caused a veridical increase in the amount of biomass by 9.3% compared to control ($p = 0.0011$) and that of 30 mg/L—a decrease by 4.6% ($p = 0.0076$).

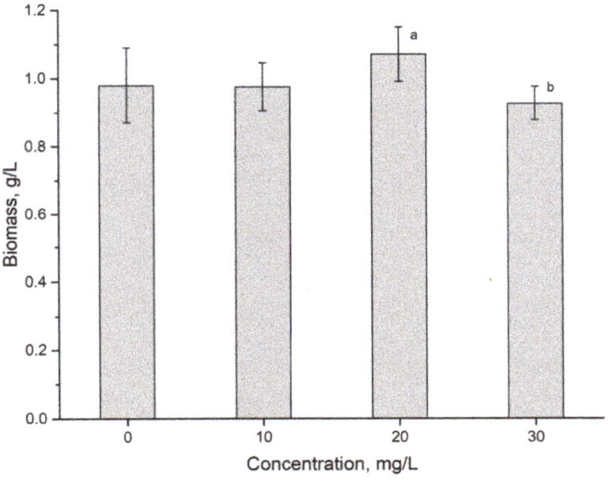

Figure 3. The influence of Y(III) applied in concentrations range of 10–30 mg/L on the amount of biomass of *Arthrospira platensis* (a—$p < 0.005$, b—$p < 0.01$ for the difference between experimental samples and control).

At the same time, it should be mentioned that from a biotechnological point of view, the amount of biomass in both control and experimental samples was within the physiological norm characteristic for spirulina. Thus, the amount of biomass at the end of the cultivation cycle was 0.927–1.071 g/L.

The stimulating effect of rare earth elements on spirulina productivity was reported in other studies as well. For example, dysprosium in concentrations of 10–30 mg/L stimulated biomass accumulation by 19.3% compared to the control and lanthanum—by 18.1% [37]. A pattern revealed for Y(III) was also observed in the case of Yb, when a metal concentration of 20 mg/L led to a slight increase in the amount of biomass and a concentration of 30 mg/L resulted in its decrease by 26.8% [38]. Such response of spirulina biomass is probably an expression of the phenomenon of hormesis, when the dose-response effect is biphasic, with stimulation at low doses and inhibition at higher doses. The stimulation of certain processes at low doses of xenobiotics is generated by an overcompensation reaction of the culture.

At the same time, spirulina appears to be much more resistant to Y(III) compared to other cyanobacteria. Thus, the growth of the *Microcystis aeruginosa* culture was inhibited by 0.3 mg/L of Y(III) and the amplitude of the effect depended on the amount of phosphorus in the medium [38]. In the case of stress caused by lead, the concentration of Y(III) of up to 0.5 mg/L stimulated the growth of the *Microcystis aeruginosa* culture, while concentrations of 5 and 10 mg/L enhanced the negative effect provoked by lead [39]. Other rare earth elements applied in concentrations similar to those used in the present study caused inhibitory effects in other cyanobacteria. For example, lanthanum at a concentration of 72 μmol/L

(10 mg/L) had a strong inhibitory effect on the cyanobacterium *Microcystis aeruginosa* [40], and cerium at a concentration of 5 mg/L on the cyanobacterium *Anabaena flosaquae* [41]. Concentrations of Nd 5 and 10 mg/L significantly inhibited the growth of *Microcystis aeruginosa* by reducing the amount of accumulated biomass two times. The stationary phase occurred more rapidly, was shorter, and was followed by a very pronounced biomass decline [42]. Thus, according to the obtained results, spirulina is a cyanobacterium with a high tolerance to rare earth elements, in particular to Y(III), with mild toxic effects on biomass accumulation being observed only at metal concentration of 30 mg/L.

Although applied Y(III) concentrations led to different responses of the spirulina culture in terms of the amount of biomass, changes in the protein content in the biomass indicate a similar effect at all concentrations (Figure 4). The content of proteins decreased in all experimental variants, which denoted an impairment of physiological processes in the presence of Y(III). The content of proteins in the control biomass was 63.65% of the dry biomass and in the experimental variants 54.4–60.05%, corresponding to a decrease of 5.714.5% compared to control. At a Y(III) concentration of 10 mg/L, a slight but statistically veridical decrease ($p = 0.0269$) in protein content was observed. At concentrations of 20 and 30 mg/L, the amount of protein decreased by 11.1 and 14.5%, respectively, compared to the control ($p = 0.0013$ and $p = 0.0002$).

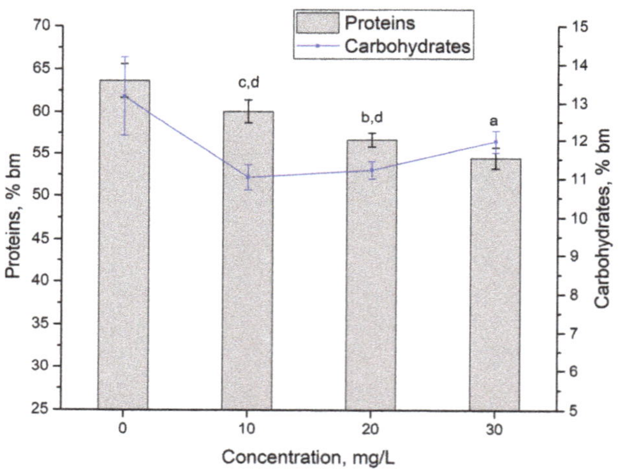

Figure 4. The content of proteins and carbohydrates in Arthrospira platensis exposed to Y(III) in concentrations of 10–30 mg/mL (a—$p < 0.0005$, b—$p < 0.005$, c—$p < 0.05$ for the difference between the experimental sample and control—proteins; d—$p < 0.05$ for the difference between the experimental sample and control—carbohydrates).

A slight decrease in the content of protein was observed for other rare earth elements applied in the same range of concentrations. Thus, the concentration of Nd, Tb, and Sm of 30 mg/L caused a statistically significant decrease in the content of protein in the spirulina biomass. The most pronounced inhibitory effect was provoked by Sm at a concentration of 30 mg/L, when the content of proteins in the biomass decreased by 16% compared to the control [37]. Eu(III) also had a negative effect on protein content, reducing it by 10.2% at an Eu(III) concentration of 20 mg/L and by 17.7% at a concentration of 30 mg/L [10].

The decrease in the amount of protein in spirulina biomass under conditions of stress is already a proven fact and it can be caused by both physical and chemical factors [11–13]. At the addition of Y(III) in concentrations of 10–30 mg/L in a cultivation medium, the content of protein has not decreased below 54% of dry biomass, which ensures an adequate adaptation of the culture to the created conditions of growth.

The content of carbohydrates in the spirulina biomass in the present study varied between 13.18 and 11.03%. The highest level of this parameter was determined in the control sample and in experimental variants its content decreased by 17–16.25%. At Y(III) concentrations of 10 and 20 mg/L, the decrease in the content of carbohydrates is statistically significant ($p = 0.013$ and $p = 0.016$, respectively), while at the concentration of 30 mg/L, it only can be said regarding a decreasing trend.

It was previously shown that rare earth elements can differently affect the content of carbohydrates in spirulina biomass. Thus, in the presence of Sm, Tb, La, and Dy in the cultivation medium, the content of carbohydrates increased by up to 48% compared to the control, the increase being dependent on the element and its concentration. On the other hand, Nd and Yb significantly reduced the content of carbohydrates in spirulina—up to 21.9% compared to the control [37]. In the case of Eu, only a concentration of 30 mg/L led to a decrease in carbohydrates content by 27.4% with respect to the control [10].

Phycobiliproteins in cyanobacterial cells have a dual function. Firstly, there are secondary photosynthetic pigments, which absorb solar energy with a wavelength of 495–650 nm and transfer it to chlorophyll in the reactive centres of the photosynthetic apparatus, thus increasing the efficiency of the photosynthesis process. Secondly, these molecules have the function of antioxidant protection. As effective antioxidants, phycobiliproteins manifest themselves in the process of neutralizing free radicals. Thus, it was established that these compounds eliminate oxyl, hydroxyl, and peroxyl radicals. Their protective effect is expressed by the protection of physiologically active membranes from the peroxidation process [43]. Spirulina cells contain two phycobilin pigments—phycocyanin and allophycocyanin. The change in their content in Y-loaded spirulina biomass is presented in Figure 5.

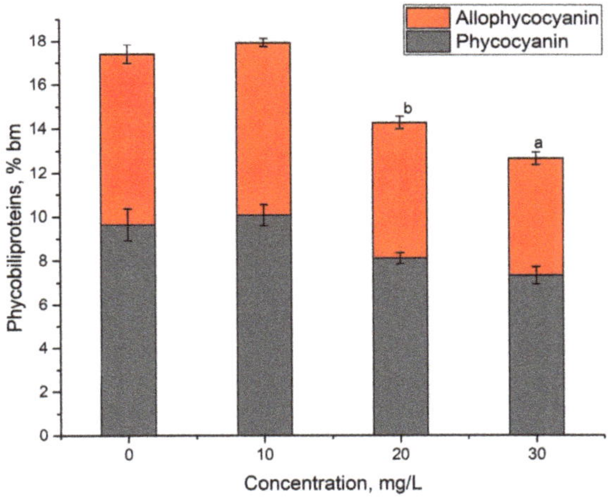

Figure 5. The content of phycobiliproteins in the biomass of *Arthrospira platensis* exposed to Y(III) in a concentration of 10–30 mg/L (a—$p < 0.005$, b—$p < 0.01$, for the difference between the sample and the control—the content of phycobiliproteins).

The content of phycocyanin in the control biomass was 9.64 ± 0.72% of the dry biomass, and of allophycocyanin 0.76 ± 0.41%. The sum of phycobiliproteins in the control biomass was 17.41 ± 1.14%, very close to values of 17.93 ± 0.65% obtained at the addition of 10 mg/L of Y(III) in the cultivation medium. In this case, the difference between control and experimental samples was statistically insignificant. Instead, the other two applied Y(III) concentrations reduced the phycobilin content by 18.17% at a concentration of 20 mg/L ($p = 0.0059$) and by 27.44% at a concentration of 30 mg/L ($p = 0.0015$). It is known that phycobiliproteins react very quickly to changes in the state of the spirulina culture, and intense or moderate stress can result in a very pronounced decrease in their content.

In the case of other rare earth elements, the same pattern was observed. Thus, concentrations of 30 mg/L of Nd and Yb caused a reduction in the content of phycobiliproteins by 10.7–19.0% compared to the control, and the same concentration of La, Dy, Sm, and Tb reduced the content of phycobiliproteins twice [37]. In the present study, the reduction in the content of phycobiliproteins in the biomass at Y(III) concentrations of 20 and 30 mg/L could be associated with the decrease in the efficiency of photosynthesis (which is most likely valid only for the concentration of 30 mg/L at which the amount of biomass was reduced) and with a decrease in the antioxidant capacity of the spirulina biomass.

The change in the content of basic photosynthetic pigments under the action of Y(III) is illustrated in Figure 6.

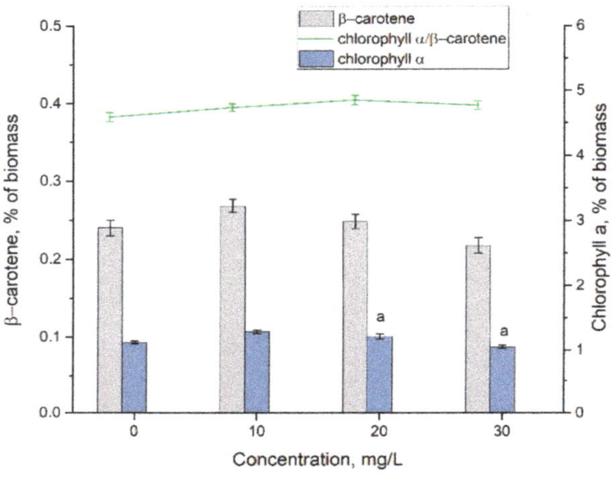

Figure 6. The content of pigments in the biomass of *Arthrospira platensis* exposed to Y(III) at a concentration of 10–30 mg/L (a—$p < 0.01$ for the difference between sample and control—chlorophyll).

In the control biomass, the amount of α-chlorophyll was 1.07 ± 0.021% and of β-carotene 0.24 ± 0.01%. The ratio of chlorophyll α/β–carotene, a parameter that can be considered as an indicator of photosynthetic activity, was also taken into account. The content of β-carotene in the experimental samples was very close to control values and felt within the limits of 0.22–0.27% of the biomass, which are characteristic physiological values for spirulina.

Similar results were obtained for other rare earth elements—Sm, La, Dy, Nd, and Yb—in concentrations similar to those applied in the present study ensured in most cases the maintenance of the amount of β-carotene at the control level or a moderately high one [37]. Only Eu in a concentration of 30 mg/L led to a slight decrease in the content of β-carotene [10].

The content of α-chlorophyll in the experimental samples varied between 1.05 and 1.28% of the dry spirulina biomass. Y(III) concentrations of 10 and 20 mg/L caused an increase in the content of this pigment by 19.92 and 13.34, respectively. Cultivation of spirulina in the presence of other rare earth elements did not influence significantly the content of α-chlorophyll, except for Sm (concentrations 10–30 mg/L) and Eu (concentration 30 mg/L), when a significant decrease in the content of α-chlorophyll in the biomass was observed [10,37].

The chlorophyll α/β-carotene ratio indicates not only photosynthetic activity but also a possible state of stress. A decrease in the ratio is usually associated with stress. According to values in Figure 7, the ratio did not change significantly in the presence of Y(III) in the medium, which also correlates with the results obtained for the amount of biomass accumulated during the spirulina growth.

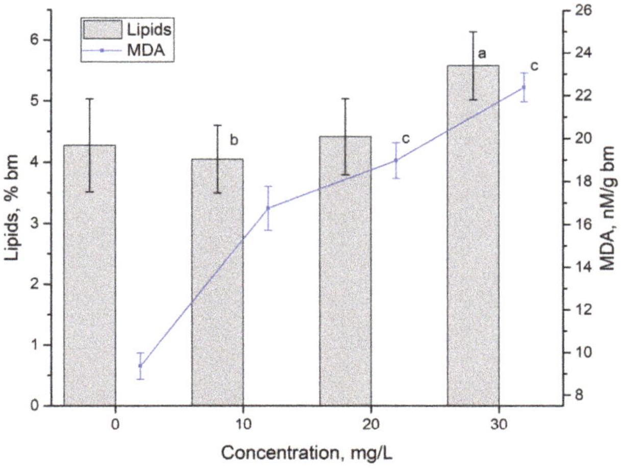

Figure 7. The content of lipids and malonic dialdehyde in the biomass of *Arthrospira platensis* exposed to Y(III) in a concentration of 10–30 mg/L (a—$p < 0.005$ for the difference between the sample and the control—lipids; b—$p < 0.005$ and c—$p < 0.0005$ for the difference between sample and control—MDA).

Thus, the results obtained for primary and secondary photosynthetic pigments in the spirulina biomass show that they remained at a level characteristic of the good physiological state of the culture, which allowed maintaining the productivity of the cyanobacterium at a level acceptable for spirulina.

The content of lipids in the biomass and the level of malonic dialdehyde—a product of the oxidative degradation of lipids—are two parameters that can indicate a state of stress caused by different types of factors on the spirulina culture. The results obtained for these parameters are shown in Figure 7.

The content of lipids in the control sample was 4.27 ± 0.20% of the dry biomass and in experimental variants, at Y(III) concentrations of 10 and 20 mg/L, it was almost on the same level (4.04 ± 0.43 and 4.42 ± 0.16%, respectively). The concentration of 30 mg/L of Y(III) led to an increase in the content of lipids in the biomass of up to 5.58 ± 0.29% of the dry biomass, which is 30.69% higher compared to the control.

The data obtained for other rare earth elements showed that inhibition of lipid accumulation occurred more often compared to the stimulation effect. Thus, Tb, Dy, Yb, and Eu, in concentrations similar to those applied in the present study, led to a decrease in lipid content in spirulina biomass by up to 35% compared to the control. Some of the rare earth elements, for example, Sm and Nd, did not affect the content of lipids in the biomass. A lipid-enhancing effect was observed at La concentrations of 20 and 30 mg/L, when lipids content in the dry spirulina biomass was 14.7 and 29.1% higher compared to the control [10,37–42].

The amount of MDA in the control biomass was 9.35 ± 1.48 nM per gram of dry biomass, and in the experimental variants the amount of this marker of oxidative stress was 1.78–2.38 times higher. The amount of MDA increased proportionally to the increase in the Y(III) concentration in the medium. In this case, a clear dose-effect type dependence was obtained, which proved the toxic action of Y(III) toward spirulina.

An increase in the level of MDA in spirulina biomass grown on a medium with the addition of rare earth elements is a common phenomenon. Thus, La, Dy, Sm, Nd, Yb, and Eu in concentrations of 10–30 mg/L provoked a moderate to a very pronounced increase in the MDA level [10,37]. In the case of other cyanobacteria, the doubling of the MDA content under the influence of rare earth elements occurred at much lower concentrations. For example, in the case of *Microcystis aeruginosa*, the concentration of 1 mg/L of Y(III) increased the MDA content in biomass two times [38].

The activity of the ethanolic and water extracts from spirulina biomass grown on a medium with the addition of Y(III) can be seen in Figure 8. In the experimental variants, the water and ethanolic extracts had a higher activity compared to those obtained for control samples. Thus, the water extract obtained from the biomass grown on the medium with 10 mg/L of Y(III) was 19.97% more active compared to the control, for a concentration of 20 mg/L—36.62% more active—and at the concentration of 30 mg /L—41.61% more active.

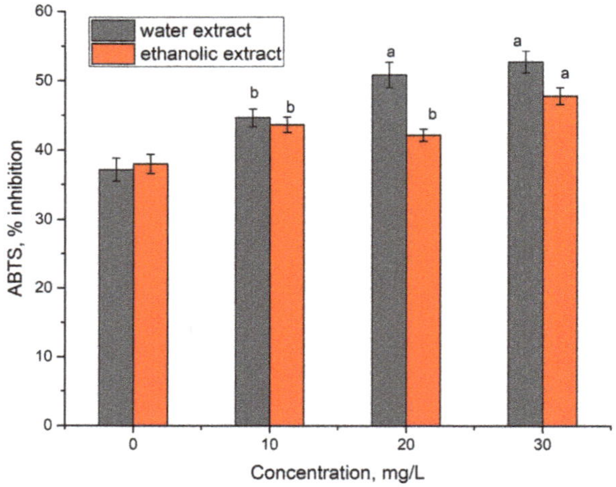

Figure 8. The antioxidant activity of extracts from *A. platensis* biomass exposed to Y(III) at concentrations 10–30 mg/L (a—$p < 0.0005$, b—$p < 0.005$ for the difference between sample and control).

The ethanolic extracts in the experimental variants were 10.63–25.53% more active compared to the control.

In previously performed studies, it was also observed a pronounced change in the level of activity of the water and ethanolic extracts obtained from spirulina biomass grown on media with the addition of rare earth elements in concentrations of 10–30 mg/L. The antioxidant activity of the extracts from Eu-loaded spirulina increased by 10–30% compared to the control [10]. Similarly, an increase in the antioxidant activity of the water and ethanolic extracts was observed when La, Dy, Sm, and Tb were added to the cultivation medium in the range of concentrations applied in the present study. Nd and Yb, on the contrary, caused a considerable decrease in the inhibition activity of the ABTS radical [37].

4. Conclusions

The potential of *A. platensis* to recover Y(III) by applying biosorption and bioaccumulation was shown. The process of Y(III) biosorption was dependent on the pH of the solution, adsorption time, Y(III) concentration, and temperature. A maximum theoretical sorption capacity of 719.8 mg/g can be achieved at a pH of 3, temperature of 20 °C, and adsorption time of 3 min. The equilibrium of the biosorption was described using the Langmuir and Freundlich isotherm models, while the kinetic of the biosorption was better presented by the pseudo-first-order model. From the thermodynamic point of view, Y(III) biosorption was a spontaneous and exothermic process. The interaction of Y(III) with living biomass in bioaccumulation experiments resulted in the removal of up to 70% of Y(III) from the media. At applied concentrations, Y(III) caused serious changes in the composition of the spirulina biomass. Thus, in all experimental variants, a decrease in the content of proteins, phycobiliproteins, and carbohydrates, and an increase in MDA content and antioxidant activity were observed. An adequate level of spirulina productivity in Y-loaded biomass was ensured by maintaining the content of photosynthetic pigments and the chlorophyll α/β-carotene ratio on the level characteristic for the control biomass. According to high values of spirulina adsorption capacity, biosorption seems to be a more efficient process for the Y(III) removal from wastewater. The sorption of Y(III) on dead spirulina biomass showed to be more efficient in acidic environments, while Y(III) is prone to hydrothermal mobilization in alkaline solutions. Thus, depending on the experimental conditions, both processes are promising to be realized. The results obtained using the cyanobacteria *Arthrospira platensis* can serve as a basis for further research using other cyanobacteria with higher productivity and lower biotechnological values, thus making it possible to reduce considerably the costs of bioremediation technologies.

A study of the Y(III) effect on the biomass composition allows for the consideration of the present work also as a study of the toxic effects produced by yttrium ions in the aquatic environment.

Author Contributions: Conceptualization, N.Y., I.Z. and L.C.; methodology, N.Y., T.C. and L.R.; software, D.G.; formal analysis, D.G.; investigation, N.Y., I.Z., L.C., T.C. and L.R.; data curation, I.Z. and L.C.; writing—original draft preparation, I.Z. and L.C.; writing—review and editing, all authors; visualization, I.Z. and N.Y. All authors have read and agreed to the published version of the manuscript.

Funding: The study was carried out with the financial support of the Association of Young Scientists and Specialist of the JINR Grant No. 22-402-11.

Institutional Review Board Statement: Not applicable.

Informed Consent Statement: Not applicable.

Data Availability Statement: Not applicable.

Conflicts of Interest: The authors declare no conflict of interest.

Appendix A

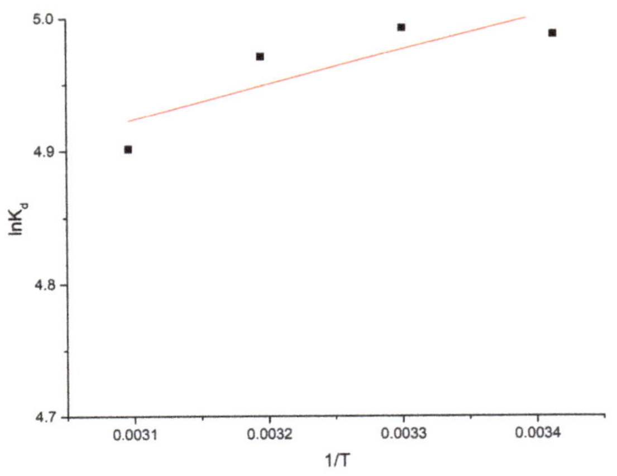

Figure A1. Graph of $\ln K_d$ versus 1/T.

References

1. Wang, W.; Xu, C.; Jin, Y.; Zhang, Z.; Yan, R.; Zhu, D. The accumulation of rare-earth yttrium ions by Penicillium sp. ZD28. *AMB Express* **2020**, *10*, 25. [CrossRef] [PubMed]
2. Liang, C.L.; Shen, J.L. Removal of yttrium from rare-earth wastewater by Serratia marcescens: Biosorption optimization and mechanisms studies. *Sci. Rep.* **2022**, *12*, 4861. [CrossRef] [PubMed]
3. Du, X.; Graedel, T.E. Global in-use stocks of the rare earth elements: A first estimate. *Environ. Sci. Technol.* **2011**, *45*, 4096–4101. [CrossRef] [PubMed]
4. Dubey, S.S.; Grandhi, S. Sorption studies of yttrium(III) ions on surfaces of nano-thorium(IV) oxide and nano-zirconium(IV) oxide. *Int. J. Environ. Sci. Technol.* **2019**, *16*, 59–70. [CrossRef]

5. Cardon, P.Y.; Triffault-Bouchet, G.; Caron, A.; Rosabal, M.; Fortin, C.; Amyot, M. Toxicity and Subcellular Fractionation of Yttrium in Three Freshwater Organisms: Daphnia magna, Chironomus riparius, and Oncorhynchus mykiss. *ACS Omega* **2019**, *4*, 13747–13755. [CrossRef] [PubMed]
6. Ding, Y.; Tian, Y.; Zeng, Z.; Shuai, P.; Lan, H.; Zhu, X.; Zhong, Y.; Wu, L.; Fan, X. YCl3 promotes neuronal cell death by inducing apoptotic pathways in rats. *Biomed Res. Int.* **2017**, *2017*, 2183658. [CrossRef]
7. Nakamura, Y.; Tsumura, Y.; Tonogai, Y.; Shibata, T.; Ito, Y. Differences in Behavior among the Chlorides of Seven Rare Earth Elements Administered Intravenously to Rats. *Toxicol. Sci.* **1997**, *37*, 106–116. [CrossRef]
8. Hamdan, A.M.; Abd-El-Mageed, H.; Ghanem, N. Biological treatment of hazardous heavy metals by Streptomyces rochei ANH for sustainable water management in agriculture. *Sci. Rep.* **2021**, *11*, 9314. [CrossRef]
9. Razzak, S.A.; Faruque, M.O.; Alsheikh, Z.; Alsheikhmohamad, L.; Alkuroud, D.; Alfayez, A.; Hossain, S.M.Z.; Hossain, M.M. A comprehensive review on conventional and biological-driven heavy metals removal from industrial wastewater. *Environ. Adv.* **2022**, *7*, 100168. [CrossRef]
10. Yushin, N.; Zinicovscaia, I.; Cepoi, L.; Chiriac, T.; Rudi, L.; Grozdov, D. Biosorption and Bioaccumulation Capacity of Arthospiraplatensis toward Europium Ions. *Water* **2022**, *14*, 2128. [CrossRef]
11. Cepoi, L. Environmental and Technological Stresses and Their Management in Cyanobacteria. In *Cyanobacteria: From Basic Science to Applications*; Elsevier: Amsterdam, The Netherlands, 2018; pp. 217–244. ISBN 9780128146682.
12. Cepoi, L.; Zinicovscaia, I.; Rudi, L.; Chiriac, T.; Miscu, V.; Djur, S.; Strelkova, L.; Grozdov, D. Spirulina platensis as renewable accumulator for heavy metals accumulation from multi-element synthetic effluents. *Environ. Sci. Pollut. Res.* **2020**, *27*, 31793–31811. [CrossRef] [PubMed]
13. Cepoi, L.; Zinicovscaia, I.; Rudi, L.; Chiriac, T.; Miscu, V.; Djur, S.; Strelkova, L.; Vergel, K.; Nekhoroshkov, P. Growth and heavy metals accumulation by Spirulina platensis biomass from multicomponent copper containing synthetic effluents during repeated cultivation cycles. *Ecol. Eng.* **2020**, *142*, 105637. [CrossRef]
14. Gunasundari, E.; Kumar, P.S. Higher adsorption capacity of Spirulina platensis alga for Cr(VI) ions removal: Parameter optimisation, equilibrium, kinetic and thermodynamic predictions. *IET Nanobiotechnol.* **2017**, *11*, 317–328. [CrossRef]
15. Zinicovscaia, I.; Safonov, A.; Ostalkevich, S.; Gundorina, S.; Nekhoroshkov, P.; Grozdov, D. Metal ions removal from different type of industrial effluents using Spirulina platensis biomass. *Int. J. Phytoremediation* **2019**, *21*, 1442–1448. [CrossRef] [PubMed]
16. Almomani, F.; Bohsale, R.R. Bio-sorption of toxic metals from industrial wastewater by algae strains Spirulina platensis and Chlorella vulgaris: Application of isotherm, kinetic models and process optimization. *Sci. Total Environ.* **2021**, *755*, 142654. [CrossRef]
17. Cepoi, L.; Zinicovscaia, I. Spirulina platensis as a model object for the environment bioremediation studies. In *Handbook of Algal Science, Technology and Medicine*; Academic Press: Cambridge, MA, USA, 2020; pp. 629–640.
18. Lowry, O.H.; Rosebrough, N.J.; Farr, A.L.; Randall, R.J. Protein measurement with the Folin phenol reagent. *J. Biol. Chem.* **1951**, *193*, 265–275. [CrossRef]
19. Park, J.; Jeong, H.J.; Yoon, E.Y.; Moon, S.J. Easy and rapid quantification of lipid contents of marine dinoflagellates using the sulpho-phospho-vanillin method. *Algae* **2016**, *31*, 391–401. [CrossRef]
20. Carra, P.O. Algal biliproteins. *Biochem. J.* **1970**, *119*, 2P–3P. [CrossRef]
21. Hodges, D.M.; DeLong, J.M.; Forney, C.F.; Prange, R.K. Improving the thiobarbituric acid-reactive-substances assay for estimating lipid peroxidation in plant tissues containing anthocyanin and other interfering compounds. *Planta* **1999**, *207*, 604–611. [CrossRef]
22. Re, R.; Pellegrini, N.; Proteggente, A.; Pannala, A.; Yang, M.; Rice-Evans, C. Antioxidant activity applying an improved ABTS radical cation decolorization assay. *Free Radic. Biol. Med.* **1999**, *26*, 1231–1237. [CrossRef]
23. Fialkovsky, I.S.; Lutskiy, D.S.; Litvinova, T.E.; Oleijnik, O.I. Effect of anion composition on the extraction of cerium (III) and yttrium (III) by oleic acid. *ARPN J. Eng. Appl. Sci.* **2018**, *13*, 3152–3161.
24. Kazak, E.S.; Kalitina, E.G.; Kharitonova, N.A.; Chelnokov, G.A.; Elovskii, E.V.; Bragin, I.V. Biosorption of Rare-Earth Elements and Yttrium by Heterotrophic Bacteria in an Aqueous Environment. *Moscow Univ. Geol. Bull.* **2018**, *73*, 287–294. [CrossRef]
25. Karavaiko, G.I.; Kareva, A.S.; Avakian, Z.A.; Zakharova, V.I.; Korenevsky, A.A. Biosorption of scandium and yttrium from solutions. *Biotechnol. Lett.* **1996**, *18*, 1291–1296. [CrossRef]
26. Umar, A.; Khan, M.S.; Alam, S.; Zekker, I.; Burlakovs, J.; dC Rubin, S.S.; Bhowmick, G.D.; Kallistova, A.; Pimenov, N.; Zahoor, M. Article synthesis and characterization of Pd-Ni bimetallic nanoparticles as efficient adsorbent for the removal of acid orange 8 present in wastewater. *Water* **2021**, *13*, 1095. [CrossRef]
27. Rahman, N.U.; Ullah, I.; Alam, S.; Khan, M.S.; Shah, L.A.; Zekker, I.; Burlakovs, J.; Kallistova, A.; Pimenov, N.; Vincevica-Gaile, Z.; et al. Activated ailanthus altissima sawdust as adsorbent for removal of acid yellow 29 from wastewater: Kinetics approach. *Water* **2021**, *13*, 2136. [CrossRef]
28. Treto-Suárez, M.A.; Prieto-García, J.O.; Mollineda-Trujillo, Á.; Lamazares, E.; Hidalgo-Rosa, Y.; Mena-Ulecia, K. Kinetic study of removal heavy metal from aqueous solution using the synthetic aluminum silicate. *Sci. Rep.* **2020**, *10*, 10836. [CrossRef]
29. Nithya, R.; Gomathi, T.; Sudha, P.N.; Venkatesan, J.; Anil, S.; Kim, S.K. Removal of Cr(VI) from aqueous solution using chitosan-g-poly(butyl acrylate)/silica gel nanocomposite. *Int. J. Biol. Macromol.* **2016**, *87*, 545–554. [CrossRef]
30. Zinicovscaia, I.; Yushin, N.; Abdusamadzoda, D.; Grozdov, D.; Shvetsova, M. Efficient Removal of Metals from Synthetic and Real Galvanic Zinc–Containing Effluents by Brewer's Yeast Saccharomyces cerevisiae. *Materials* **2020**, *13*, 3624. [CrossRef]

31. Liu, J.; Zeng, L.; Liao, S.; Liao, X.; Liu, J.; Mao, J.; Chen, Y.; Qiu, T.; Ren, S. Highly efficient enrichment and adsorption of rare earth ions (yttrium(III)) by recyclable magnetic nitrogen functionalized mesoporous expanded perlite. *Chin. Chem. Lett.* **2020**, *31*, 2849–2853. [CrossRef]
32. Sakr, A.K.; Cheira, M.F.; Hassanin, M.A.; Mira, H.I.; Mohamed, S.A.; Khandaker, M.U.; Osman, H.; Eed, E.M.; Sayyed, M.I.; Hanfi, M.Y. Adsorption of yttrium ions on 3-amino-5-hydroxypyrazole impregnated bleaching clay, a novel sorbent material. *Appl. Sci.* **2021**, *11*, 10320. [CrossRef]
33. Li, Y.; Xia, B.; Zhao, Q.; Liu, F.; Zhang, P.; Du, Q.; Wang, D.; Li, D.; Wang, Z.; Xia, Y. Removal of copper ions from aqueous solution by calcium alginate immobilized kaolin. *J. Environ. Sci.* **2011**, *23*, 404–411. [CrossRef]
34. He, C.; Salih, K.A.M.; Wei, Y.; Mira, H.; Abdel-Rahman, A.A.H.; Elwakeel, K.Z.; Hamza, M.F.; Guibal, E. Efficient recovery of rare earth elements (Pr(iii) and tm(iii)) from mining residues using a new phosphorylated hydrogel (algal biomass/pei). *Metals* **2021**, *11*, 294. [CrossRef]
35. Moldoveanu, G.A.; Papangelakis, V.G. Recovery of rare earth elements adsorbed on clay minerals: I. Desorption mechanism. *Hydrometallurgy* **2012**, *117–118*, 71–78. [CrossRef]
36. Pinto, J.; Henriques, B.; Soares, J.; Costa, M.; Dias, M.; Fabre, E.; Lopes, C.B.; Vale, C.; Pinheiro-Torres, J.; Pereira, E. A green method based on living macroalgae for the removal of rare-earth elements from contaminated waters. *J. Environ. Manag.* **2020**, *263*, 110376. [CrossRef]
37. Zinicovscaia, I.; Cepoi, L.; Rudi, L.; Chiriac, T.; Grozdov, D.; Pavlov, S.; Djur, S. Accumulation of dysprosium, samarium, terbium, lanthanum, neodymium and ytterbium by Arthrospira platensis and their effects on biomass biochemical composition. *J. Rare Earths* **2021**, *39*, 1133–1143. [CrossRef]
38. Wang, Y.; Li, Y.; Luo, X.; Ren, Y.; Gao, E.; Gao, H. Effects of yttrium and phosphorus on growth and physiological characteristics of Microcystis aeruginosa. *J. Rare Earths* **2018**, *36*, 781–788. [CrossRef]
39. Wu, Y.; Wang, Y.; Du, J.; Wang, Z.; Wu, Q. Effects of yttrium under lead stress on growth and physiological characteristics of Microcystis aeruginosa. *J. Rare Earths* **2016**, *34*, 747–756. [CrossRef]
40. Jin, X.; Chu, Z.; Yan, F.; Zeng, Q. Effects of lanthanum(III) and EDTA on the growth and competition of Microcystis aeruginosa and Scenedesmus quadricauda. *Limnologica* **2009**, *39*, 86–93. [CrossRef]
41. Wang, Y.; Li, J.; Lü, Y.; Jin, H.; Deng, S.; Zeng, Y. Effects of cerium on growth and physiological characteristics of Anabaena flosaquae. *J. Rare Earths* **2012**, *30*, 1287–1292. [CrossRef]
42. Wang, Y.; Jin, H.; Deng, S.; Chen, Y.; Yu, Y. Effects of neodymium on growth and physiological characteristics of Microcystis aeruginosa. *J. Rare Earths* **2011**, *29*, 388–395. [CrossRef]
43. Sekar, S.; Chandramohan, M. Phycobiliproteins as a commodity: Trends in applied research, patents and commercialization. *J. Appl. Phycol.* **2008**, *20*, 113–136. [CrossRef]

metals

MDPI

Article

The Role of Mycorrhizal-Assisted Phytomining in the Recovery of Raw Materials from Mine Wastes

Adalgisa Scotti [1,2,*,†], Vanesa Analía Silvani [3,†], Natalia Andrea Juarez [1], Alicia Margarita Godeas [3] and Stefano Ubaldini [2]

1 Bioenvironmental Laboratory, International Center for Earth Sciences, National Atomic Energy Commission, Technological University National-FRSR, San Rafael Mendoza 5600, Argentina
2 Institute of Environmental Geology and Geoengineering, National Research Council of Italy, Research Area of Rome 1, 00015 Montelibretti, Italy
3 Institute of Biodiversity and Applied and Experimental Biology, Faculty of Exact and Natural Science, National Scientific and Technical Research Council—University of Buenos Aires, Buenos Aires 1428, Argentina
* Correspondence: scotti@cnea.gov.ar
† These authors contributed equally to this work.

Abstract: In recent years, critical and secondary raw materials (CRMs and SRMs, respectively) have received great interest within the circular economy model. In this work, the mycorrhizal-assisted phytomining (MAP) system, composed of *Helianthus annuus*–arbuscular mycorrhizal fungus *Rhizophagus intraradices*–Zn-volcanic ashes, was applied in bioreactors for the recovery of CRMs (Sr, P) and SRMs (Cr, Zn, Cu, Mn, Rb, Ni) from mining wastes of the *Los Cóndores* mine (Argentina). Our results showed high bioaccumulation of Sr, P, Mn, and Zn in the aerial tissues, and a high root-to-shoot translocation for Mn (4.02) > Sr > P > Rb > Zn (0.84). Mycorrhization treatment increased the root-to-leaf translocation for Cr and P and prevented translocation towards flower tissues in most elements. The estimated bioextracting potential of the MAP system (290 plants) in a vegetable depuration module (VDM) ranged from 158 mg/m^3 P > Zn > Mn > 15.1 mg/m^3 Sr. We demonstrated a promising and cost-effective biotechnology applicable in agronomical practices, given the exclusion of toxic elements in flower parts, as well as for the recovery of CRMs and SRMs by hydrometallurgy from plant biomass.

Keywords: mycorrhizal-assisted phytomining; bioreactors; hydrometallurgy; resource recovery

Citation: Scotti, A.; Silvani, V.A.; Juarez, N.A.; Godeas, A.M.; Ubaldini, S. The Role of Mycorrhizal-Assisted Phytomining in the Recovery of Raw Materials from Mine Wastes. *Metals* **2022**, *12*, 1828. https://doi.org/10.3390/met12111828

Academic Editor: Antonije Onjia

Received: 22 September 2022
Accepted: 24 October 2022
Published: 27 October 2022

1. Introduction

The circular economy represents a completely new concept for the life cycle of a product, which involves sustainable activities such as recycling and reuse. Nowadays, this model is replacing the traditional linear economy of "take—make—dispose" model given that the circular economy brings benefits not only to the environment through waste reduction but also to the economy through savings in raw materials [1,2]. Nowadays, the exploitation of new raw materials for the development of high-value-added products is not profitable, but these could be gained if efficient waste recycling technologies are applied. Specifically, the major anthropogenic sources of critical and secondary raw materials (CRMs and SRMs, respectively) in the environment include wastes from metalliferous mining or smelting industries [2]. Many chemical elements were declared of special interest in the fourth EU list [3] due to the growing demand in industry and the limited availability in nature, highlighting the abandoned mining sites as a source of SRMs and CRMs [2].

Phytomining biotechnology involves plant species capable of extracting chemical elements from subeconomic deposits or mineral wastes and accumulating them in plant biomass to subsequently recover the element in value [4]. Currently, phytomining is applied for the recovery of CRMs and SRMs from solid metal wastes, resulting in an excellent option for a circular economy [5,6]. The application of phytomining benefits the environment by reduction of toxic effects [7], and it could be improved by incorporating beneficial microorganisms associated with the rhizosphere of target plants. In particular, the phytoextraction and phytoaccumulation capacities could be enhanced by the mutualistic association of these plants with root symbionts, such as the arbuscular mycorrhizal (AM) fungi [8]. These fungi generate a direct relationship between soil and roots, and therefore, they contribute to the bioavailability of chemical elements in soil [9]. Various reports have mentioned the abilities of AM fungi to uptake different elements from the soil, sequester them in their structures, or translocate them to their host plant. The mechanisms of bioaccumulation in the mycorrhizal association are highly dependent on the fungal and plant partner involved, as well as some environmental factors [10]. Previously, we obtained an efficient extraction of CRMs and SRMs from mining soil by using the mycorrhizal-assisted phytomining (MAP) system, comprised of "*Helianthus annuus–Rhizophagus intraradices*–Zn 350-volcanic ash" (Patent 130100620) [11] under laboratory conditions at Technological Readiness Level (TRL) 2 [5]. Also in this work, the bioextracting potential (BP) values were estimated after calibration of the MAP system in an engineering module depuration (vegetable depuration module, VDM) at TRL 6 [8], obtaining values of 2.417 g (K) > BP > 0.14 g (As) per m^3 of contaminated soil. Finally, the recovery of SRMs and CRMs by hydrometallurgical techniques was suggested as a viable and cost-effective option. However, the scaling-up of a pilot test or prototype technology at TRL 4 before TRL 5-6, to reach the maturity levels towards a real environment, is urgently needed [12].

In the present work, we used an innovative automatized technological prototype, called a bioreactor, which allowed us to validate the MAP system at TRL 4 [13] for the decontamination of polluted soil, and simultaneously, recovering useful CRMs and SRMs. The bioreactor permits the calibration of hydraulic variables of entry and exit of effluents and other physicochemical parameters. After the MAP system performance in the bioreactor, it can be followed with the harvest of chemical elements from plant biomass and purification with bio-hydrometallurgical technology for the recovery of SRMs and CRMs with high extraction efficiency and high purity degree. The microbial-assisted phytomining linked to hydrometallurgy techniques is an innovative and economical technology to mine waste treatment [1,7].

The aim of this work was to calibrate and validate the MAP system applied in the bioreactor at TRL 4 using mining wastes from the Los Cóndores mine (San Luis province, Argentina) for the recovery of SRMs and CRMs, and subsequent decontamination of polluted soil. Finally, a prediction of the performance of the MAP system in the VDM at TRL 6 containing mine residues was estimated.

2. Materials and Methods

2.1. Sampling Sites

The sampling campaign in the Los Cóndores mine (Concarán, San Luis province, Argentina) was carried out within the framework of the Horizon 2020 ERA MIN BioCriticalMetals project 'Recognition of microbial functional communities and assessment of the mineralizing potential (bioleaching) for high-tech critical metals' (https://www.uc.pt/en/org/biocriticalmetals, accesed on 20 July 2022). The Los Cóndores mine has a relatively long mining history from late 1800 to 1985 by exploitation of wolfram, bismuth, and copper ore deposits [14,15].

In October and November 2016, different kinds of soil samples and mine wastes were collected (approximately 2 kg each), but for this study, only four samples were used from two tailings and two dumps of the Los Cóndores mine (Figure 1; Supplementary Table S1).

The quartering and homogenization corresponding to the four samples were carried out in the mineral processing laboratory at Universidad Nacional de San Luis (Argentina). Subsamples were analyzed by Total X-Ray Fluorescence (TXRF), and the remaining samples were kept in the Bioenvironmental Laboratory (CNEA FRSR-UTN San Rafael, Mendoza province, Argentina) until their use in the bioreactors.

Figure 1. Geographical location of the Los Cóndores Mine in San Luis province (Argentina). Black points indicate the sampling sites under study.

2.2. Mycorrhizal-Assisted Phytomining (MAP) System

The MAP system consisted of sunflower plants (*Helianthus annuus* L., hybrid cultivar DK4045, Syngenta seeds) colonized by the AM fungus *Rhizophagus intraradices* GA5 strain grown in pots (1.2 L) with a homogeneous mixture of contaminated mine soil substrate (CS) or blank soil substrate (B), mixed with volcanic ashes (50:50, v/v) and supplemented with $ZnSO_4$ (as a catalyst). The B treatment consisted of commercial soil substrate (Fertile Soil Arhumus®).

The AM fungus GA5 strain, provided by Banco de Glomeromycota in vitro (Faculty of Exact and Natural Sciences of Buenos Aires University, Argentina, https://bgiv.com.ar/strains/Rhizophagus-intraradices/ga5, accesed on 20 July 2022) was propagated as described in Silvani et al. [16]. Four sunflower plants from each pot with the CS and B substrate were inoculated with 5 gr of general inocula of GA5 strain (containing mycorrhizal root fragments, external spores, and mycelia). The other four sunflower plants remained uninoculated with the AM fungus. The MAP systems were maintained for 30 days at controlled temperature (23 C day/16 C night) and natural light conditions until transplantation to the bioreactors.

2.3. Bioreactors at TRL 4

Two bioreactors, placed in the Bioenvironmental Laboratory (CNEA FRSR, Mendoza), were made with polycarbonate containers (30 cm × 60 cm and 80 cm in height and with a slope of 6%) (Figure 2) [13]. Each bioreactor was filled with three layers of stones of different granulometry. The first deeper 10 cm layer consisted of coarse gravel (average diameter = 10 cm), then a second 15 cm layer of medium gravel (average diameter = 5 cm), a third 20 cm layer of fine gravel (average diameter = 1 cm), and the upper 15 cm layer was composed of mine soil samples, volcanic ashes 1:1 (v/v) and 350 ppm $ZnSO_4$ (mine treatment, CS) (Figure 2). The control treatment was performed with the commercial soil substrate (blank soil treatment, B). Each container was connected to a collection chamber (10 cm × 30 cm and 80 cm deep) to collect percolated water. These bioreactors at TRL 4 were made to a 1:10 scale with the vegetable depuration module (VDM) at TRL 6 [8], taking into account the dimensions, slope, filter layers, and physicochemical (pH, Eh) and hydraulic parameters (hydraulic constant, hydraulic retention time, and vertical income flow).

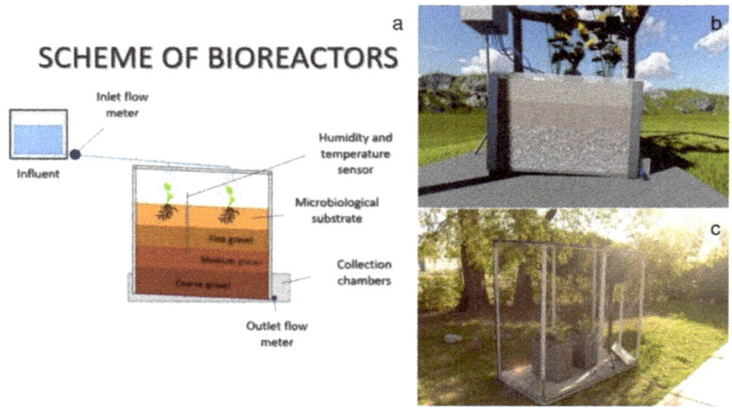

Figure 2. Schemes of the bioreactor (TRL 4). Transversal section of the bioreactor (**a**,**b**) and general view of the bioreactor at the Bioenvironmental Laboratory (**c**).

2.4. Hydraulic Calibration

The bioreactors and the VDM behave as a modified subsurface artificial wetland with inlet vertical flow and outlet flow collected in a collecting chamber (Figure 2). The volume of effluent obtained in the collecting chamber was not significant, given that irrigation along the assay was carried out considering the humidity of the substrate registered with sensors immersed into the upper top layer (CS layer).

To model the hydraulic flow in the bioreactors and the VDM, Darcy's law was used [13]. This law considers the inlet and outlet flow when field capacity is reached. The hydraulic constant (Ks) was also estimated. The hydraulics in both followed the Darcy model:

$$Ks = Q/(Ac \times s) \quad (1)$$

where:

Ks: hydraulic constant (m/days);

Q: average flow rate at which it enters and which it exits (m^3/days);

Ac = perpendicular area to the flow (m^2) = 0.3 m × 0.5 m;

s = slope (m/m) = 0.03 m/0.5 m.

Furthermore, the hydraulic retention time (th) was measured by registering the time it took the influent to cross down the different layers and to exit towards the collecting chamber when a 2 cm film of the fluent covered the last surface layer of stone in the bioreactor.

2.5. Experimental Design and Measurement of Parameters

Two bioreactors were performed as described above under two treatments: blank soil (B) and contaminated mine soil substrate (CS). In each bioreactor, 8 sunflower plants in total were grown: 4 plants were inoculated with the AM fungus GA5 strain and 4 were not inoculated. The plants were contained in separate pots in each bioreactor to avoid AM root colonization in uninoculated treatments. Previously, a complete hole in the bottom of each pot was made to let drainage occur. After 6 months, sunflower plants were harvested from each bioreactor. The shoots were separated into leaves and flowers, and the roots were carefully rinsed with distilled water to eliminate substrate particles. Fresh weights of leaves, flowers, and roots were registered, then dried in an oven at 80 °C for 72 h to obtain dry biomass.

To assess the mycorrhizal colonization, subsamples of roots were stained with trypan blue following the modified technique of Phillips and Hayman [17]. The frequency of mycorrhizal colonization was calculated as the percentage of root segments containing

AM fungal structures according to Giovanneti and Mosse [18]. All measurements were performed under a Nikon Optiphot-2 light binocular microscope at 100× magnification. Finally, the chemical elements of dried plant tissues (roots, leaves, and flowers) and soil substrate from pots were analyzed by the Total X-Ray Fluorescence (TRX) technique. The bioconcentration factors (calculated as the ratio between the concentration of chemical elements in plant tissue and soil substrate, BC) and translocation factors (TF, calculated as the ratio between the concentration of chemical elements in leaves and roots and between leaves and flowers) were determined for the following chemical elements: Cr, Zn, P, Ni, Rb, Sr, Cu, and Mn [8,12].

2.6. TXRF Analysis

The chemical determination of plant biomass and soil substrate from each pot was carried out by the TXRF analysis. For that, each plant sample was weighed and placed in a Teflon beaker with 3 mL sub-boiling HNO_3 and 1 mL H_2O_2, and then microwave digested. After digestion, the solution was transferred to a 10 mL volumetric flask (or 5 mL in cases where the sample was smaller than 0.05 g) and made up to volume with Milli-Q water. A total of 10 µL of Ga 100 mg/L (Merck, Munich, Germany) was added as an internal standard to 1 mL of sample solution. Then, µ5 L of the solution was deposited on a quartz reflector and measured for 300 s in the TXRFS2 Picofox spectrometer (Bruker, Buenos Aires, Argentina) with a Mo tube. For soil samples, a 28 mm diameter pressed pellet was prepared with 3 g of dry sample. Elemental determination was carried out with a WDXRF S8 Tiger spectrometer (Bruker, Munro, Buenos Aires, Argentina), using the Quant Express application.

2.7. Bioextracting Potential (BP) in VDM

The estimation of BP of CRMs (P and Sr) and SRMs (Cr, Cu, Mn, Zn, Rb, and Ni) by using the MAP system in the VDM at TRL 6 containing the same contaminated mine soil, was carried out as detailed in Scotti et al. [5]. The VDM is located at the Bioenvironmental Laboratory CNEA FRSR (San Rafael, Mendoza, Argentina) and a description of its performance is detailed in Scotti et al. [8]. The estimation of BP of SRMs and CRMs by the VDM and the MAP system was calculated considering the concentration of each element in plant biomass obtained at the TRL 4 experiment, and the total biomass grown in the VDM with 1 m^3 of contaminated mine soil substrate, as follows:

$$BP\ (mg) = [(Wtot\ (L) \times Cp(L)) + (Wtot\ (R) \times Cp(R)) + (Wtot\ (F) \times Cp(F))]/1000 \quad (2)$$

where:

Wtot (L, R, F): Total dry weight (g) of aerial (leaves, L and flowers, F) and radical (roots, R) plant tissue;

Cp (L, R, F): concentration (ppm) of chemical elements in the aerial (Leaves, L and flowers, F) or radicular (roots, R) plant tissue.

3. Results

After calibration, we registered a flow rate (Q) of 7.3 m^3/day and the hydraulic retention time resulted in 10 min. After application of the Darcy's law equation, we obtained a hydraulic constant value (Ks) of 811.11 m/day for both bioreactors.

At the end of the experiment, all sunflower plants had survived, and no visible signs of toxicity were observed during the assay in the bioreactor. The dry biomass of sunflower plants grown in the contaminated mine soil substrate (CS) and blank soil (B), inoculated (M+) and uninoculated (M−) with the AM fungus Rhizophagus intraradices GA5 strain after 6 months is shown in Table 1. No statistical variation in plant biomass was registered between M+ and M− in both B and CS substrates. The root biomass from the CS treatment was lower than those values from the B treatment, whilst the leaf biomass of mycorrhized plants growing in the CS substrate had more biomass than those uninoculated plants.

Table 1. Plant dry biomass of Helianthus annuus grown in the contaminated mine soil substrate (CS) and blank soil (B) and inoculated (M+) and uninoculated (M−) with the arbuscular mycorrhizal fungus Rhizophagus intraradices GA5 strain.

		Biomass (g)	
Soil Substrate	**Plant Tissue**	**M+**	**M−**
CS	F	0.43 (0.02)	0.40 (0.03)
	L	0.29 (0.03)	0.20 (0.02)
	R	0.02 (0.01)	0.010 (0.006)
B	F	0.39 (0.03)	0.41 (0.03)
	L	0.31 (0.03)	0.30 (0.02)
	R	0.16 (0.01)	0.17 (0.01)

The Fisher test did not show significant differences between M+ and M− in B and CS treatments. Values are mean and standard deviation is reported in brackets (n = 4). F: Flower tissue; L: Leaf tissues; R: Root tissues.

The AM fungus R. intraradices GA5 strain colonized the sunflower roots in both substrates after 6 months (Figure 3a,b), but the percentage (%) of mycorrhizal root colonization of sunflowers grown in the CSM+ treatment was 45 ± 7.5 % (mean ± error standard), while the uninoculated plants showed a 3 ± 0.5 % in the same substrate (CSM−). The inoculated plants in the B treatment reached 88 ± 3 % of root colonization (BM+), and no colonization of roots by any AM fungi was observed in the roots of sunflowers uninoculated and developed in the same substrate (Figure 3c).

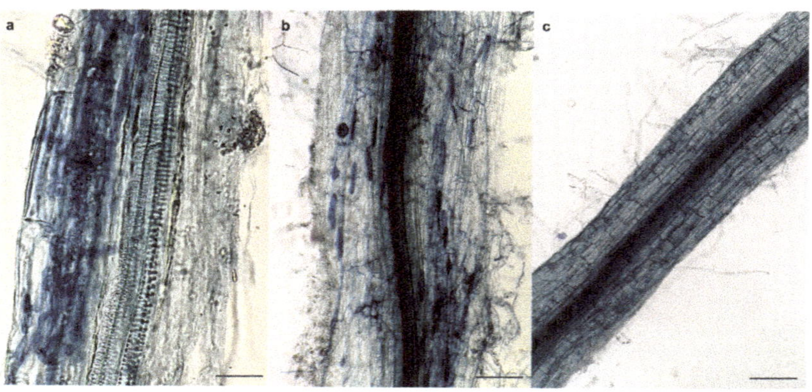

Figure 3. (**a**) Sunflower root fragments colonized by the arbuscular mycorrhizal fungus *Rhizophagus intraradices* GA5 strain after 6 months of growth in contaminated mine soil + volcanic ash + Zn (CSM+), and (**b**) in blank substrate (BM+); and (**c**) uncolonized root fragment of sunflower plant without inoculation of AM fungi and grown blank substrate (BM-). Bars: (**a**) 100 μm; (**b**) 200 μm; (**c**) 400 μm.

Table 2 shows the concentration of the raw materials (ppm) in different plant tissues of sunflowers inoculated (CSM+) and uninoculated (CSM−) with the AM fungus when grown in the CS and B substrate. The concentration of Zn, Cu, and Rb in the CS was markedly superior to the B substrate, while the P concentration resulted lower in the mining substrate. No differences were recorded between inoculated and uninoculated treatments for these elements. Zn accumulated more in the different plant tissues in both CSM− and CSM+ treatments. For Cr, no statistical differences were found among treatments, but a slight increase was observed in flower and leaves tissues and a decrease in roots when sunflowers were inoculated and developed in the mining substrate. In addition, more concentration of Cr was registered in the mining substrate when AM fungus was present. The P concentration was markedly different in B and CMS+ treatment in leaves of sunflowers. For Ni, a higher concentration was found in the leaves and roots of inoculated plants grown in a contaminated substrate. No statistical differences were found in Cu concentration between CSM− and CSM− treatment but a tendency of increasing Cu in leaves and roots of sunflowers under CSM+ treatment was observed. Mn concentration was higher in the leaves and flower tissues of plants under CS treatment but without statistical differences between inoculated and uninoculated plants. For Rb and Sr, no significant differences were found between treatments. However, the Rb concentration in aerial parts of sunflower plants was slightly higher in the CSM+ treatment, and more Sr was concentrated in leaves with a decrease in roots in the CSM− treatment.

Table 2. Raw materials concentration (ppm) in the different plant tissues of *Helianthus annuus* inoculated (CSM+) and uninoculated (CSM−) with the arbuscular mycorrhizal fungus *Rhizophagus intraradices* GA5 strain, when grown in contaminated mine soil + volcanic ash + Zn (CS) and control blank soil (B).

Chemical Elements	Plant Tissues and Soil	B	CSM−	CSM+
Zn	Flowers	46 (4.24) **a**	756 (76) **b**	600 (60) **b**
	Leaves	41 (0) **a**	1200 (100) **b**	930 (93) **b**
	Roots	38.5 (12.0) **a**	1200 (100) **b**	1100 (100) **b**
	Substrate	137.5 (2.1) **a**	6000 (90) **b**	5900 (90) **b**
Cr	Flowers	3.15 (0.64) **a**	3.50 (0.70) **a**	4.02 (0.80) **a**
	Leaves	5.4 (0.6) **a**	4.7 (0.70) **a**	7.4 (0.7) **a**
	Roots	17.9 (14.2) **a**	29 (3) **a**	13 (1) **a**
	Substrate	64 (1) **a**	83 (8) **a**	95 (19) **a**
P	Flowers	3250 (353) **a**	1700 (200) **b**	1500 (200) **b**
	Leaves	2050 (353) **a**	615 (61) **b**	1500 (200) **a**
	Roots	1250 (212) **a**	584 (58) **b**	543 (54) **b**
	Substrate	3350 (70) **a**	2200 (300) **b**	2200 (300) **b**
Ni	Flowers	0.5 (0) **a**	0.7 (0.1) **a**	0.5 (0.1) **a**
	Leaves	0.5 (0) **a**	0.80 (0.02) **b**	1.4 (0.02) **c**
	Roots	1.75 (0.92) **a**	2.2 (0.3) **a**	5.2 (0.5) **b**
	Substrate	30.5 (0.7) **a**	33 (3) **a**	31 (6)a
Cu	Flowers	5 (0.7) **a**	15 (1) **b**	15 (1) **b**
	Leaves	8.4 (0.6) **a**	19 (2) **b**	25 (3) **b**
	Roots	11.25 (3.61) **a**	244 (24) **b**	254 (25) **b**
	Substrate	51.00 (1.41) **a**	4500 (700) **b**	4600 (700) **b**
Mn	Flowers	19.5 (3.5) **a**	118 (12) **b**	120 (12) **b**
	Leaves	32.5 (0.7) **a**	241 (24) **b**	338 (34) **b**
	Roots	51.5 (19.1) **a**	60 (6) **a**	84 (8) **a**
	Substrate	690 (19) **a**	1000 (200) **a**	1000 (200) **a**
Rb	Flowers	12.5 (0.7) **a**	15 (2) **a**	16 (2) **a**
	Leaves	10.7 (1.8) **a**	16 (2) **a**	16 (2) **a**
	Roots	21 (3) **a**	12 (1) **b**	14 (1) **ab**
	Substrate	82 (58) **a**	283 (28) **b**	280 (56) **b**
Sr	Flowers	95.5 (7.8) **a**	98 (10) **a**	97 (10) **a**
	Leaves	156.5 (6.4) **a**	220 (22) **a**	208 (21) **a**
	Roots	123.5 (13.4) **a**	93 (9) **a**	103 (10) **a**
	Substrate	138.5 (31.8) **a**	181 (18) **a**	177 (35) **a**

LSD Fisher—different letters in the rows indicate significant differences. The standard deviation is reported in brackets.

Table 3 shows the bioaccumulation coefficient (BC) values obtained in the different tissues of sunflower plants inoculated with the AM fungus *R. intraradices* GA5 strain and grown under contaminated mine substrate (CSM+). The BC values for flower biomass were 0.003 Cu < 0.016 Ni < 0.04 Cr < 0.06 Rb < 0.10 Zn < 0.12 Mn < 0.55 Sr < 0.68 P, for leaf biomass covered the range of 0.005 Cu < 0.045 Ni < 0.06 Rb < 0.08 Cr < 0.16 Zn < 0.34 Mn <

0.68 P < 1.17 Sr, and for root biomass resulted in 0.05 Rb < 0.06 Cu < 0.08 Mn < 0.14 Cr < 0.17 Ni < 0.19 Zn < 0.25 P < 0.58 Sr. The BC values in the CS treatment for flower biomass of sunflower plants without mycorrhizal inoculation (CSM-) were 0.003 Cu < 0.0021 Ni < 0.04 Cr < 0.05 Rb < 0.13 Zn < 0.12 Mn < 0.54 Sr < 0.77 P, whilst BC values for leaf biomass were 0.004 Cu < 0.024 Ni < 0.06 Cr = Rb < 0.20 Zn < 0.24 Mn < 0.27 P < 1.21 Sr, and for root biomass were 0.04 Rb < 0.05 Cu < 0.06 Mn < 0.07 Ni < 0.20 Zn < 0.26 P < 0.35 Cr < 0.51 Sr (Table 3). The BC values in the control treatment (B) for flower biomass resulted in 0.016 Ni < 0.03 Mn < 0.05 Cr < 0.08 Rb < 0.1 Cu < 0.33 Zn < 0.69 Sr < 0.97 P, for leaf biomass from 0.016 Ni < Mn < Cr < Rb < Cu < Zn < P< 1.13 Sr, while BC values for root covered the range 0.06 Ni < 0.07 Mn < 0.11 Rb < 0.22 Cu < 0.28 Cr = Zn < 0.37 P < 0.89 Sr (Table 3). The highest values of BC for P and Mn were registered in the aerial part when sunflower plants were inoculated and grown in the mine substrate (CSM+), while Sr showed BC > 1 without significant differences between treatments. Meanwhile, the highest values of BC were observed in roots for Cr, Cu, and Ni in comparison with leaves and flower tissues, and similar BC values were obtained for Zn in roots and leaf tissues of sunflower plants for treatments with the mine substrate (CSM+ and CSM−). The highest translocation factors (TF) (root to leaf) in mycorrhizal plants were recorded for Mn (4.02) > Sr > P > Rb > Zn (0.84) (Table 3). The mycorrhizal treatment decreased Cr input to the root but increased the TF for this element. Furthermore, mycorrhization increased the translocation of Cu, Cr, and P from root to leaf, but no variation occurred for Mn, and for Zn, Rb, Ni, and Sr the TF values were lower. In the case of translocation from leaf to flower tissues, the TF factor only increased for Rb in mycorrhizal treatments.

Table 3. Values of the bioaccumulation coefficients (BC) and translocation factors (TF) from roots to leaves and from leaves to flowers of the chemical elements under study in *Helianthus annuus* plants growing in blank soil (B) or contaminated soil substrate (CS), inoculated (CSM+) and uninoculated (CSM-) with the arbuscular mycorrhizal fungus *Rhizophagus intraradices* GA5 strain. The comparative effects of TF in mycorrhization and control treatment are shown as greater (+), less (−), or equal (=).

Elements	Treatment	BC Flower	BC Leaves	BC Roots	TF Leaves/Roots	Effect M+	TF Flower/Leaves	Effect M+
Cr	CSM+	0.04 (0.02)	0.08 (0.02)	0.14 (0.04)	0.57	(+)	0.50	(−)
	CSM−	0.04 (0.01)	0.06 (0.01)	0.35 (0.07)	0.16		0.67	
	B	0.05 (0.01)	0.09 (0.01)	0.28 (0.23)	0.30		0.57	
Zn	CSM+	0.10 (0.01)	0.16 (0.02)	0.19 (0.02)	0.84	(−)	0.63	(−)
	CSM−	0.13 (0.02)	0.20 (0.02)	0.20 (0.02)	1.00		0.65	
	B	0.33 (0.04)	0.30 (0.01)	0.28 (0.10)	1.06		1.10	
P	CSM+	0.68 (0.18)	0.68 (0.18)	0.25 (0.06)	2.76	(+)	1.00	(−)
	CSM−	0.77 (0.20)	0.28 (0.07)	0.26 (0.06)	1.05		2.75	
	B	0.97 (0.13)	0.61 (0.12)	0.37 (0.07)	1.64		1.59	
Cu	CSM+	0.003 (0.0007)	0.005 (0.001)	0.06 (0.01)	0.10	(+)	0.60	(−)
	CSM−	0.003 (0.0007)	0.004 (0.001)	0.05 (0.01)	0.08		0.75	
	B	0.10 (0.02)	0.16 (0.02)	0.22 (0.08)	0.75		0.63	
Mn	CSM+	0.12 (0.04)	0.34 (0.10)	0.08 (0.02)	4.02	(=)	0.35	(−)
	CSM−	0.12 (0.04)	0.24 (0.07)	0.06 (0.02)	4.02		0.50	
	B	0.03 (0.006)	0.05 (0.002)	0.07 (0.03)	0.63		0.60	
Rb	CSM+	0.06 (0.02)	0.06 (0.02)	0.05 (0.01)	1.14	(−)	1.00	(+)
	CSM−	0.05 (0.01)	0.06 (0.01)	0.04 (0.01)	1.33		0.83	
	B	0.15 (0.12)	0.13 (0.11)	0.26 (0.22)	0.51		1.15	
Sr	CSM+	0.55 (0.16)	1.17 (0.35)	0.58 (0.17)	2.02	(−)	0.47	(−)
	CSM−	0.54 (0.11)	1.21 (0.24)	0.51 (0.10)	2.37		0.45	
	B	0.69 (0.21)	1.13 (0.30)	0.89 (0.30)	1.26		0.61	
Ni	CSM+	0.016 (0.006)	0.045 (0.01)	0.17 (0.05)	0.27	(−)	0.36	(−)
	CSM−	0.021 (0.005)	0.024 (0.003)	0.07 (0.02)	0.36		0.88	
	B	0.016 (0.0004)	0.016 (0.0004)	0.06 (0.03)	0.29		1.00	

Standard deviation is reported in brackets.

Table 4 shows the estimation of bioextracting potential (BP) for each chemical element when using the MAP system in the VDM containing 2 m^3 of the substrate (50:50, v/v consisted of contaminated soil: volcanic ash). The BP (mg) varied in 0.21 Ni < 1.2 Cr < 3.4 Rb < 5.45 Cu < 30.2 Sr < 43.9 Mn < 159 Zn < 316 P. The highest BP estimated value considering root and leaf biomass was obtained for P, Zn, Mn, and Sr.

Table 4. Estimation of bioextracting potential (BP) for each chemical element when using the MAP system in the VDM containing 2 m^3 of substrate (50:50, v/v consisted of contaminated soil: volcanic ash).

Parameter	Zn (ppm)	Cr (ppm)	P (ppm)	Ni (ppm)	Cu (ppm)	Mn (ppm)	Rb (ppm)	Sr (ppm)
CpF × WtF	74,820 (10,962)	501 (123)	187,050 (33,640)	62.3 (15.4)	1870 (211)	14,964 (2192)	1995 (342)	12,096 (1810)
CpR × WtR	6380 (3770)	75 (43)	3149 (1888)	30 (18)	1473 (882)	487 (290)	81 (46)	597 (357)
CpL × WtL	78,213 (15,912)	622 (123)	126,150 (29,870)	118 (14)	2102 (470)	28,425 (5800)	1346 (307)	17,492 (3576)
BP (mg)	159 (30)	1.2 (0.3)	316 (65)	0.21 (0.05)	5.4 (1.6)	43.9 (8.3)	3.4 (0.7)	30.2 (11.8)

Standard deviation is reported in brackets. Cp (L, R, F): concentration (ppm) of chemical elements in leaves (L), flowers (F), or roots (R) tissues; Wtot (L, R, F): total dry weight (g) of leaves (L), flower (F) or roots (R) tissues.

4. Discussion

In this work, the mycorrhizal-assisted phytomining (MAP) system, composed of *Helianthus annuus*–arbuscular mycorrhizal fungus *Rhizophagus intraradices*–Zn-volcanic ashes was evaluated in bioreactors at TRL 4 for the recovery of CRMs (Sr, P) and SRMs (Cr, Zn, Cu, Mn, Rb, Ni) from Los Cóndores mining wastes. The mine substrate showed high concentrations of Cu, Zn, Rb, and low P concentration in comparison with the blank substrate and other mining-contaminated soils previously reported [2,5]. The variation in composition and concentration of chemical elements in the substrate might cause differences in the behavior of the MAP system. It is known that AM symbiosis functions as a 'buffer' to protect the host plant against damage caused by heavy metals in the soil [19,20]. The different behavior against heavy metals in excess may be due to competition for membrane receptors or passive diffusions under pH and Eh values in the experiment, as well as to the activation of certain fungal and/or plant enzymes/glycoproteins that are involved in chelation, speciation (redox processes), sequestration and immobilization of chemical elements (by phytochelatins, glomalin, or fungal metallothioneins) [19]. Some substances such as Zn ions act as enzyme cofactors that activate/deactivate enzymes involved in the tolerance and uptake/exclusion of chemical elements at high concentrations [8,11]. In this sense, we highlighted that the concentration of Zn in the MAP system affected the entry of different chemical elements in mycorrhizal plants. In addition, the inoculation with AM fungus *R. intraradices* GA5 strain favored phytoextraction processes by increasing translocation from roots to leaves of Cr, P, and Cu. On the other hand, there were no differences in Mn behavior, and phytostabilization occurred for Ni, Sr, Rb, and Zn by decreasing its translocation to aerial plant tissues. The same mycorrhizal effect on translocation from roots to leaves of Cr, Cu, and Ni, and the opposite mycorrhizal effect for Zn, P, Sr, Rb, and Mn were found with the MAP system developed in mine substrates from Odisha (India). These differences could be attributed to the different physicochemical characteristics of the mine substrates under study (Odisha and Los Cóndores mines) [2,5].

When mycorrhized, plants grown in the blank soil substrate increased the concentration of P in leaves, Ni in leaves and roots, and Rb in roots. An increase in the translocation of P from roots to leaves in M+ treatments was observed. It is widely known that AM fungi improve the nutritional status of colonized plants, mainly by P and N uptake [21], and also contribute to the enhancement of Ni and Cr in sunflower biomass [5,22].

No differences in Sr bioaccumulation were found between inoculated and uninoculated treatments, but a very significant difference was observed in the bioaccumulation of this element in leaves compared to roots (TF > 2). Many authors have confirmed that Sr can be accumulated in leaves, leaf trichomes, and stems in several plant species [2,5,23,24].

Rb also presented a higher bioconcentration value in leaves (with a TF > 1) of sunflower plants grown in the mine substrate in comparison with those in the blank soil substrate.

As in previous works, at considerable amounts of Rb in the substrate (>280 ppm) and after reaching a bioconcentration threshold in leaves, the mycorrhizal treatment exerts an exclusion system and slows down the translocation of Rb to the aerial plant tissues. This effect had not been observed in our previous works probably due to the low concentrations of Rb in a substrate. Rb ion is an analogue for K ion, which is taken up along the same pathways as K and typically occurs at very low levels in soils and plant tissues [25].

A low plant biomass production in the mining substrate was recorded. This result could be due to the high concentration of some toxic elements in the mine soil that affected sunflower plant development. Likewise, mycorrhizal root colonization is generally reduced at high heavy metal concentrations in soil, and the potential uptake of Zn and other nutrients may be reduced [26,27]. For this reason, it is possible that we have not found a significant difference in plant biomass and bioconcentration values for many metals between inoculated and uninoculated treatments.

We detected a decrease of translocation in most heavy metals from the leaves to the flower parts when plants were colonized by AM fungi, except for Rb (which did not present significant differences). De María and Rivelli [28] found differences in the accumulation and distribution of Cd, Zn, and Cu in diverse phenological stages; these metal concentrations increased in the stems and leaves, particularly in the old ones, whereas decreased in bud flowers. The storage of some heavy metals in roots and their low translocation along the plant tissues during the growth stage could be considered a strategy of mycorrhizal sunflower plants to preserve young metabolically active leaves and reproductive organs from toxic metal concentrations [28]. Our results demonstrate a promising biotechnology applicable to agronomical practices given the exclusion of toxic elements in flower parts.

The highest BP of the MAP system in the VDM at TRL 6 containing mine residues was for P, Zn, Mn, and Sr (considering root and leaves). These values are much lower than those found in Scotti et al. [5] and those obtained in Guglietta et al. [2] with the same MDV system. The different mine substrates could be the cause of performance variation of the MAP system in the VDM. For that reason, calibration of those variables in the bioreactor at TRL 4 is needed before scaling up at a higher TRL maturity. The BP encourages a recovery of SRMs and CRMs by hydrometallurgical techniques [1], with subsequent purification by selective electrodeposition, permitting a sustainable and selective metals recovery at a high degree of purity (95%) and determining commercial reuse. By application of a process circuit with leaching/purification of the SRMs and CRMs accumulated and concentrated on plant biomass, a recovery of 90-95% of purified metals has been demonstrated [1,29].

5. Conclusions

Interesting aspects in the behavior of the MAP system regarding the extraction, translocation, and exclusion of chemical elements were found, associated with the composition of the substrate.

The MAP system allowed the extraction of the elements under study, being tolerant to high concentrations of heavy metals.

A high efficiency of the MAP system for the exclusion of heavy metals in flower tissues was registered.

The BP was effective for Sr and P more than other elements. These results are of great importance given that P and Sr are valuable CRMs.

The results encourage the application of this methodology but improve plant biomass; this is a critical issue for the sustainable exploitation of mining waste within a circular economy context.

Supplementary Materials: The following supporting information can be downloaded at: https://www.mdpi.com/article/10.3390/met12111828/s1, Table S1. Additional information about sampling sites (Source Verónica Saavedra, Universidad Nacional de San Luis, Horizon 2020 ERA MIN BioCriticalMetals project).

Author Contributions: Conceptualization, A.S. and V.A.S.; methodology, A.S.; validation, A.S. and V.A.S.; formal analysis, V.A.S.; investigation, A.S., V.A.S., and S.U.; resources, A.S. and S.U.; writing—original draft preparation, N.A.J., V.A.S., S.U., and A.M.G.; writing—review and editing, V.A.S. and A.M.G.; supervision, A.S. and V.A.S.; project administration, A.S.; funding acquisition, A.S. and S.U. All authors have read and agreed to the published version of the manuscript.

Funding: This research was funded by PID 5252 MSUTNSR0005252 Technological University National-FRSR and the biogeochemical research line of the CNEA Bioenvironmental Laboratory.

Institutional Review Board Statement: Not applicable.

Informed Consent Statement: Not applicable.

Data Availability Statement: Not applicable.

Acknowledgments: We acknowledge the technical collaboration of Veronica Saavedra, Juan Cerioni, Sol Cerioni, Ana Rosa Castaño Gañan, Martín Mengarelli, Matías Benavidez, Mauricio Visciglia, Roberto Biondi, Juliana Turano, Camila Quiroga, and Gabriela Coria. Furthermore, we acknowledge Paola Babay, Florencia Gonzalez, and Roxana Leguizamon for their support with the X-ray fluorescence chemistry laboratory of the National Atomic Energy Commission for chemical analysis.

Conflicts of Interest: The authors declare no conflict of interest.

References

1. Ubaldini, S.; Guglietta, D.; Vegliò, F.; Giuliano, V. Valorization of Mining Waste by Application of Innovative Thiosulphate Leaching for Gold Recovery. *Metals* **2019**, *9*, 274. [CrossRef]
2. Guglietta, D.; Belardi, G.; Cappai, G.; Casentini, B.; Godeas, A.; Milia, S.; Passeri, D.; Salvatori, R.; Scotti, A.; Silvani, V.; et al. Toward a Multidisciplinary Strategy for the Classification and Reuse of Iron and Manganese Mining Wastes. *Chem. J. Mold.* **2020**, *15*, 21–30. [CrossRef]
3. European Commission, COM 2020/474 Final. Available online: https://eur-lex.europa.eu/legal-content/EN/TXT/?uri=CELEX:52020DC0474 (accessed on 16 September 2022).
4. Peuke, A.D.; Rennenberg, H. Phytoremediation. Molecular biology, requirements for application, environmental protection, public attention and feasibility. *EMBO J.* **2005**, *6*, 497–501. [CrossRef]
5. Scotti, A.; Milia, S.; Silvani, V.; Cappai, G.; Guglietta, D.; Trapasso, F.; Tempesta, E.; Passeri, D.; Godeas, A.; Gómez, M.; et al. Sustainable Recovery of Secondary and Critical Raw Materials from Classified Mining Residues Using Mycorrhizal-Assisted Phytoextraction. *Metals* **2021**, *11*, 1163. [CrossRef]
6. Dinh, T.; Dobo, Z.; Kovacs, H. Phytomining of noble metals—A review. *Chemosphere* **2022**, *286*, 131805. [CrossRef]
7. Ubaldini, S.; Guglietta, D.; Trapasso, F.; Carloni, S.; Passeri, D.; Scotti, A. Treatment of Secondary Raw Materials by Innovative Processes. *Chem. J. Mold.* **2019**, *14*, 32–46. [CrossRef]
8. Scotti, A.; Silvani, V.; Cerioni, J.; Visciglia, M.; Benavidez, M.; Godeas, A.A. Pilot testing of a bioremediation system of water and soils contaminated with heavy metals: Vegetable Depuration Module. *Int. J. Phytoremediation* **2019**, *21*, 899–907. [CrossRef]
9. Colombo, R.P.; Benavidez, M.E.; Fernandez Bidondo, L.; Silvani, V.; Bompadre, M.J.; Statello, M.; Scorza, M.V.; Scotti, A.; Godeas, A.M. Arbuscular mycorrhizal fungi in heavy metal highly polluted soil in the Riachuelo river basin. *Rev. Arg. Microbiol.* **2019**, *52*, 145–149. [CrossRef]
10. Bompadre, M.J.; Benavidez, M.; Colombo, R.P.; Silvani, V.A.; Godeas, A.M.; Scotti, A.; Pardo, A.G.; Fernandez Bidondo, L. Mycorrhizal stress alleviation in *Senecio bonariensis* Hook & Arn growing in urban polluted soils. *J. Environ. Qual.* **2021**, *50*, 589–597.
11. Scotti, A.; Godeas, A.; Silvani, V. Procedimiento para Aumentar la Capacidad Biorremediadora de Plantas Hiperacumuladoras a Través de Hongos Formadores de Micorrizas Arbusculares (HMA) para Tratamiento de Suelos y/o Aguas Contaminados. Patent AR090183 B1130100620, 2022.
12. Scotti, A.; Silvani, V.; Milia, S.; Cappai, G.; Ubaldini, S.; Ortega, V.; Colombo, R.; Godeas, A.; Gómez, M. Scale-up of Mycorrhizal-Assisted Phytoremediation system from Technology Readiness Level 6 (Relevant Environment) to 7 (Operational Environment): Cost-benefits within a Circular Economy Context. In *Soil Science—Emerging Technologies, Global Perspectives and Applications*; Aide, M.T., Braden, I., Eds.; IntechOpen: London, UK, 2022. [CrossRef]
13. Scotti, A.; Cerioni, J.; Reviglio, H.; Visciglia, M.; Cerioni, S.; Biondi, R.; Saavedra, V.; Litter, M.; Silvani, V.; Godeas, A.; et al. Scaling to technological readiness levels 6 in the bio-environmental laboratory. *Robot. Autom. Eng. J.* **2019**, *4*, 5555637.
14. Brodtkorb, M.K.; Ametrano, S. Estudio mineralógico de la mina "Los Cóndores", provincia de San Luis. In Proceedings of the 8° Congreso Geológico Argentino 3, San Luis, Argentina, 20–26 September 1981; pp. 259–302.
15. Romano, E.; Saavedra, V. UNSL Mineralogical Report for WP1 Technical Report; Bucharest, Romania, May 2018. Available online: https://www.uc.pt/en/org/biocriticalmetals# (accessed on 16 September 2022).

16. Silvani, V.A.; Fernández Bidondo, L.; Bompadre, M.J.; Pérgola, M.; Bompadre, A.; Fracchia, S.; Godeas, A.M. Growth dynamics of geographically different arbuscular mycorrhizal fungal isolates belonging to the 'Rhizophagus clade' under monoxenic conditions. *Mycologia* **2014**, *106*, 963–975. [CrossRef] [PubMed]
17. Phillips, J.M.; Hayman, D.S. Improved procedures for clearing roots and staining parasitic and vesicular-arbuscular mycorrhizal fungi for rapid assessment of infection. *Trans. Br. Mycol. Soc.* **1970**, *55*, 158–160. [CrossRef]
18. Giovannetti, M.; Mosse, M. An evaluation of techniques for measuring vesicular arbuscular infection in roots. *New Phytol.* **1980**, *84*, 589–600. [CrossRef]
19. Ferrol, N.; Tamayo, E.; Vargas, P. The heavy metal paradox in arbuscular mycorrhizas: From mechanisms to biotechnological applications. *J. Exp. Bot.* **2016**, *67*, 6253–6265. [CrossRef]
20. Luthfiana, N.; Inamura, N.; Tantriani; Sato, T.; Saito, K.; Oikawa, A.; Chen, W.; Tawaraya, K. Metabolite profiling of the hyphal exudates of Rhizophagus clarus and Rhizophagus irregularis under phosphorus deficiency. *Mycorrhiza* **2021**, *31*, 403–412. [CrossRef]
21. Smith, S.E.; Read, D.J. *Mycorrhizal Symbiosis*, 3rd ed.; Academic Press: New York, NY, USA, 2008.
22. Ker, K.; Charest, C. Nickel remediation by AM colonized sunflower. *Mycorrhiza* **2010**, *20*, 399–406. [CrossRef]
23. Burger, A.; Lichtscheidl, I. Strontium in the environment: Review about reactions of plants towards stable and radioactive strontium isotopes. *Sci. Total Environ.* **2019**, *653*, 1458–1512. [CrossRef] [PubMed]
24. Hanaka, A.; Dresler, S.; Wójciak-Kosior, M.; Strzemski, M.; Ková˘cik, J.; Latalski, M.; Zawi'slak, G.; Sowa, I. The Impact of Longand Short-Term Strontium Treatment on Metabolites and Minerals in Glycine max. *Molecules* **2019**, *24*, 3825. [CrossRef]
25. Hawkes, C.V.; Casper, B.B. Lateral root function and root overlap among mycorrhizal and non mycorrhizal herbs in Florida shrubland, measured using rubidium as a nutrient analog. *Am. J. Bot.* **2002**, *89*, 1289–1294. [CrossRef]
26. Watts-Williams, S.J.; Smith, F.A.; McLaughlin, M.J.; Patti, A.F.; Cavagnaro, T.R. How important is the mycorrhizal pathway for plant Zn uptake? *Plant Soil* **2015**, *390*, 157–166. [CrossRef]
27. Watts-Williams, S.J.; Patti, A.F.; Cavagnaro, T.R. Arbuscular mycorrhizas are beneficial under both deficient and toxic soil zinc conditions. *Plant Soil* **2013**, *371*, 299–312. [CrossRef]
28. De Maria, S.; Rivelli, A.R. Trace element accumulation and distribution in sunflower plants at the stages of flower bud and maturity. *Ital. J. Agron.* **2013**, *8*, e9. [CrossRef]
29. Ubaldini, S.; Povar, I.; Lupascu, T.; Spinu, O.; Trapasso, F.; Passeri, D.; Carloni, S.; Guglietta, D. Application of Innovative Processes for Gold Recovery from Romanian Mining Wastes. *Chem. J. Mold.* **2020**, *15*, 29–37. [CrossRef]

metals MDPI

Article

Leaching of Gold and Copper from Printed Circuit Boards under the Alternating Current Action in Hydrochloric Acid Electrolytes

Vera Serga [1], Aleksej Zarkov [2,3], Ervins Blumbergs [4,5,6], Andrei Shishkin [4,7], Janis Baronins [4,7,8,*], Edgars Elsts [2] and Vladimir Pankratov [2]

1 Institute of Materials and Surface Engineering, Faculty of Materials Science and Applied Chemistry, Riga Technical University, 3/7 P. Valdena Street, LV-1048 Riga, Latvia
2 Institute of Solid State Physics, University of Latvia, 8 Kengaraga Street, LV-1063 Riga, Latvia
3 Institute of Chemistry, Vilnius University, 24 Naugarduko Street, LT-03225 Vilnius, Lithuania
4 ZTF Aerkom SIA, 32 Miera Str., LV-2169 Salaspils, Latvia
5 Institute of Physics, University of Latvia, 32 Miera Street, LV-2169 Salaspils, Latvia
6 Faculty of Civil Engineering, Riga Technical University, 21/1 Azenes Street, LV-1048 Riga, Latvia
7 Rudolfs Cimdins Riga Biomaterials Innovations and Development Centre of RTU, Institute of General Chemical Engineering, Faculty of Materials Science and Applied Chemistry, Riga Technical University, 3 Pulka Street, LV-1007 Riga, Latvia
8 Latvian Maritime Academy, 12 k-1 Flotes Street, LV-1016 Riga, Latvia
* Correspondence: janis.baronins@gmail.com

Abstract: Modern technologies for recycling electronic waste (e-waste) have high economic efficiency and environmental safety requirements. Among the existing technologies, hydrometallurgy is considered to be the most promising technology for e-waste recycling. Increasing attention paid to the chlorination method is associated with the complex recycling of low-grade ores containing noble metals and the raw materials of secondary polymetallic. In this paper, we propose a new scheme for leaching metals from computer printed circuit boards (PCBs) pre-crushed in a disintegrator: The processes of chlorine production and hydrochlorination are implemented in one reactor under the action of an alternating current (AC) of industrial frequency (50 Hz). Three fine fractions of raw material powders with particle size d < 90 μm, d = 90–180 μm, and d = 180–350 μm were used as research objects and the finest fraction of the raw material (d < 90 μm) was studied in more detail. It was found that complete leaching of gold is achieved from fractions of raw materials with a particle size d = 90–180 μm and d = 180–350 μm, containing 277 ppm and 67 ppm of the gold, respectively, at an experiment duration (t_{ex}) of 2 h, a current density (i) of 0.66 $A \cdot cm^{-2}$, and a solid/liquid (S/L) ratio of 8.6 $g \cdot L^{-1}$. Under the same conditions of the electrochemical leaching process from the fraction of raw materials with a particle size of d < 90 μm and a gold content of 824 ppm, the degree of metal leaching is 80.5%. At the same time, with an increase in particle size in the raw material fractions from d < 90 μm to d = 180–350 μm and a copper content in the raw material from 1.40% to 6.13%, an increase in the degree of its leaching from 81.6% to 95.2%, respectively, is observed. In the framework of the preliminary study presented in this work, for the finest raw material fraction with d < 90 μm the highest gold leaching degree (86.3%) was achieved under the following experimental conditions: t_{ex}= 4 h, C_{HCl} = 6 M, i = 0.88 $A \cdot cm^{-2}$, S/L ratio—8.6 $g \cdot L^{-1}$ and the highest copper leaching degree (94.2%) was achieved under the following experimental conditions: t_{ex} = 2 h, C_{HCl} = 6 M, i = 0.64 $A \cdot cm^{-2}$, and S/L ratio—2.9 $g \cdot L^{-1}$.

Keywords: printed circuit boards; gold; copper; electrochemical leaching; alternating current

Citation: Serga, V.; Zarkov, A.; Blumbergs, E.; Shishkin, A.; Baronins, J.; Elsts, E.; Pankratov, V. Leaching of Gold and Copper from Printed Circuit Boards under the Alternating Current Action in Hydrochloric Acid Electrolytes. *Metals* **2022**, *12*, 1953. https://doi.org/10.3390/met12111953

Academic Editor: Stefano Ubaldini

Received: 4 October 2022
Accepted: 11 November 2022
Published: 15 November 2022

1. Introduction

The volume of electronic waste (e-waste) suitable for recycling in order to recover precious and non-ferrous metals is rapidly growing worldwide. Personal computers is the largest source of electronic waste [1]. Computer printed circuit boards (PCBs) are manufactured using complex technologies. The composition of PCBs includes a non-conductive substrate (glass fibre/paper reinforced with epoxy/phenolic resin) or laminate, polymer and ceramic materials [1], and metals, including copper, aluminium, tin, lead, zinc, nickel, iron and noble metals (gold, silver and palladium [2–4]). The content of individual metals in PCBs, given in the literature [3,5], varies widely, and is associated not only with the manufacturer and the date of production of computers, but also with the methods of preliminary grinding of the raw material, sampling, and leaching method for transferring metals into a solution for quantitative analysis.

There are three leading technologies for recycling electronic waste: Pyrometallurgy, hydrometallurgy, and bio-metallurgy [2,6–10]. Pyrometallurgy is mainly used industrially for recycling large volumes of primary (ore) and secondary raw materials [9]. Biotechnological techniques are commercially applied to process mineral raw materials [10]. The use of bioleaching for the recovery of valuable metals from e-waste is currently one of the rapidly developing areas [8]. However, hydrometallurgy is mainly associated with research, development, and application potential. Therefore, hydrometallurgical electronic scrap recycling possibilities have been actively studied [11–16]. Leaching is the initial stage of the hydrometallurgical process and the most critical stage in recovering valuable metals from waste PCBs. Usually, for the enrichment of raw materials, a preliminary acid treatment (H_2SO_4, HCl or HNO_3) is carried out [14,17,18]. Within the framework of hydrometallurgical technology, there are many methods of gold leaching, including cyanide, aqua regia, halogen-halide, thiosulfate, thiourea, among others. [9]. Cyanidation is the primary method of hydrometallurgical recovery of gold and silver from ores, which consists of the leaching of metals using an alkaline cyanide solution (NaCN and KCN) in the presence of atmospheric oxygen [19]. The use of cyanidation for leaching gold from PCBs showed that this method is only effective for dissolving gold and/or silver from the surface of PCBs. In addition, this process has a low leaching rate and generates a large amount of wastewater containing cyanide [20].

Gold dissolves in aqua regia to form chloroauric acid $H[AuCl_4]$. Similarly, gold dissolves in hydrochloric acid solutions in the presence of Fe^{3+} [21,22], acting as an oxidizing agent. The dissolution reaction examples can be expressed, as in Equations (1) and (2):

$$Au + 2Cl^- + Fe^{3+} \rightarrow [AuCl_2]^- + Fe^{2+} \quad (1)$$

$$[AuCl_2]^- + 2Cl^- + 2Fe^{3+} \rightarrow [AuCl_4]^- + 2Fe^{2+} \quad (2)$$

Leaching with the use of chlorine, bromine, and iodine is characterized by a high rate [23,24]. In this case, gold forms both Au^+ and complexes with halide ions, depending on the chemical composition of the solution. The reaction of dissolving metallic gold in a chlorine-saturated solution of hydrochloric acid (hydrochlorination process) proceeds according to Equations (3) and (4) [21,25]:

$$Au + 0.5Cl_2 + Cl^- \rightarrow [AuCl_2]^- \quad (3)$$

$$Au + 1.5Cl_2 + Cl^- \rightarrow [AuCl_4]^- \quad (4)$$

Chlorination of gold and copper in chloride solutions is characterized by the high solubility of metal complexes with chlorine and a high leaching rate [26].

Chlorination processes for recovering gold from ores and concentrates have been used since the second half of the 19th century and preceded cyanide leaching technology. A significant release of chlorine characterizes industrial methods of hydrochlorination into the atmosphere [21]. Therefore, reducing chlorine consumption and improving maintenance personnel's safety is crucial when developing a hydrochlorination technology for recovering valuable metals from secondary raw materials.

New technological solutions are proposed to minimize the risks of chlorine/chloride leaching systems, including in situ chlorine production (electrogeneration). Thus, to study the leaching of copper, zinc, lead, and tin from PCBs, the authors [27] proposed two reactor designs: A combined reactor (with an anion-exchange membrane) with simultaneous formation of chlorine under the action of direct current and metal leaching, and a separate metal leaching reactor connected with the anode compartment of the chlorine gas generator. The non-magnetic fraction of crushed PCBs was used as a dispersed phase in a hydrochloric acid solution. The results obtained showed that the efficiency of copper leaching in a combined reactor is lower than in a separate reactor. In the separate reactor at 50 °C and experiment duration of 240 min, the degree of Zn, Sn, Pb and Cu leaching in a 2M HCl solution is 98, 96, 96 and 71%, respectively. The impact of copper ions in solution on the rate of copper leaching using electrogenerated chlorine in a combined reactor showed that with an increase in the concentration of cuprous ions in the solution, the share of current for the release of chlorine decreases. For the oxidation of cuprous ions, it increases [28]. In this case, the following reactions of leaching in hydrochloric acid solution (Equations (5)–(8)) on the anode took place [29]:

$$Cu + Cl_{2(aq)} \rightarrow Cu^{2+} + 2Cl^- \quad (5)$$

$$Cu^{2+} + 4Cl^- + Cu \rightarrow 2[CuCl_2]^- \quad (6)$$

$$2Cl^- \rightarrow Cl_{2(g)} + 2e^-,\ E^0 = 1.358\ V_{SHE} \quad (7)$$

$$[CuCl_2]^- \rightarrow Cu^{2+} + 2Cl^- + e^-,\ E^0 = 0.457\ V_{SHE} \quad (8)$$

The gaseous chlorine released at the anode Equation (7) and dissolves in hydrochloric acid solutions follows the reactions in Equations (9)–(11) [30]:

$$Cl_{2(g)} \leftrightarrow Cl_{2(aq)} \quad (9)$$

$$Cl_{2(aq)} + H_2O \leftrightarrow HCl + HOCl \quad (10)$$

$$Cl_{2(aq)} + Cl^- \leftrightarrow Cl_3{}^- \quad (11)$$

The product distribution depends on the HCl concentration, pH, and temperature. At electrolyte temperatures of 50 °C, 60 °C, and 70 °C, with an increase in the HCl concentration from 3 mol·L^{-1} to 7 mol·L^{-1}, a decrease in the content of $Cl_{2(aq)}$ in the solution and an increase in the content of $Cl_3{}^-$ are observed. At the same time, the content of HClO remains insignificant [30].

The results of the two-stage leaching process of gold from waste mobile phone PCBs in a separate leaching reactor are also presented in ref. [31]. It was demonstrated that, at the first stage of the process, 97% copper and 5% gold are leached into a 2 M HCl solution, and in the second stage, 93% of gold is leached into a 0.1 M HCl solution. In this case, the reactions of gold dissolution in the reactor are presented in the form of the following reactions in Equations (12) and (13):

$$Au + Cl^- + 3/2Cl_2 \rightarrow [AuCl_4]^-{}_{(aq)} \quad (12)$$

$$Cl_{2(aq)} + H_2O \leftrightarrow HCl + HOCl \quad (13)$$

A separate reactor connected with the anode compartment of a chlorine gas generator was also used in ref. [32] to leach gold from waste PCBs. The leaching stage was preceded by treatment in a 4 M HCl with the addition of H_2O_2. It is shown that, with over 75 min of the experiment at a temperature of 50 °C, more than 99% of the gold passes into a 2.0 M NaCl solution (pH 1.0) at a feed rate of electrogenerated Cl_2 of 0.62 mmol·min^{-1}.

An electrochemical process is also proposed to separate Cu, Sn, Ag, Pd, and Au in two stages. In the second stage, Pd and Au are dissolved by chlorine generated on the anode under the action of direct current [33].

Thus, the above-mentioned studies have shown that the electrogenerated chlorine leaching process can be successfully used to leach metals from electronic waste.

At the same current densities, the voltage using alternating current (AC) is almost two times lower than when using direct current, which makes the process energetically more favorable. The AC of various shapes contributes to the intensification of the processes of anodic dissolution/dispersion of metals and the elimination of the passivation of the metal electrode [34]. The authors [35] demonstrated that the electrochemical dissolution of platinum group metals in hydrochloric acid solutions using AC is an effective and promising way to obtain pure salts from noble metals. A method for the dissolution of noble metals (Au, Ir, Pd, Pt, and Rh) and metals resistant to hydrochloric acid (Cd, Cu, In, Ta, and W) under the action of a half-wave or a full-wave AC was also proposed in ref. [36]. The study showed that the rate of dissolution of metals noticeably increases with an increasing current density and HCl concentration in the electrolyte solution from 3.2 M to 12 M.

This work aims to study some regularities of the process of copper and gold leaching from disintegrator-crushed PCBs under the action of an alternating current in hydrochloric acid electrolytes. In the proposed scheme, the processes of producing an oxidizing agent—chlorine and leaching of metals—occur in one electrochemical cell. The production of chlorine occurs on the electrodes under the action of AC.

2. Materials and Methods

The scheme of studies presented in this paper is shown in Figure 1.

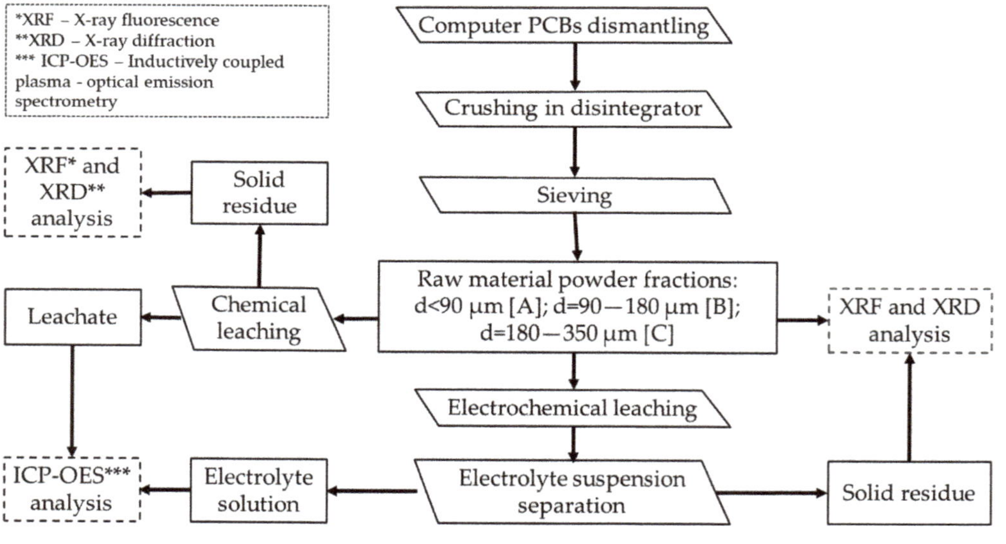

Figure 1. Schematic diagram of the process investigated.

Preliminary preparation of raw materials included only stages of PCB dismantling, double crushing in a hammer mill (DS-A) and single crushing using the high-energy semi-industrial disintegration-milling system (disintegrator DSL-350, Tallinn University of Technology, Tallinn, Estonia) and subsequent sieving using FRITSCH ANALYSETTE 3 PRO Vibratory Sieve Shaker (FRITSCH GmbH, Weimar, Germany). As a result, the three finest fractions of the obtained powders with different particle sizes (d) were selected as raw research materials: <90 μm, 90–180 μm and 180–350 μm.

The electrical circuit and the electrochemical cell used are shown in Figure 2. The electrochemical cell was a square-shaped diaphragm-less polypropylene reactor with a tight-fitting lid on which two cylindrical graphite electrodes (for spectral analysis, d = 0.8 cm) were mounted. The lid also had two holes for a glass paddle stirrer, loading the raw material, filling the electrolyte, and installing a chromel–alumel thermocouple to measure the electrolyte temperature (T_{el}). A water-jacketed glass cell was only used in separate experiments to study the effect of electrolyte temperature on the efficiency of target component leaching. Powders of the raw material were used in the form of a dispersed (solid) phase in the hydrochloric acid electrolyte (liquid phase). In all experiments, the volume of the electrolyte was 350 mL, and the concentration of the hydrochloric acid was 6 M. The choice of the electrolyte concentration was based on the electrical conductivity data of hydrochloric acid solutions: Specific electrical conductivity of 18–20% (~5–6 M) solutions at 65 °C are 1.2 $Ohm^{-1}{\cdot}cm^{-1}$, and at 90 °C is 1.5 $Ohm^{-1}{\cdot}cm^{1}$ [37].

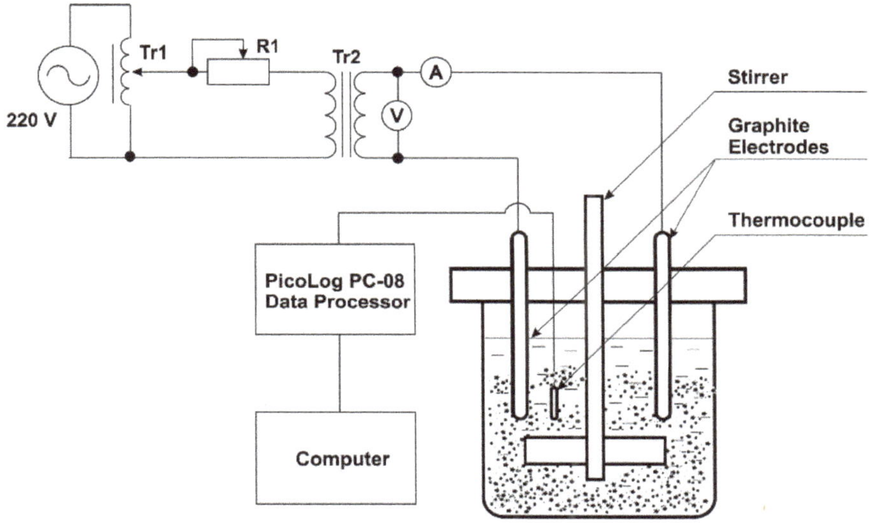

Figure 2. Scheme of the experimental setup: AC circuit (Tr1—laboratory autotransformer, R1—rheostat, Tr2—step-down transformer, V—AC voltmeter, A—AC ammeter) and electrochemical cell.

After filling the reactor with the initial components to create a suspension, the mixture was stirred at a speed of 1200 rpm for 5 min, then the stirring speed was reduced to 700 rpm, and the power was turned on. The solid-to-liquid (S/L) ratio in the suspension was 2.9 $g{\cdot}L^{-1}$, 5.7 $g{\cdot}L^{-1}$, 8.6 $g{\cdot}L^{-1}$, 11.4 $g{\cdot}L^{-1}$ and 14.3 $g{\cdot}L^{-1}$. To produce chlorine, an alternating current (AC) of industrial frequency (50 Hz) with a density (i) of 0.66 $A{\cdot}cm^{-2}$ and 0.88 $A{\cdot}cm^{-2}$ was used. The duration of the experiments (t_{ex}) varied from 0.5 to 5 h. Solid residues after leaching were washed with distilled water to pH 5–6, dried at a temperature of 105 °C and weighed.

Quantitative determination of metals in solutions resulting from chemical (sequentially boiling in 6 M HCl solution and aqua regia solution) and electrochemical (in 6 M HCl solution) leaching of raw materials was carried out by inductively coupled plasma optical emission spectrometry (ICP-OES, Perkin Elmer Optima 7000 DV ICP-OES, Perkin Elmer Inc., Waltham, MA, USA). X-ray diffraction (XRD, diffractometer Rigaku—MiniFlex 600 with CuK_{α} radiation with λ = 1.5418 Å, Rigaku Corp., Tokyo, Japan) and X-ray fluorescence (XRF, analyser EAGLE III XPL, EDAX Inc., Mahwah, NJ, USA) analyses were used to characterize raw materials and solid residues after leaching. The disintegrator-crushed fractions of a raw material microscope (KEYENCE VHX-2000, Keyence Corporation, Osaka, Japan) were used for optical imaging. For the quantitative determination of free chlorine in the electrolyte solution, the method of titration with methyl orange was used [38].

The efficiency of electrochemical leaching of metals from raw materials was evaluated by the degree of metal leaching (R), which was defined as the ratio of the quantity of metal transferred into the electrolyte solution (M_{el}) to the quantity of metal contained in the raw material (M_{PCBs}), as demonstrated in Equation (14):

$$R(\%) = (M_{el}/M_{PCBs}) \times 100\% \quad (14)$$

The content of metals in the raw material (MPCB) was calculated from the results of the ICP-OES analysis of the leachates obtained from the chemical leaching of a representative sample of the raw material.

In the proposed process organization scheme (Figure 2), the oxidizing agent—chlorine—is produced during the anodic half-cycles of the passage of AC through the cell on graphite electrodes, and raw material leaching occurs mainly in the volume of the electrolyte solution. This process does not require special conditions or additional reagents, and the process can be carried out in batch mode. This dramatically simplifies chlorination technology and makes it more environmentally friendly.

3. Results and Discussion

3.1. Raw Material Treatment and Characterisation

The high-energy semi-industrial disintegration-milling system, specially designed for processing mechanically durable materials, was used to grind fragments of PCBs into finer particles. The detailed disintegration-milling system is described in the works in refs. [39–41]. The device crushes materials by collisions. Supplied particles collide with the fingers of the milling device and with other particles, which have a moving counter direction. The resulting intensive wave of pressure begins to propagate, and the resulting values of stresses exceed the material strength. The previous work in ref. [42] described a novel single and two-stage PCBs disintegration-milling approach.

Mixed computer PCBs (Figure 3) were used as raw material to imitate wasted PCBs from real landfilled damp, which is essential for commercial technology development. This mixture of PCBs was treated as described in the previous work in ref. [42] and was used as an object of this research. As a result of crushing and subsequent sieving, three fine powder fractions of the raw material were obtained, and the morphology of these fractions was studied using optical microscopy (Figure 4a–c) and designated [A], [B] and [C], respectively.

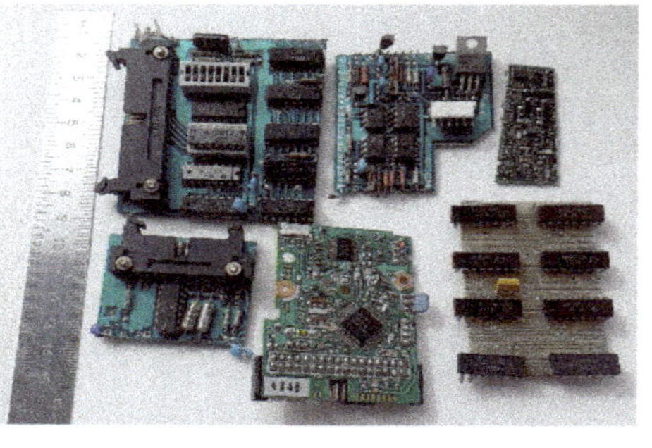

Figure 3. Image of computer PCBs used for the study.

From the presenting data, it can be seen that the obtained powders are heterogeneous and consist of irregularly shaped particles (Figure 4). It should be noted that these raw material powders, without any additional physical and/or chemical pre-treatment, were used in the study of Au and Cu electrochemical leaching.

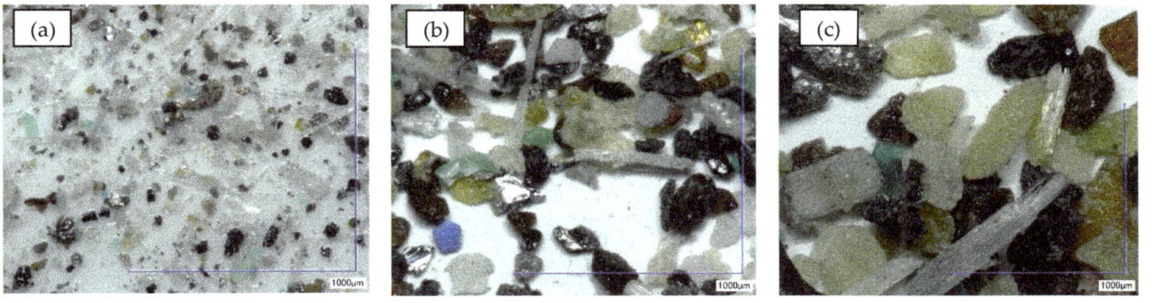

Figure 4. Optical images at magnification ×200 times, of disintegrator-crushed PCBs fractions: (**a**) is d < 90 μm; (**b**) is d = 90–180 μm; and (**c**) is d = 180–350 μm and designated [A], [B] and [C], respectively.

Complex Characterisation of Raw Material

From the obtained fractions of [A], [B] and [C], the finest fraction [A] was chosen as the main object of the research. The choice of this faction is due to several reasons. First, the scheme for studying the leaching process involves the use of the reaction mixture in the form of a suspension, and finer particle suspension is more stable than coarse particles (according to Stock's Law). Second, a finer particle powder has a higher specific surface area, which is crucial for leaching reaction intensity. Third, "open" laminated PCBs, small fragments with Cu intermediate layers, are essential. Finally, this fraction has the highest gold content.

The quantitative determination of the elements included in the composition of PCBs is complicated by both the presence of a wide range of elements present in various chemical forms (metals, alloys, and their corrosion products) and the range of their concentrations and by the significant content of the base. Therefore, atomic spectral analysis methods such as X-ray fluorescence (XRF), optical emission analysis with inductively coupled plasma (ICP-OES), as well as X-ray diffraction (XRD), were used for the complex investigation of raw material.

In this work, XRF analysis was used to qualitatively/semi-quantitatively determine the composition of the raw material sample. According to the results obtained (Figure 5), metals such as Fe, Ca, Zn, Cu, Ni, Al, Ti, Mn, and Si are present in the sample. However, based on the obtained spectrum, it is impossible to unequivocally judge the presence of gold in the sample under study in the region of the gold (AuL) peak (about 9.50 keV), overlapping with a second zinc peak (ZnK), indicated by the arrow in Figure 5.

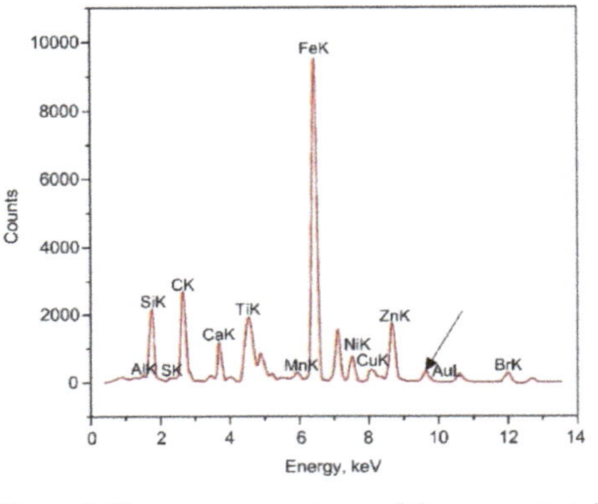

Figure 5. Fluorescence spectrum of the raw material fraction [A].

In turn, the X-ray phase analysis data (Figure 6, curve 1) demonstrate the presence of such basic crystalline phases as SiO_2, Si, Al_2O_3 and Fe_3O_4. Established compounds are included in PCB materials, such as fiberglass, ceramics, and semiconductor materials [43]. The phase composition of the solid residue obtained after the chemical leaching of a raw material sample was also studied (Figure 6, curve 2). A comparison of the obtained results (Figure 6) shows that the chemical leaching of the raw material has practically no effect on the identified phase composition, except the crystalline phase of the magnetite

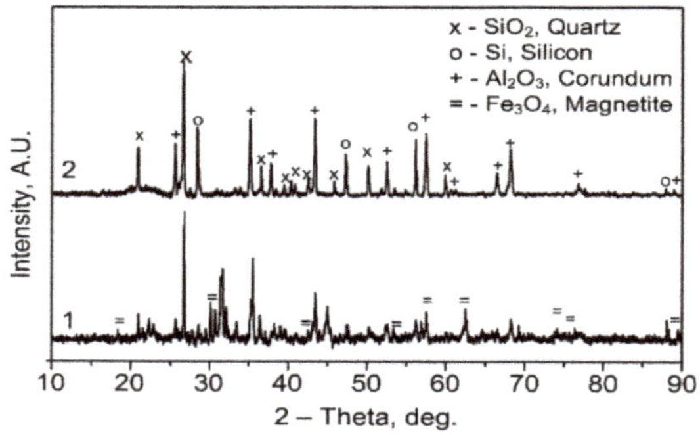

Figure 6. XRD patterns: 1—raw material fraction [A]; 2—solid residue after chemical leaching of a raw material fraction [A].

The solution obtained after chemical leaching of the raw material sample—leachate—was analyzed by ICP-OES. A quantitative determination was carried out not only on metals identified in a solid sample using XRF analysis (Figure 5), but also on some others, usually included in PCBs, including tin, lead, antimony, nickel, and cobalt. The results are presented in Figure 7. According to the results of weighing, the content of the solid residues after chemical leaching raw material fractions [A], [B] and [C] is 45.76 wt.%, 61.28 wt.% and 68.20 wt.% of the mass of raw material, respectively. It should be noted that 26.53 wt.% are elements of a fraction [A], 19.08 wt.% of a fraction [B] and 15.81 wt.% of a fraction [C], whose quantitative determination in the solution was not carried out and/or could not be identified by the ICP-OES method. A comparison of the results of ICP-OES obtained for three fractions shows that, with an increase in the particle size of the raw material, a decrease in the content of all analyzed elements is observed, except the copper. For this metal, an inverse relationship is established. More significant amounts of copper are most likely present in larger disintegrator-milled fractions (d > 350 μm). Thus, the content of target metals in fraction A was established: Gold—824 ppm and copper—1.40 wt.%.

According to the results of the ICP-OES analysis of the leachates obtained after chemical leaching of the raw material fractions B and C, the gold content is 277 ppm, 67 ppm, and the copper content is 2.67 wt.%, 6.13 wt.%, respectively.

3.2. Impact of Electrochemical Leaching Process Parameters on the Leaching Degree of Gold and Copper

3.2.1. Solid/Liquid Ratio

The results presented in Figure 8 show that the degree of gold leaching slightly increased from 77.3 wt.% at S/L = 2.9 g·L^{-1}, reaching a maximum of 80.5 wt.% at S/L = 8.6 g·L^{-1}. With a further increase in the S/L ratio, there is a significant decrease in R to 46.9 wt.% (S/L = 14.3 g·L^{-1}). This could be because, at a higher S/L ratio under these experimental conditions, chlorine concentration in the electrolyte solution is insufficient for complete metal dissolution. Since, in the process under study, chlorine ions are not only a precursor for the production of an oxidizing agent—chlorine—but also a complexing agent that forms soluble complex ions with metal ions [30]. To increase the degree of gold leaching by an increasing S/L ratio, it is also necessary to raise the concentration of the oxidizing agent. The chlorine concentration in the electrolyte solution can be raised by increasing the chloride ions' concentration and the current density [44]. At the same time, with an increase in the S/L ratio from 2.9 $g \cdot L^{-1}$ to 14.3 $g \cdot L^{-1}$, the degree of copper leaching increased linearly from 82.1 wt.% to 87.3 wt.%, respectively.

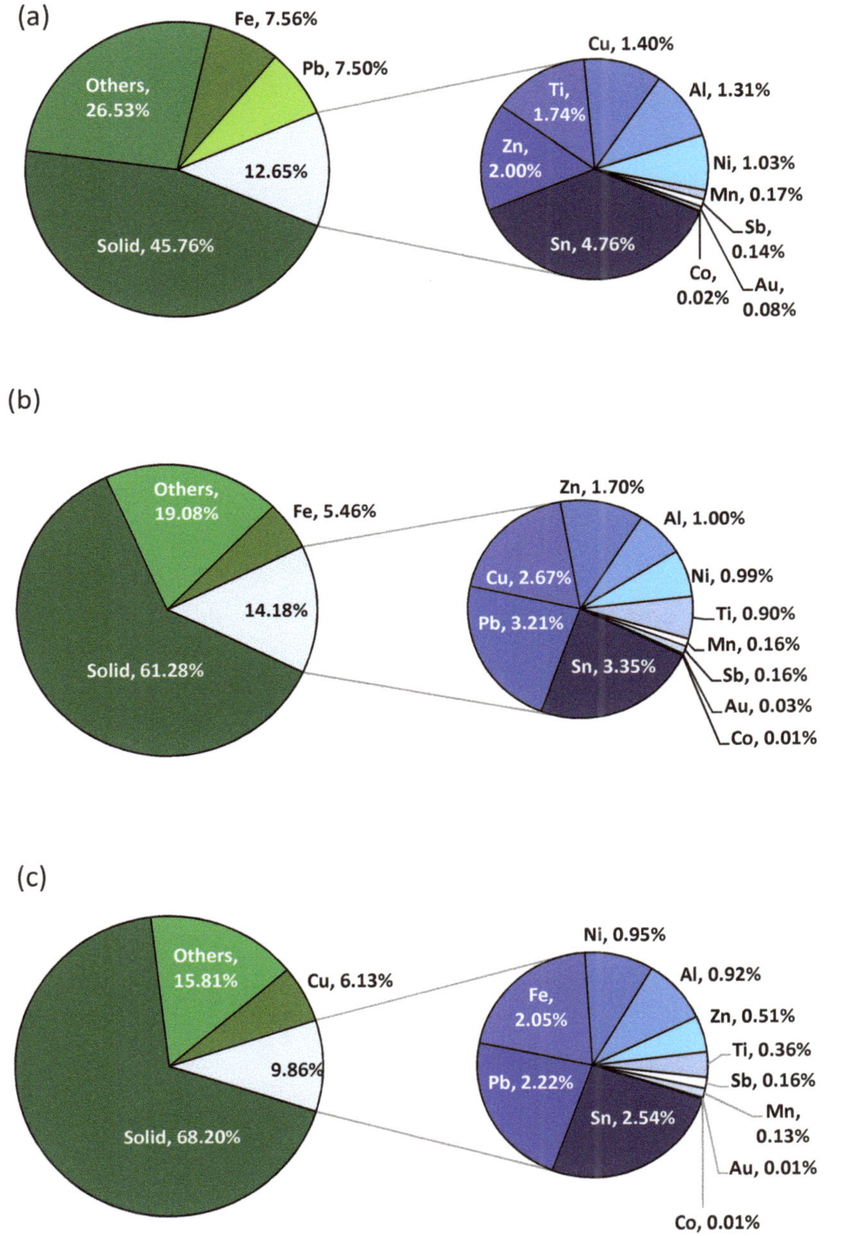

Figure 7. Content of metals in raw materials according to the results of ICP-OES analysis of the leachates prepared by chemical leaching: (**a**)—Fraction [A]; (**b**)—fraction [B]: and (**c**)—fraction [C].

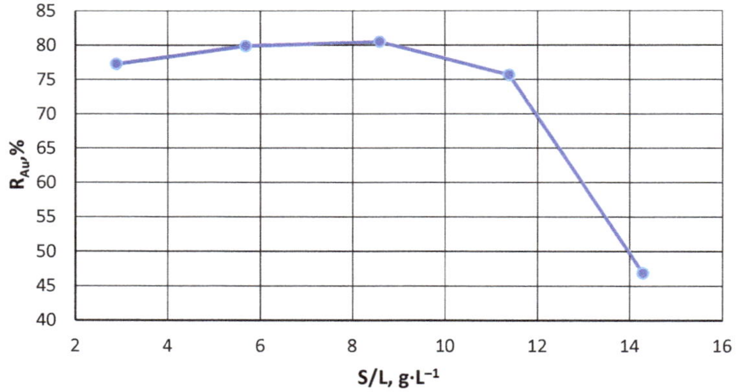

Figure 8. Impact of S/L ratio on the degree of gold leaching. Experimental conditions: Raw material fraction [A], C_{HCl} = 6 M, i = 0.66 A·cm^{-2}, and t_{ex}= 2 h.

3.2.2. Raw Material Particle Size

The particle size of the PCBs obtained after crushing is one of the essential factors determining the efficiency of metal leaching, but this effect is not the same for all elements [45]. According to Figure 9, complete leaching of gold is achieved at a larger particle size of raw material and lower gold content (fractions [B] and [C], Figure 4b,c). The degree of copper leaching also increases with the particle size, but complete leaching was not achieved.

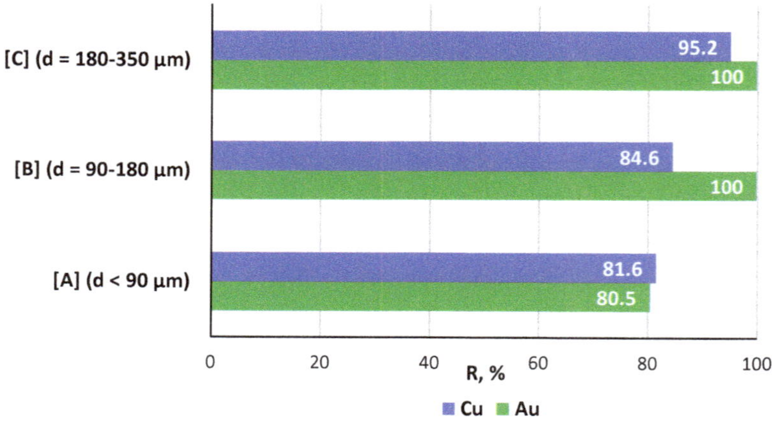

Figure 9. Impact of raw material particle size on the degree of gold and copper leaching. Experimental conditions: i = 0.66 A·cm^{-2}, t_{ex} = 2 h, S/L = 8.6 g·L^{-1}, and C_{HCl} = 6 M.

In the case of Au, the efficiency of metal leaching depends more on the content of the metal in the raw material, but not on raw material particle size. An increase in the degree of copper leaching with an increase in the particle size of the raw material is probably due to an increase in the content of not only copper, but also the product of its corrosion formed during the storage of electronic waste in humid air—dicopper dihydroxo carbonate $(CuOH)_2CO_3$—which reacts with an HCl solution [46]. Unfortunately, the results obtained are complicated to compare with the literature data since, as a rule, larger fractions of raw materials, a different method of preparing raw material and crushing, and other leaching agents are used in the studies.

3.2.3. Alternating Current

Applying AC has resulted in a noticeable improvement in the degree of gold leaching compared to the experiment without AC. In contrast, the degree of copper leaching does not seem to significantly benefit from AC (Figures 9 and 10).

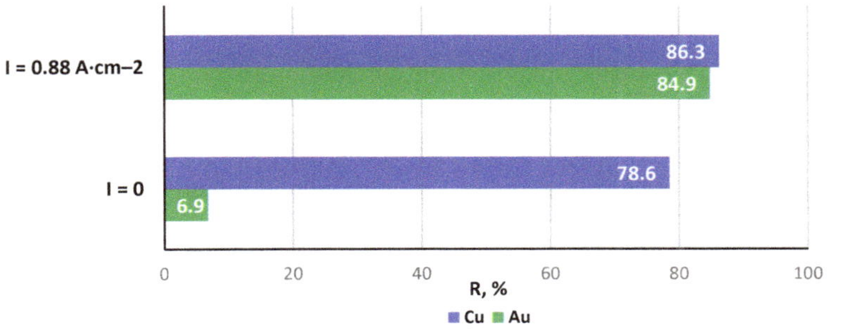

Figure 10. Effect of AC applying on gold and copper leaching degree. Experimental conditions: Raw material—fraction [A], t_{ex}= 2 h, S/L = 8.6 g·L^{-1}, and C_{HCl} = 6 M.

Increasing the current density from 0.66 A·cm^{-2} to 0.88 A·cm^{-2} leads to a slight increase in the gold and copper degree of leaching (above 4%).

From the presented data (Figure 10), it can be seen that 6.9% of gold and 78.6% of copper at ambient temperature passes into a hydrochloric acid solution that does not contain chlorine. On the one hand, this phenomenon is due to the reaction of oxidation of these metals by iron ions (Equations (1) and (2)) [21,22,47], which are formed as a result of the dissolution of iron or its compounds in a solution of hydrochloric acid. On the other hand, such a significant transition of copper into the solution under these conditions is also probably associated with substantial amounts of $(CuOH)_2CO_3$ [POD]. Thus, these results show that acid pre-treatment of finely dispersed raw material with hydrochloric acid solutions, which are used in certain hydrometallurgical technologies for pre-leaching of active metals, can lead not only to losses of copper but also of gold.

3.2.4. Electrolyte Temperature

With the passage of alternating current during the first 30 min of the process, a noticeable increase in the temperature of the electrolyte and a decrease in the voltage on the cell are observed. A further increase in the duration of the experiment leads to a gradual stabilisation of these parameters (Figure 11).

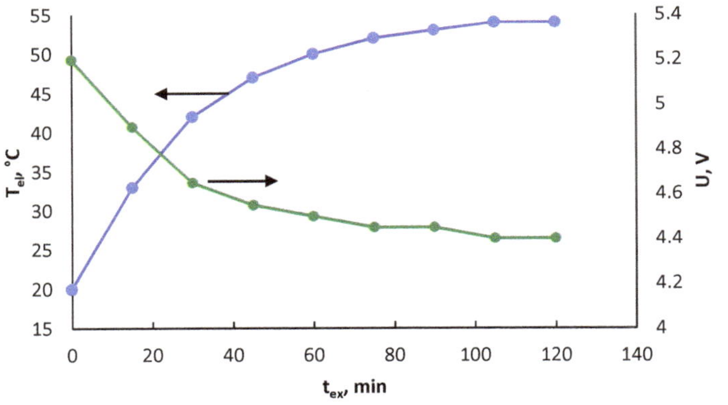

Figure 11. Electrolyte temperature and cell voltage versus the experiment duration. Experimental conditions: Raw material—fraction [A], C_{HCl} = 6 M, i = 0.66 A·cm^{-2}, and S/L = 8.6 g·L^{-1}.

To study the effect of electrolyte temperature on the efficiency of metal leaching, the processes were carried out with forced air cooling of the electrolyte while maintaining a constant electrolyte temperature in a polypropylene reactor and a jacketed glass cell, respectively (Figure 12).

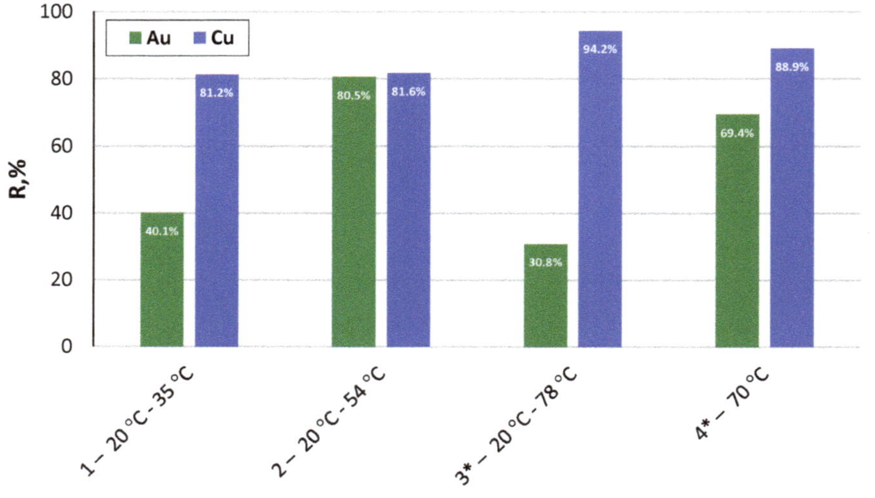

Figure 12. Effect of the electrolyte temperature on the degree of gold and copper leaching from the raw material fraction [A]: 1—Forced electrolyte air cooling; 2—without cooling; 3*—without thermostatic; and 4*—thermostatic at 70 °C. Experimental conditions: 1, 2, 3* and 4*—C_{HCl} = 6 M, 1, 2—I = 0.66 A·cm^{-2}, t_{ex} = 2 h, S/L = 8.6 g·L^{-1}; 3*, 4*—i = 0.64 A·cm^{-2}, t_{ex} = 2 h, and S/L = 2.9 g·L^{-1}. * Water-jacketed electrochemical cell.

The results showed that the electrolyte temperature practically does not affect the copper leaching efficiency in the polypropylene cell. However, the degree of copper leaching increases in the water-jacketed cell when the electrolyte heats during the electrochemical process compared to the process while maintaining a constant temperature of 70 °C by 5%. Such an impact is probably because, in the first case, the final temperature of the electrolyte exceeds 70 °C. In turn, increasing the electrolyte temperature twice improves the gold leaching efficiency.

3.2.5. Experiment Duration

The dependence of all metal leaching on the experiment duration (Figure 13a) was investigated, as well as studies on the effect of the experiment duration on the efficiency of gold and copper leaching were studied and shown in Figure 13b. The results of the iron leaching are also presented for comparison. The choice of this metal is because iron (III) ions in hydrochloric acid solutions also act as an oxidizing agent for gold [21,22] and copper [47]. The presented results show that the most significant increase in the degree of gold leaching is observed in the range of 0.5–1 h and reaches 85.1 wt.% after 1 h. In this case, after 0.5 h from the beginning of the experiment, the concentration of iron ions in the solution reaches 251.1 mg·L^{-1}. With a further increase in the experiment duration, periodic changes in the content of gold ions in the solution are observed, reaching a maximum of 86.3% within 4 h of the experiment. This phenomenon may be associated with the formation of so-called "secondary" gold, formed on the surface of gold particles as a result of the reaction of disproportionation (15) that occurs in the leaching solution with a decrease in the concentration of dissolved chlorine [21,25].

$$3[AuCl_2]^- \rightarrow [AuCl_4]^- + 2Au + 2Cl^- \quad (15)$$

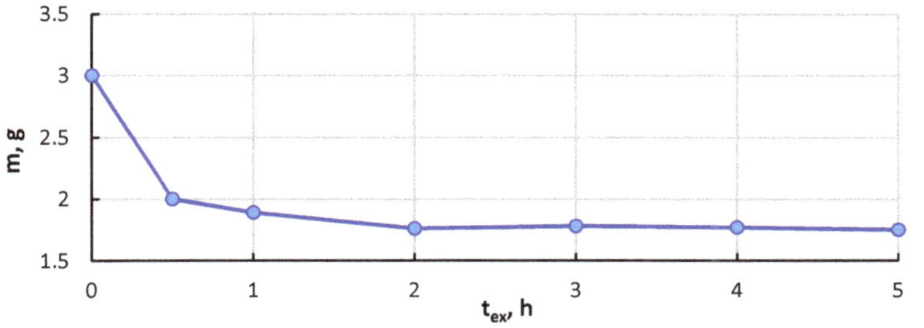

(**a**)

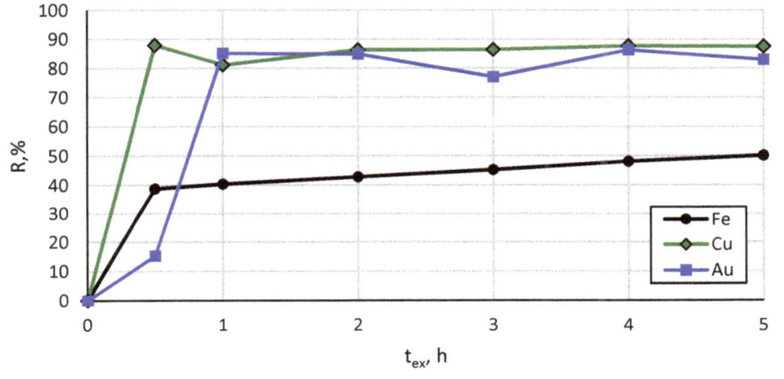

(**b**)

Figure 13. Impact of the experiment duration on the weight of solid residue (**a**) and the degree of gold, copper, and iron leaching (**b**). Experimental conditions: raw material fraction [A], C_{HCl} = 6 M, i = 0.88 $A \cdot cm^{-2}$, and S/L = 8.6 $g \cdot L^{-1}$.

To eliminate this phenomenon, it is necessary to adjust the concentration of chloride ions in an electrolyte solution during the process.

It should be noted that, after the completion of the four-hour experiment, the degree of copper leaching was 87.7%. In addition, 48.0% Fe, 86.4% Mn, 28.2% Zn, 41.0% Ni, and 97.0% Ti also pass into the electrolyte solution. Additionally, for metals such as Al, Pb and Sn, complete leaching from the raw material is achieved.

Solid residues from the experiments at different durations: 0.5 h, 2 h, 4 h and 5 h (Figure 13a) were studied using the XRD method (Figure 14).

The presented data show that the $Pb(TiO_3)$ crystalline phase is present only in the solid residue obtained after 0.5 h of electrochemical treatment of the raw material (Figure 14, curve 1). With a further increase in the duration of the experiment, the phase composition of solid residues practically does not change (Figure 14, curves 2–4). It is a mixture of SiO_2, Al_2O_3, Fe_3O_4 and Si crystalline phases. Thus, the results of the XRD analysis showed that chemical leaching could dissolve all magnetite (Figure 6, curve 2), not electrochemical leaching.

The XRD analysis of solid residues does not contradict the results of the ICP-OES analysis of the electrolyte solutions (Figure 13b).

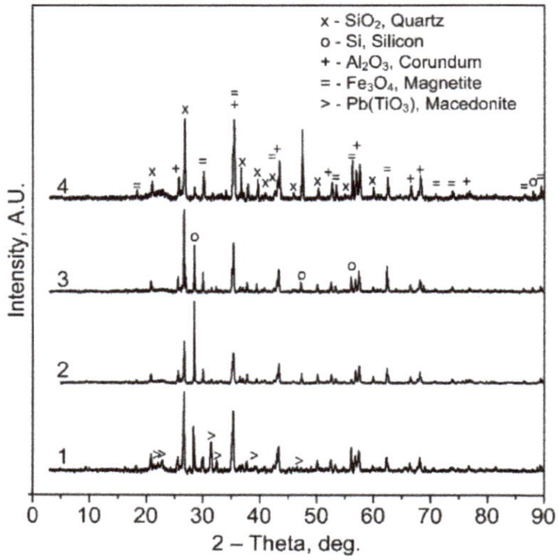

Figure 14. XRD patterns of solid residues from the experiments at different experiment durations: 1–0.5 h, 2–2 h, 3–4 h and 4–5 h. Other experimental conditions are provided in the caption of Figure 13.

To study the kinetics of the dissolution of chlorine in a solution of electrolyte in the conditions of an experiment corresponding to Figure 13, a series of "idle" experiments (without adding the raw material to the electrolyte) was carried out (Figure 15).

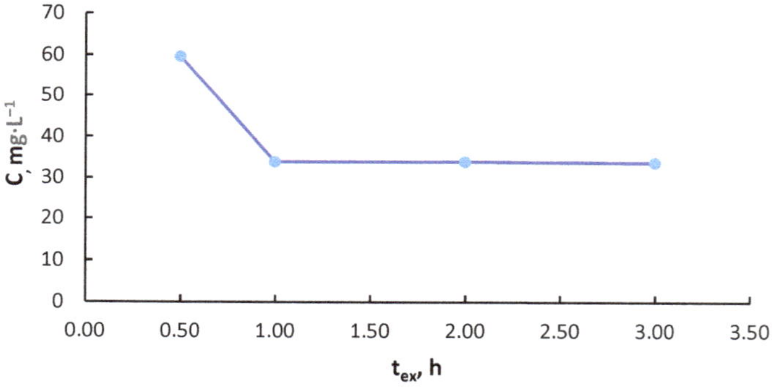

Figure 15. Change of free chlorine concentration in electrolyte solution on the experiment duration. Experimental conditions: C_{HCl} = 6 M, i = 0.88 A·cm^{-2}.

It can be seen from the presented data that, after 0.5 h of the experiment, the concentration of chlorine in the electrolyte solution reaches 59.72 mg·L^{-1}. A further increase in the duration of the investigation to 1 h leads to a decrease in the chlorine concentration in the electrolyte solution by 1.7 times. It remains constant for the next two hours. It has been established that an increase in the duration of the experiment from 0.5 to 3 h leads to additional heating of the electrolyte from 52 °C to 76 °C, which causes a decrease in the solubility of chlorine in the hydrochloric acid electrolyte solution [48]. It should be noted that these data are in good agreement with the results presented in Figure 13b, namely that active dissolution of gold begins 30 min after the beginning of the experiment.

Thus, in the framework of the presented study, for the finest raw material fraction with d < 90 µm the highest gold leaching degree (86.3%) was achieved under the following experimental conditions: t_{ex} = 4 h, C_{HCl} = 6 M, i = 0.88 A·cm^{-2}, S/L ratio—8.6 g·L^{-1} and the highest copper leaching degree (94.2%) was achieved under the following experimental conditions: t_{ex} = 2 h, C_{HCl} = 6 M, i = 0.64 $A \cdot cm^{-2}$, and S/L ratio—2.9 $g \cdot L^{-1}$.

4. Conclusions

The results showed that the developed process of implementing the method applies to leaching gold and copper from the disintegrator-crushed waste PCBs in fine fractions < 90 µm, 90–180 µm, and 180–350 µm. According to the results of the ICP-OES analysis, with increasing particle size in the fractions of raw material, the gold content decreases from 824 ppm to 67 ppm and the copper content increases from 1.40% to 6.13%. Under the same conditions of the electrochemical leaching process, namely, an experiment duration of 2 h, a current density of 0.66 $A \cdot cm^{-2}$, and a solid/liquid ratio of 8.6 $g \cdot L^{-1}$ complete leaching of gold is achieved from raw materials with a particle size of d = 90–180 µm and d = 180–350 µm containing 277 ppm and 67 ppm of gold, respectively. At the same time, increasing the particle size in raw material fractions from d < 90 µm to d = 180–350 µm and the copper content means an increase from 1.40% to 6.13%, respectively, and an increase in the degree of copper leaching from 81.6% to 95.2% is observed. Determine that applying AC (i = 0.66 $A \cdot cm^{-2}$) has resulted in a noticeable improvement in the degree of gold leaching up to 80.5% from the finest fraction of raw material with d < 90 µm compared to the experiment in the absence of AC—6.9%. In contrast, the degree of copper leaching does not seem to benefit from AC significantly. Applying AC leads to an increase in the degree of metal leaching by only 3%. The increase in the electrolyte temperature also has a more significant effect on the gold leaching efficiency than copper. These results could be used as a foundation for developing a complex technology for recovering valuable metals from PCBs.

Author Contributions: Conceptualization, V.S., V.P. and E.B.; methodology, V.S., A.Z. and A.S.; validation, A.S., E.E. and V.P.; formal analysis, E.E. and A.S.; investigation, V.S., V.P., E.E. and A.S.; resources, E.B. and V.P.; data curation, J.B. and A.Z.; writing—original draft preparation, V.S. and A.S.; writing—review and editing, A.S. and J.B.; visualization, A.S. and J.B.; supervision, E.B. and V.P.; project administration, E.B.; funding acquisition, E.B. and V.P. All authors have read and agreed to the published version of the manuscript.

Funding: This research was supported by ERDF project no. 1.1.1.1/20/A/139 "Development of sustainable recycling technology of electronic scrap for precious and non-ferrous metals extraction". The project was co-financed by REACT-EU funding to mitigate the effects of the pandemic crisis. The article was published with the financial support from the Riga Technical University Research Support Fund. This research was also supported by the Institute of Solid State Physics, University of Latvia, as the Center of Excellence has received funding from the European Union's Horizon 2020 Framework Program H2020-WIDESPREAD-01-2016-2017-TeamingPhase2 under grant agreement No. 739508, project CAMART2.

Data Availability Statement: Not applicable.

Acknowledgments: The authors would also like to mention the support from the "Innovation Grants for Maritime Students" performed at Latvian Maritime Academy (Project No: 1.1.1.3/18/A/006, funded by the European Regional Development Fund—ERDF, Republic of Latvia). The article is published with the financial support from the Riga Technical University Research Support Fund.

Conflicts of Interest: The authors declare no conflict of interest. The funders had no role in the design of the study; in the collection, analyses, or interpretation of data; in the writing of the manuscript; or in the decision to publish the results.

References

1. Vidyadhar, A. A Review of Technology of Metal Recovery from Electronic Waste. In *E-Waste in Transition—From Pollution to Resource*; Intech: Rijeka, Croatia, 2016. [CrossRef]
2. Paola, M. Recycling of Printed Circuit Boards. In *Integrated Waste Management—Volume II*; Intech: Rijeka, Croatia, 2011. [CrossRef]
3. Bizzo, W.A.; Figueiredo, R.A.; De Andrade, V.F. Characterization of Printed Circuit Boards for Metal and Energy Recovery after Milling and Mechanical Separation. *Materials* **2014**, *7*, 4555–4566. [CrossRef] [PubMed]

4. Gao, Z.; Li, J.; Zhang, H.-C. Printed circuit board recycling: A state-of-the-art survey. *IEEE Trans. Electron. Packag. Manuf.* **2004**, *27*, 33–42. [CrossRef]
5. Van Yken, J.; Cheng, K.Y.; Boxall, N.J.; Sheedy, C.; Nikoloski, A.N.; Moheimani, N.R.; Kaksonen, A.H. A Comparison of Methods for the Characterisation of Waste-Printed Circuit Boards. *Metals* **2021**, *11*, 1935. [CrossRef]
6. Xolo, L.; Moleko-Boyce, P.; Makelane, H.; Faleni, N.; Tshentu, Z. Status of recovery of strategic metals from spent secondary products. *Minerals* **2021**, *11*, 673. [CrossRef]
7. Işıldar, A.; van de Vossenberg, J.; Rene, E.R.; van Hullebusch, E.D.; Lens, P.N. Two-step bioleaching of copper and gold from discarded printed circuit boards (PCB). *Waste Manag.* **2016**, *57*, 149–157. [CrossRef]
8. Baniasadi, M.; Vakilchap, F.; Bahaloo-Horeh, N.; Mousavi, S.M.; Farnaud, S. Advances in bioleaching as a sustainable method for metal recovery from e-waste: A review. *J. Ind. Eng. Chem.* **2019**, *76*, 75–90. [CrossRef]
9. Krishnan, S.; Zulkapli, N.S.; Kamyab, H.; Taib, S.M.; Din, M.F.B.M.; Majid, Z.A.; Chaiprapat, S.; Kenzo, I.; Ichikawa, Y.; Nasrullah, M.; et al. Current technologies for recovery of metals from industrial wastes: An overview. *Environ. Technol. Innov.* **2021**, *22*, 101525. [CrossRef]
10. Zhou, H.-B.; Zeng, W.-M.; Yang, Z.-F.; Xie, Y.-J.; Qiu, G.-Z. Bioleaching of chalcopyrite concentrate by a moderately thermophilic culture in a stirred tank reactor. *Bioresour. Technol.* **2009**, *100*, 515–520. [CrossRef]
11. Tuncuk, A.; Stazi, V.; Akcil, A.; Yazici, E.Y.; Deveci, H. Aqueous metal recovery techniques from e-scrap: Hydrometallurgy in recycling. *Miner. Eng.* **2012**, *25*, 28–37. [CrossRef]
12. Zhang, L.; Xu, Z. A review of current progress of recycling technologies for metals from waste electrical and electronic equipment. *J. Clean. Prod.* **2016**, *127*, 19–36. [CrossRef]
13. Feng, F.; Liu, J.; Zhao, M.; Yu, L.; Wang, H.; Lu, C.; Zhang, Q.; Zhao, J.; Sun, Y.; Cen, J.; et al. Study of an Environmentally Friendly Method for the Dissolution of Precious Metal with Ionic Liquid and Iodoalkane. *Metals* **2021**, *11*, 919. [CrossRef]
14. Zhang, Y.; Liu, S.; Xie, H.; Zeng, X.; Li, J. Current Status on Leaching Precious Metals from Waste Printed Circuit Boards. *Procedia Environ. Sci.* **2012**, *16*, 560–568. [CrossRef]
15. Akcil, A.; Erust, C.; Gahan, C.S.; Ozgun, M.; Sahin, M.; Tuncuk, A. Precious metal recovery from waste printed circuit boards using cyanide and non-cyanide lixiviants—A review. *Waste Manag.* **2015**, *45*, 258–271. [CrossRef] [PubMed]
16. Birich, A.; Stopic, S.; Friedrich, B. Kinetic Investigation and Dissolution Behavior of Cyanide Alternative Gold Leaching Reagents. *Sci. Rep.* **2019**, *9*, 1–10. [CrossRef]
17. Shah, M.B.; Tipre, D.; Dave, S.R. Chemical and biological processes for multi-metal extraction from waste printed circuit boards of computers and mobile phones. *Waste Manag. Res. J. A Sustain. Circ. Econ.* **2014**, *32*, 1134–1141. [CrossRef] [PubMed]
18. Masavetas, I. Production of copper powder form printed circuit boards by electrodeposition. *Glob. Nest J.* **2009**, *11*, 241–247. [CrossRef]
19. Ghorbani, Y.; Franzidis, J.-P.; Petersen, J. Heap leaching technology-current state, innovations and future directions: A review. *Miner. Process. Extr. Met. Rev.* **2016**, *37*, 73–119. [CrossRef]
20. Zhou, Q.F.; Zhu, W. Recovery of gold from waste computer and its accessories. China Resources Comprehensive Utilization. *China Resour. Compr. Util.* **2003**, *7*, 31–35.
21. Mineev, G.G.; Panchenko, A.F. *Solvents of gold and silver in hydrometallurgy*; Metallurgy: Moscow, Russia, 1994; p. 244.
22. Seisko, S.; Lampinen, M.; Aromaa, J.; Laari, A.; Koiranen, T.; Lundström, M. Kinetics and mechanisms of gold dissolution by ferric chloride leaching. *Miner. Eng.* **2018**, *115*, 131–141. [CrossRef]
23. Filmer, A.O.; Lawrence, P.R.; Hoffmann, W. Comparison Of Cyanide, Thiourea And Chlorine As Lixiviants For Gold. In *Symposia Series—Australasian Institute of Mining and Metallurgy*; Australasian Institute of Mining and Metallurgy: Melbourne, VIC, Australia, 1984; pp. 279–287.
24. Sahin, M.; Akcil, A.; Erust, C.; Altynbek, S.; Gahan, C.S.; Tuncuk, A. A Potential Alternative for Precious Metal Recovery from E-waste: Iodine Leaching. *Sep. Sci. Technol.* **2015**, *50*, 2587–2595. [CrossRef]
25. Diaz, M.; Kelsall, G.; Welham, N. Electrowinning coupled to gold leaching by electrogenerated chlorine. *J. Electroanal. Chem.* **1993**, *361*, 25–38. [CrossRef]
26. Winand, R. Chloride hydrometallurgy. *Hydrometallurgy* **1991**, *27*, 285–316. [CrossRef]
27. Kim, E.-Y.; Kim, M.-S.; Lee, J.-C.; Jeong, J.; Pandey, B. Leaching kinetics of copper from waste printed circuit boards by electro-generated chlorine in HCl solution. *Hydrometallurgy* **2011**, *107*, 124–132. [CrossRef]
28. Kim, E.-Y.; Kim, M.-S.; Lee, J.-C.; Jha, M.K.; Yoo, K.; Jeong, J. Effect of cuprous ions on Cu leaching in the recycling of waste PCBs, using electro-generated chlorine in hydrochloric acid solution. *Miner. Eng.* **2008**, *21*, 121–128. [CrossRef]
29. Pilone, D.; Kelsall, G. Metal Recovery from Electronic Scrap by Leaching and Electrowinning IV. *Miner. Eng.* **2003**, *2*, 1565–1575. [CrossRef]
30. Kim, M.-S.; Lee, J.-C.; Park, S.-W.; Jeong, J.; Kumar, V. Dissolution behaviour of platinum by electro-generated chlorine in hydrochloric acid solution. *J. Chem. Technol. Biotechnol.* **2013**, *88*, 1212–1219. [CrossRef]
31. Kim, E.-Y.; Kim, M.-S.; Lee, J.-C.; Pandey, B. Selective recovery of gold from waste mobile phone PCBs by hydrometallurgical process. *J. Hazard. Mater.* **2011**, *198*, 206–215. [CrossRef]
32. Ilyas, S.; Srivastava, R.R.; Kim, H. Gold recovery from secondary waste of PCBs by electro-Cl2 leaching in brine solution and solvo-chemical separation with tri-butyl phosphate. *J. Clean. Prod.* **2021**, *295*, 126389. [CrossRef]

33. Lister, T.E.; Wang, P.; Anderko, A. Recovery of critical and value metals from mobile electronics enabled by electrochemical processing. *Hydrometallurgy* **2014**, *149*, 228–237. [CrossRef]
34. Shulgin, L.P. Overvoltage of electrode reactions in solutions during the passage of a symmetrical alternative current. *J. Phys. Chem.* **1979**, *3*, 2048–2051. (In Russian)
35. Styrkas, A.; Styrkas, D. Electrochemical dissolution of metals of the platinum group by alternating current. *J. Appl. Electrochem.* **1995**, *25*, 490–494. [CrossRef]
36. Box, W.D. Dissolution of Metals by AC Electrolysis. *Nucl. Appl.* **1966**, *2*, 299–303. [CrossRef]
37. Alabyshev, A.F.; Vyacheslavov, P.M.; Galnbek, A.A.; Zhivotinsky, P.B.; Rotinyan, A.L.; Fedotiev, N.P. *Applied Electrochemistry*; Khimija: Lningrad, Russia, 1974. (In Russian)
38. Taras, M. The Micro-Titration of Free Chlorine With Methyl Orange. *J. Am. Water Work. Assoc.* **1946**, *38*, 1146–1150. [CrossRef]
39. Zimakov, S.; Goljandin, D.; Peetsalu, P.; Kulu, P. Metallic powders produced by the disintegrator technology. *Int. J. Mater. Prod. Technol.* **2007**, *28*, 226. [CrossRef]
40. Peetsalu, P.; Goljandin, D.; Kulu, P.; Mikli, V. Micropowders Producted By Disintegrator Milling. *Powder Metall.* **2003**, *3*, 99–110.
41. Goljandin, D.; Sarjas, H.; Kulu, P.; Käerdi, H.; Mikli, V. Metal-Matrix Hardmetal/Cermet Reinforced Composite Powders for Thermal Spray. *Mater. Sci.* **2012**, *18*, 84–89. [CrossRef]
42. Blumbergs, E.; Serga, V.; Shishkin, A.; Goljandin, D.; Shishko, A.; Zemcenkovs, V.; Markus, K.; Baronins, J.; Pankratov, V. Selective Disintegration–Milling to Obtain Metal-Rich Particle Fractions from E-Waste. *Metals* **2022**, *12*, 1468. [CrossRef]
43. Xian, Y.; Tao, Y.; Ma, F.; Zhou, Y. Recovery of Metals from Heat-Treated Printed Circuit Boards via an Enhanced Gravity Concentrator and High-Gradient Magnetic Separator. *Materials* **2021**, *14*, 4566. [CrossRef]
44. Mazanko, A.F. *Industrial Membrane Electrolysis*; Chemistry: Moscow, Russia, 1989; p. 240.
45. Oliveira, P.C.; Taborda, F.C.; Nogueira, C.A.; Margarido, F. The Effect of Shredding and Particle Size in Physical and Chemical Processing of Printed Circuit Boards Waste. *Mater. Sci. Forum* **2012**, *730–732*, 653–658.
46. Podchainova, V.N.; Simonova, L.N. *Analytical Chemistry of the Elements. The Cooper*; Nauka: Moscow, Russia, 1990; p. 279. (In Russian)
47. Bryce, C.; Berk, D. Kinetics of the dissolution of copper in iron(III) chloride solutions. *Ind. Eng. Chem. Res.* **1995**, *34*, 1412–1418. [CrossRef]
48. Alkan, M.; Oktay, M.; Kocakerim, M.; Copur, M. Solubility of chlorine in aqueous hydrochloric acid solutions. *J. Hazard. Mater.* **2005**, *119*, 13–18. [CrossRef] [PubMed]

metals

MDPI

Article

Bio-Assisted Leaching of Non-Ferrous Metals from Waste Printed Circuit Boards—Importance of Process Parameters

Arevik Vardanyan [1,2], Narine Vardanyan [1], Mohamed Aâtach [2], Pierre Malavasi [2] and Stoyan Gaydardzhiev [2,*]

[1] Department of Microbiology, SPC "Armbiotechnology" of the National Academy of Sciences of Armenia, 14 Gyurjyan Str., Yerevan 0056, Armenia

[2] GeMMe—Minerals Engineering and Recycling, Faculty of Applied Sciences, ArGEnCo Department, University of Liege, Allée de la découverte 9, Bât. B52/3, Sart-Tilman, 4000 Liege, Belgium

* Correspondence: s.gaydardzhiev@uliege.be; Tel.: +32-(43)-669120

Abstract: The effect of varying process parameters during bio-catalyzed leaching of metals from end-of-life printed circuit boards (PCBs) was investigated. Fragmented PCBs (under 2 mm) were subjected to an indirect bioleaching in a stirred tank reactor while pulp density, pH and initial ferric iron content were varied. An iron oxidizing *Acidithiobacillus ferrooxidans* 61 microbial strain was used to generate the lixiviant through oxidizing Fe(II) to Fe(III). Chemically generated Fe(III) was tested as lixiviant under the same conditions as the biological one for comparative purposes. Cell enumeration during leaching and microscopic observations of the input and leached PCBs were conducted in parallel to shed light on the observed phenomena. The degree of bringing metals in solution was found to depend mainly on ferric iron concentration and pH. For the entire duration being always kept as 24 h, substantial portion of Cu (~87%) was extracted respectively at 1% pulp density (PD), 15.5 g/L Fe^{3+} and pH 1. For Zn and Ni, nearly 100% recovery was observed at 5% PD, 18 g/L Fe^{3+} and pH 1.1. The achieved results offer possibilities for further studies at higher pulp density, to ultimately render the bioleaching approach as enabling economical and environmentally friendly technology for urban mining of non-ferrous metals.

Keywords: printed circuit boards (PCB); iron oxidizing bacteria; double-stage bioleaching; metal extraction; pulp density; leaching yield

Citation: Vardanyan, A.; Vardanyan, N.; Aâtach, M.; Malavasi, P.; Gaydardzhiev, S. Bio-Assisted Leaching of Non-Ferrous Metals from Waste Printed Circuit Boards—Importance of Process Parameters. *Metals* **2022**, *12*, 2092. https://doi.org/10.3390/met12122092

Academic Editor: Stefano Ubaldini

Received: 30 October 2022
Accepted: 2 December 2022
Published: 6 December 2022

1. Introduction

Advanced electronic equipment is in high demand due to ongoing technological development, and as a consequence is characterized by relatively short lifespan [1]. This situation reflects in an unparallel accumulation of end-of-life electronic devices, commonly classified as electronic waste. According to the United Nations, the global generation of electronic waste (e-waste) streams was estimated between 20 and 50 million tons per year [2]. The main problem with e-waste is its toxicity, which is caused by the presence of hazardous metals (Ni, Cd, Se, As, Cr and Pb), brominated flame retardants, polychlorinated biphenyls and epoxy-based materials [3–5].

Printed circuit boards (PCBs), on the other hand, could turn an important source of base and precious metals, providing an economic incentive for recycling [6–11]. PCBs typically have a content of 30–40% metals, 30% plastics and 30% glass and oxides [12–17]. Thus, it is clear that there is a critical need for e-waste recycling and material valorization, which strengthen the supply chain's sustainability and support the circular economy. Several pyro- and hydrometallurgical studies have been carried out to recover metals from PCBs [2,18–21]. The application of pyrometallurgy for base metals recovery implies carrying out processes at higher temperatures, which provokes environmental concerns linked to gas emissions and slag generation [22]. On the other hand, conventional hydrometallurgical processes rely on intensive use of acids and supporting chemicals.

Organic acids, which are less corrosive and more selective than inorganic ones, can also be used for metal extraction [23]. Through acidification and complexation mechanisms they can solubilize the metallic fractions of oxidized or zero-valent metal bearing compounds [24]. Metals can also be recovered from e-waste through bio-hydrometallurgical routes, which are well-established in non-ferrous metal mining, and in some operations have proven more efficient than classical hydrometallurgical processes due to lower environmental footprint and demonstrated cost competitiveness [25,26].

Biohydrometallurgy encompasses main metallurgical steps such as leaching, solution purification and recovery of valuable metals. Bioleaching, one of the most-studied unit operations in biohydrometallurgy, enables non-negligible reduction in reagent use because they are produced in situ by microorganisms under soft operating conditions, i.e., low temperatures and mild pH [27]. On the other hand, biohydrometallurgy is less polluting and offers reduced operating costs and energy demand [28].

Various chemolithotrophic microorganisms are used in bioleaching applications thanks to their ability to facilitate metal dissolution via series of bio-oxidation reactions [29–34]. Iron exercise a central role as an electron carrier inside the bioleaching system. The oxidized form of iron (Fe(III)), which is generated by the microbial oxidation of ferrous iron (Fe(II)) substrates, acts as an oxidizing agent capable of oxidizing metal sulfides before being chemically reduced to ferrous iron [35].

Acid leaching is supported by acidophilic autotrophs, such as *Acidithiobacillus thiooxidans* [36], producing sulfuric acid in the presence of thiosulphate/sulfur/sulfide, while oxidation is performed by ferric iron produced by acidophilic iron-oxidizing autotrophs, such as *Acidithiobacillus ferrooxidans* [3,37] and *Leptospirillum ferrooxidans* [3].

Several studies have been conducted using microorganisms to extract metals from PCBs [18,31,37,38]. Recently, various chelating agents have been used in conjunction with microorganisms to improve metal extraction. Pyrite is used as an alternative source of iron in mineral processing industries, acting as an energy source for the growth of iron oxidizers. However, the majority of the studies deals with very low pulp densities and report low metal extraction efficiency and slow metal extraction rates [19,23,37,39,40].

Acidophilic heterotrophic bacteria and fungi are also used for metal leaching processes with *Acinetobacter* [41], *Acidiphilium* [42], *Aspergillus niger* [43,44], *Aspergillus fumigatus* [45], *Pseudomonas aeruginosa* [46], *Chromobacterium violaceum* [46–48] being primarily used for bioleaching, either as a pure culture or as part of a consortium.

Brandl et al., 2001 [31] presumably was the first study on published PCB bioleaching, which relied on acidophilic microorganisms. The two main reactions that occur are:

$$M^0 + 2Fe^{3+} \rightarrow M^{2+} + 2Fe^{2+} \tag{1}$$

$$Fe2^+ + O_2 + H^+ \xrightarrow{\text{bacteria}} Fe3^+ + H_2O \tag{2}$$

Metals are oxidized by Fe(III), which is reduced into Fe(II)—Equation (1). The bacteria then regenerate Fe(III) in the presence of dissolved oxygen, according to Equation (2). In the absence of bacteria, the oxidation of Fe(II) is slow, but the presence of iron oxidizing bacteria in the system significantly accelerates Fe(III) regeneration. Fe(III) is known as a low-cost oxidation agent frequently used in hydrometallurgy and is particularly suited to leaching various metals from PCBs [49].

Several published studies have focused on the feasibility of PCB bioleaching in batch mode using shake flasks and bioreactors [50–53]. However, due to their notorious heterogeneity their bioleaching should ideally be investigated using large volume stirred tank reactors (>1 L), enabling a sufficient amount of material to be processed. Such conditions guarantee efficient reactant transport and mimic the industrial situation closely [17]. Studies also show that staggered addition of PCBs with continuous biological production of an acidic lixiviant containing Fe(III) do improve bioleaching efficiency, most likely due to reduced PCB toxicity on microbial growth [17,31,40].

The toxicity of metals on microbial growth can be addressed using an indirect leaching approach in which lixiviant production and e-waste leaching are separated into two-stage processes. This concept has been demonstrated successfully in a laboratory for leaching metals from lithium-ion batteries [54] and PCBs [17,55].

On the background of the above, the presented study aimed at investigating recovery of copper, zinc, nickel and aluminum from PCBs using culture-bearing solutions obtained from an acidophilic iron-oxidizing bacteria, namely *Acidithiobacillus ferrooxidans* 61. An indirect bioleaching approach performed as either single or double-stage mode was chosen to counterbalance the potential toxicity of the PCBs on microbial growth and activity. By decoupling biolixiviant generation and bioleaching steps, an increased efficiency of metal leaching was expected.

2. Materials and Methods

2.1. PCBs

Depopulated end-of-life PCBs were fragmented down to a particle size below 22 mm by a hammer mill (Laarmann CHM4000) run at 1750 rpm. The analysis of the input PCBs revealed the following concentrations of non-ferrous metals: Cu (21.2%), Zn (1%), Ni (0.1%), Al (2.2%). Ahead of the bioleaching, the surface passivation layer (green lacquer mask) was removed by boiling the fragmented PCBs in a 10% NaOH solution for 15 min at ambient temperature. Afterwards, the samples were thoroughly washed with deionized water until neutral pH, dried at ambient temperature and preserved for leaching.

The main metal content in the prepared PCBs is shown in Table 1.

Table 1. Main metals concentration in the input PCBs.

Metal, %	Concentration
Cu	21.53
Al	6.95
Pb	3.2
Zn	0.78
Ni	5.12
Fe	3.86
Sn	1.98
Ag	0.097

2.2. Microorganisms and Culturing Conditions

The biolixiviant used for leaching was obtained on the basis of an iron-oxidizing culture of *Acidithiobacillus ferrooxidans* 61 isolated from Tandzut ore (Armenia) acid drainage water. Modified 9 K medium (0.5 g/L $(NH_4)_2SO_4$, 0.5 g/L $MgSO_4 \times 7H_2O$, 0.5 g/L K_2HPO_4, 0.05 g/L KCl, 0.01 g/L $Ca(NO_3)_2$) with $FeSO_4 \times 7H_2O$ having concentration between 44.2 and 124 g/L, depending on the target Fe^{3+} content, was used for cultivation and bioleaching experiments. 10N H_2SO_4 solution was used to adjust pH.

2.3. Bioleaching of PCBs

For generating lixiviant for the indirect bioleaching, culture of *A. ferrooxidans* 61 was cultured in the absence of PCBs using a modified 9 K medium maintained under optimum growing conditions. The growth of bacteria was carried out inside a bio-fermenter (Bionet Baby 0) coupled to control unit, to follow gas supply, stirring rate, pH, Eh and temperature, illustrated in Figure 1a. The fermenter filled in with 2 L culture medium was inoculated at 10% (*v*/*v*) and pH adjusted to 1.8–1.9. The fermenter was operated at temperature of 30 °C, stirring sped of 80 rpm and at 1 lpm air-flow. After 5–10 days of cultivation (depending on the concentration of ferrous sulfate), a reddish coloration of the culture medium was observed due to oxidation of Fe^{2+} to Fe^{3+}, with concomitant bacterial growth reaching density of 2.5×10^6–0.25×10^9 cells/mL and redox potential arriving in the range of 700 mV Ag/AgCl.

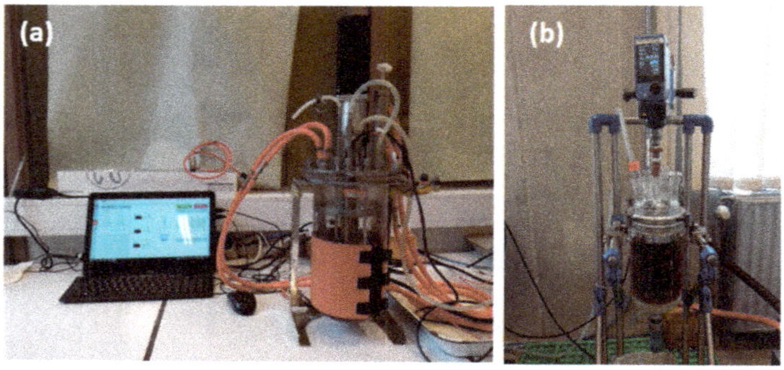

Figure 1. View of bio-fermenter (**a**) and stirred tank leaching reactor (**b**).

Indirect bioleaching of the PCBs was performed inside a 2 L jacketed reactor coupled to a circulating bath to maintain constant operating temperature of 40° needed to sustain eventual bacterial growth. The reactor was connected to a condenser to prevent excessive evaporation and a tunable compressor to supply air—Figure 1b.

Leaching tests were run for 24 h at 40 °C, stirring rate of 600 rpm and an air supply of 1 lpm. Experiments were performed under pH 1, 1.1, 1.8 and 2.1, using the following concentrations of Fe^3 as oxidizing agent (g/L): 9; 13.5, 15.5, 20 and 25. Pulp densities (PD) were varied at 1%, 3%, 5% and 10%. Sampling was performed at start-up, 1 h, 3 h, 5 h and 24 h. Quantitative determination of bacterial cells during the leaching was carried out by a serial end-point dilution methodology (e.g., "Most Probable Number" technique) [56]. The solid residues were air dried and subjected to chemical and mineralogical (XRD) analysis. To mimic industrial conditions to a maximum extent, all experiments were conducted without sterilization.

To outline the effect from the origin of Fe^{3+} used in the leaching, parallel to the biogenic ferric iron obtained from *A. ferrooxidans* 61, comparative trials using ferric iron of chemical origin ($Fe_2(SO_4)_3$) were performed. The comparative tests were realized as two different modes—as single- and double-stage batches. When double-stage leaching was tested, the first stage had 2 h duration, after elapsing of which PLS was decanted and fresh lixiviant added directly to the remaining solid. This marked the beginning of a second-stage run for further 4 h.

2.4. Analysis

Metals in the pregnant leach solution (PLS) were analyzed by atomic absorption spectrophotometry (AAS) and ICP-OES. The solid residue after leaching was dried, oven-treated at 900 °C for 240 min, grinded in an agate mortar or a ring mill and digested in aqua regia before being analyzed, similar to the liquid samples. Ferrous and ferric iron concentration was determined by an EDTA-based complexometric titration [57].

Mineralogical inspection of the leached PCBs was realized on a Bruker (Billerica, MA, USA) D8 ECO diffractometer using CuKalpha radiation (Lambda = 1.5418 Å). Representative amount of the material was pulverized in an agate mortar, and then deposited on a zero background silicon sample-holder. The X-ray powder diffraction patterns were first interpreted using the EVA 3.2 software (Bruker, 2014). This software allowed us to identify the phases and compare them to the ICDD database, version PDF-2. Then, the powder patterns were analyzed using the TOPAS 4.2 software (Bruker, 2014), which allowed quantification of the mineral phases. The quantification procedure is based on the Rietveld method.

PCBs were observed on morphology and phases liberation by a SEM-based automated mineralogy system—Zeiss sigma 300 FEG "Mineralogic"—coupled to two Brucker EDS xFlash 6 | 30 X-ray energy dispersion spectrometers (silicon drift detector). To this end, raw PCBs and samples derived after leaching were cast into 30 mm diameter resin mounts following an established procedure by [58]. The section polishing was accomplished by using polishing disks and diamond suspensions of different finesses. The SEM-EDS analyses were carried out using a probe current of 2.3 nA with an accelerating voltage of 20 kV at a working distance of 8.5 mm. A mapping mode was performed using a 3 to 5 μm step size and a dwell time of 55 ms. Analytical conditions such as contrast and brightness were set up manually to provide best possible contrast between the observed phases (plastics, composites, metals). System magnification was set to 6000X and voltage tension to 20 kV.

Leaching kinetics (g L/h) was calculated using Equation (3) below:

$$\frac{d\left[Cu^{2+}\right]}{dt} = -\frac{d\left[Fe^{3+}\right]}{2\,dt} \tag{3}$$

where: $[Cu^{2+}]$—variation in Cu concentration during the first and second hours;

dt—leaching duration (hour).

Metal recovery was calculated following Equation (4) shown below:

$$MR\ (\%) = \frac{CMl \times V}{m \times CMs} \times 100 \tag{4}$$

where MR metal recovery, %;

CMl—concentration of leached metal in the pregnant leaching solution, %;

CMs—metal concentration in the feed, %;

m—mass of used PCBs, g;

V—volume of the leachate inside the reactor, mL (in our case 2000 mL).

3. Results and Disscussion

3.1. Effect of Initial Concentration of Fe(III), Pulp Density and pH on Metals Recovery

The effect of various concentrations of oxidizing agent (Fe(III)) at 1% PD and initial pH 1 on the recovery of Cu, Zn, Al and Ni is shown in Figure 2.

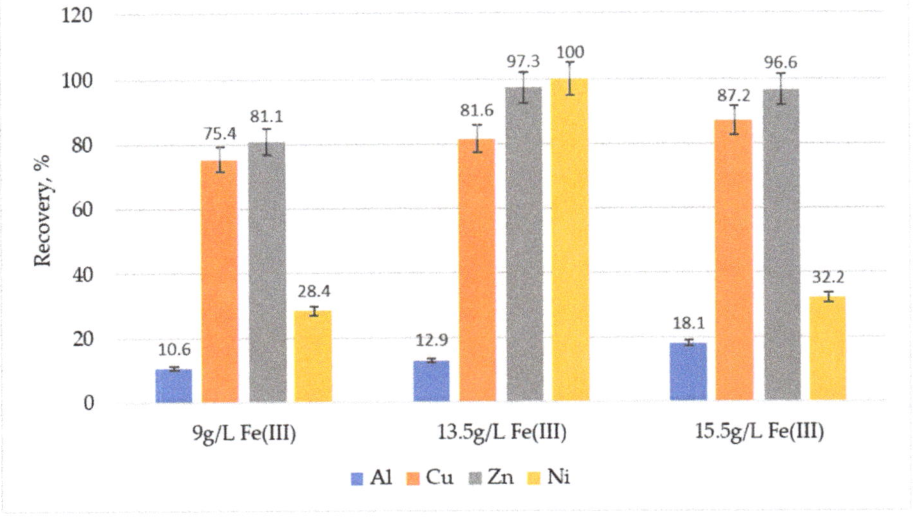

Figure 2. Recovery of metals from PCBs at various concentrations of biogenic Fe^{3+} (PD 1%; initial pH 1–1.1; 40 °C; air 1 lpm; 600 rpm; duration 24 h).

As shown in Figure 2, the ferric iron concentration exercises marked effect on copper recovery. During bioleaching at 40 °C, the increase of initial ferric iron concentration from 9 g/L to 15.5 g/L led to an increase in copper leaching rate from 0.6 g/L h to 1.5 g/L h, respectively. Thus, in the presence of 15.5 g/L Fe^{3+} the maximum amount of recovered Cu within 24 h duration was about 87.2% at initial pH 1 and pulp density (PD) of 1%.

At pH 2.1 the recovery of Cu was 81.0% in the presence of 13.5 g/L Fe^{3+}, which was roughly the same as at pH 1–81.6%, the rest conditions remained the same—Figure 3.

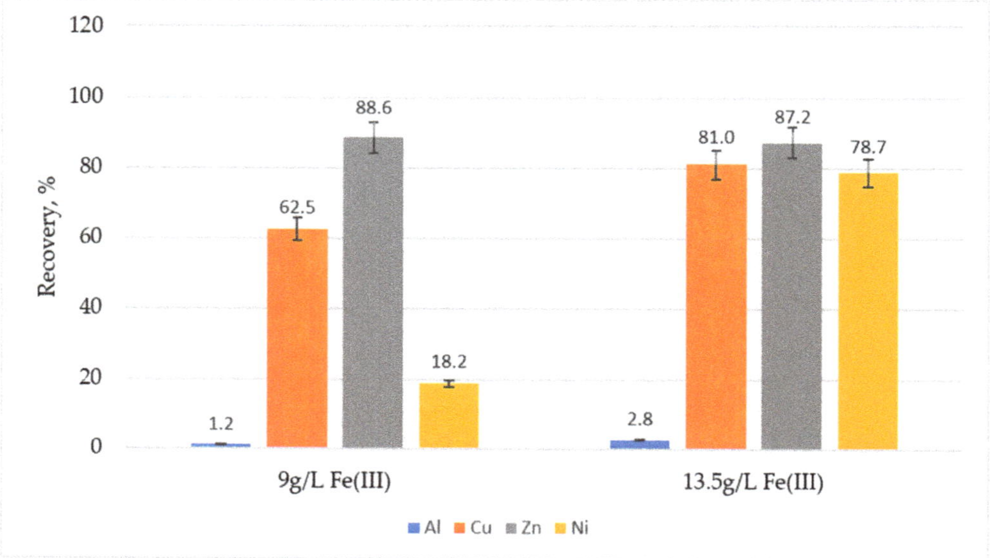

Figure 3. Recovery of metals from PCBs at 9 g/L and 13.5 g/L of biogenic Fe^{3+} (PD 1%; initial pH 2.1; 40 °C; air 1 lpm; 600 rpm; duration 24 h).

It could be inferred that the recovery of Al at 1% PD is mostly related to the low pH and Fe^{3+} concentration—Figure 2. The higher the amount of Fe^{3+}, the higher the recovery of Al. The recovery of Al at pH 1.0 in the presence of 9 g/L and 13.5 g/L Fe^{3+} was about 10.6% and 12.9%, respectively (Figure 2), while at higher pH (pH $\geq$ 2) it was about 9 and 4.6 times lower (1.2 and 2.8% respectively—Figure 3).

The recovery of Zn at 1% PD and pH 1 in the presence of 9 g/L, 13.5 g/L and 15.5 g/L Fe^{3+} was about 81.1%, 97.3% and 96.6%, respectively (Figure 2). The recovery of Zn at pH 2 in the presence of 9 g/L was slightly higher to the one at pH 1, but lower at increased ferric iron (13.5 g/L) rate—88.6% and 87.2%, respectively. Therefore, in the case of Zn, it can be inferred that the initial concentration of Fe^{3+} ions plays a secondary role and the initial pH is important.

As regarding Ni, its recovery in the presence of 9 g/L and 13.5 g/L Fe^{3+} was about 28.4% and nearly 100% (Figure 2) at pH 1, and 18.2% and 78.7% at pH 2.1 (Figure 3). Thus, under acidic conditions an increase in Ni leachability was observed with an increase in the concentration of Fe^{3+}.

It should be noted that if we consider the leaching duration, the metal recoveries reported in this paper are similar to those communicated by Hubau et al., 2020 [17]. Hubau et al. used a double-stage continuous bioleaching of PCBs shredded to 750µm (without pretreatment) at 1% PD, pH 1.5, 40 °C and biolixiviant containing 8 g/L Fe^{3+} obtained by bacterial acidophilic consortium BRGM-KCC comprising *Leptospirillum*, *Acidithiobacillus* and *Sulfobacillus* species. These conditions allowed them to recover 96% Cu, 85% Zn, 73% Ni and 54% Al over 48 h. On the other hand, Tapia et al., 2022 [28] managed to recover 58.9% Zn and 65% Cu for 3 days and 91.3% Zn and 68.5% Cu for 15 and 11 days, respectively, from PCBs shredded to ≤300µm (without pretreatment) at 1% of PD and 30 °C using biolixiviant containing 9 g/L Fe^{3+} obtained by enriched acidophilic iron oxidizing consortium comprising seven cultures.

The follow-up of metals and Fe^{3+} content in the PLS with time is shown in Figure 4. As seen Al, Cu, Ni and Zn are brought in solution predominately during the first 5 h, a period followed by a plateau trend. In parallel, it can be seen that with Fe^{3+} consumption progressing metal extraction rate tended to slow down. This observation was probably caused either by a limited availability of Fe^{3+} because of generation of chemical complexes or due to a non-complete metal liberation from the PCB matrix.

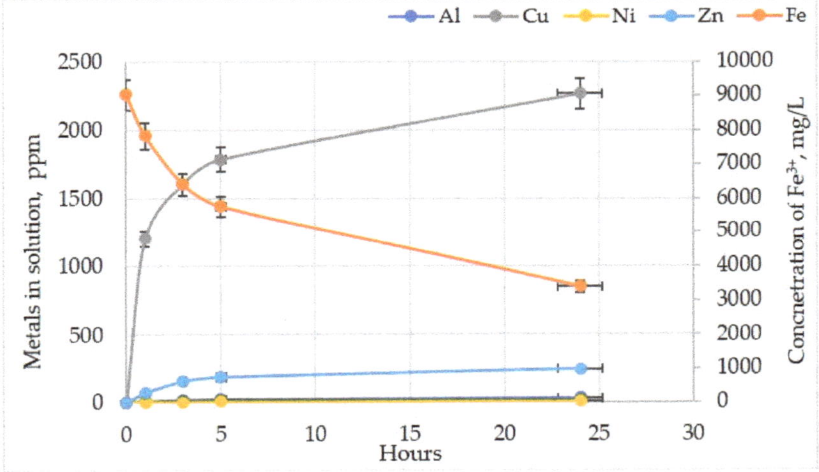

Figure 4. Evolution of metal ions and Fe^{3+} in the PLS with time (40 °C; 1% PD; pH 1.0; initial concentration of Fe^{3+} 9/L).

When we compare the effect from the Fe^{3+} origin on the single-stage leaching efficiency at 5% PD, one might note that the recoveries of Cu at pH 2 in the presence of 20 g/L biogenic Fe^{3+} and 20 g/L chemical Fe^{3+} at pH 1.1 were virtually identical (67.5% and 65.8%, respectively)—Figure 5. A very similar pattern was observed for Zn and Ni. In case of Al, the recovery was very low in all four cases regardless initial pH and type of Fe^{3+} used. As mentioned elsewhere, maintaining highly acidic conditions in leaching is essential to promoting Al dissolution [59]. However, in our case, even at very low pH, Al was virtually not dissolved, probably due to intrinsic refractoriness (e.g., present as an alloy), surface coatings or encapsulation inside the PCB matrix.

The results from the comparative two-stage leaching using bio-derived Fe^{3+} (at 19 g/L) and chemically derived Fe^{3+} (at 16 g/L) are presented in Figure 6. In this series of tests PD was increased to 10%.

Perusal of the results shown in Figure 6 suggested that the recovery of Cu is virtually the same (62 and 61.5% for Cu) in chemical and biological leaching systems. Aluminum is brought into solution to a very modest extent (3–4%). Recovery of both Zn and Ni in the biological leaching system was about 20% higher than in the chemical.

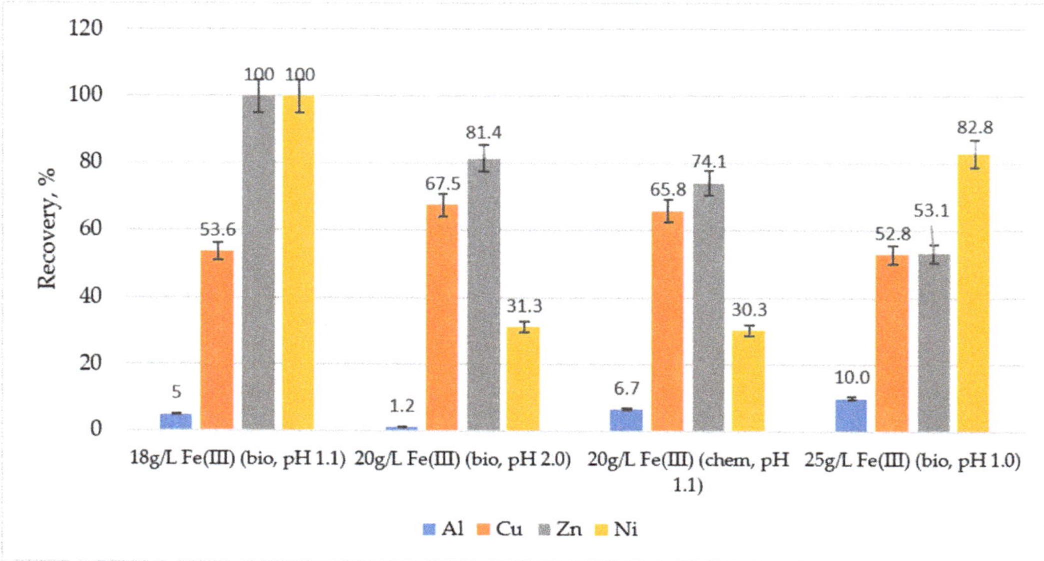

Figure 5. Metals recovery from PCBs at various concentrations of chemical and biogenic Fe^{3+} and pH (PD 5%; 40 °C, air 1 lpm; 600 rpm; duration 24 h).

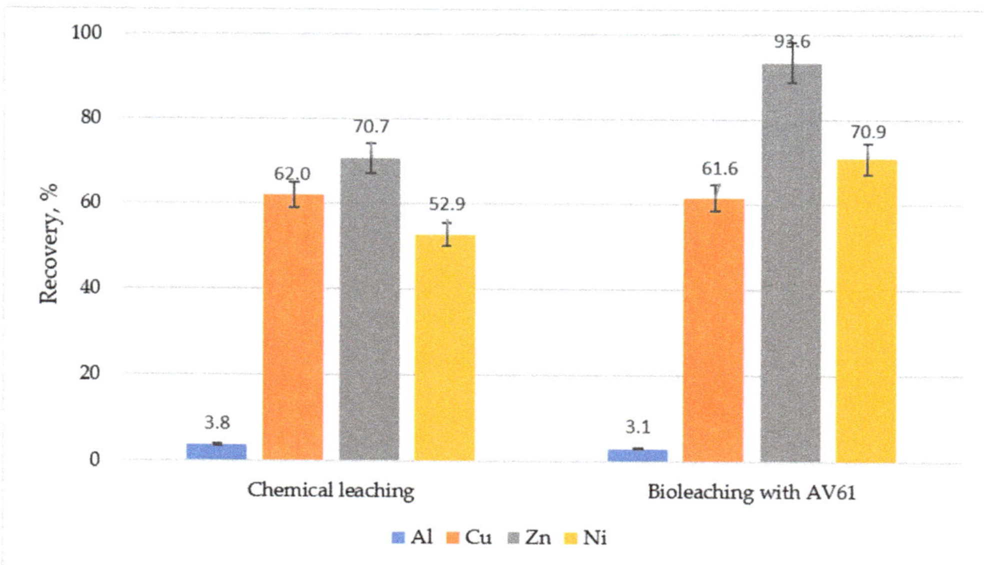

Figure 6. Recovery of metals from PCBs with two-stage leaching, 6 h duration (2 h + 4 h). (PD 10%; Fe^{3+} concentration 16 g/L (chem) and 19 g/L (bio); initial pH 1.2; 40 °C; air 1 lpm; 600 rpm.)

Cumulative metal content in the PLS during each of the two stages is plotted for each metal in Figure 7 to demonstrate the effect of ferric iron origin. It can be noted that Cu dissolution by biogenic Fe^{3+} occurred mostly during the first stage, while in case of chemical leaching, Cu dissolution rate was identical in both stages. The recovery of Zn was, on the other hand, more pronounced during the second stage in both leaching systems. The majority of Ni occurred in solution over the first stage of the chemical-assisted leaching, while in the case of biogenic Fe^{3+}, Ni dissolution rate was identical in both stages. Aluminum was brought in solution in majority during the first step.

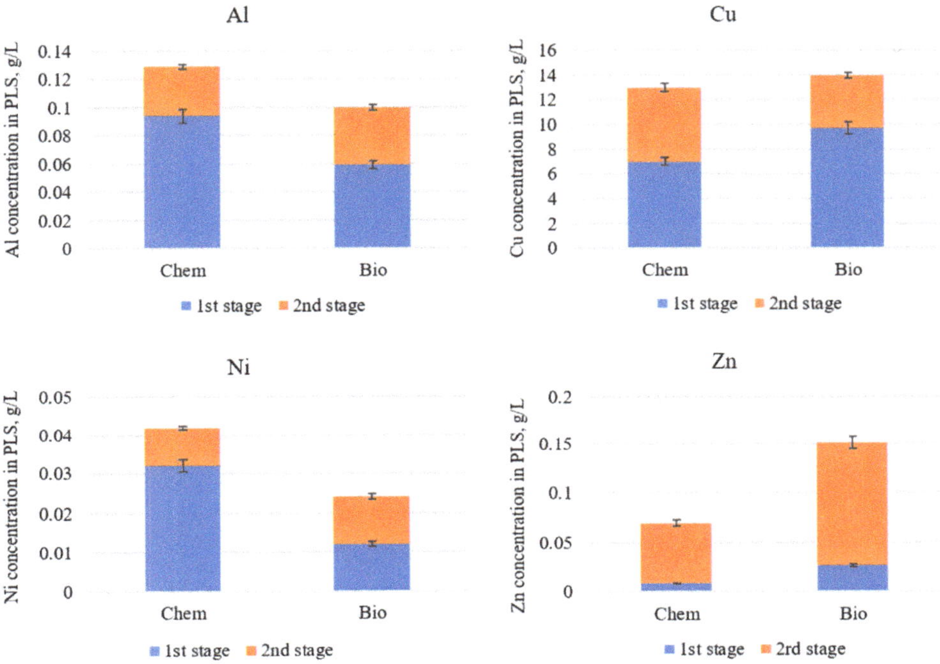

Figure 7. Cumulative metal content of the pregnant leach solution (PLS) after two-stage leaching with bio- and chemically generated Fe^{3+} (PD 10%; initial pH 1.2; 40 °C; air 1 lpm; 600 rpm).

Since the intrinsic value of the PCBs tested in the present study is to a larger extent due to Cu content, it was important to follow the kinetics of Cu bioleaching, and at the same time to perform iron speciation inside the process. Moreover, it was interesting to verify through cell enumeration if bacterial activity was maintained during leaching.

Figure 8 illustrates the copper content in PLS and iron speciation during the sampling performed at each hour of the two-stage process. During the first stage, copper seems to be brought in solution more easily thanks to the abundant availability of biogenic Fe^{3}. Ferric iron was almost consumed after two hours, dropping from 20 g/L to less than 3 g/L. During the second stage, the addition of fresh biolixiviant enabled us to bring out most of the remaining Cu in solution, however, not to a complete extent. Cu leaching almost halted after 4 h regardless of the ferric iron still being present. This was probably because part of the Cu in the PCBs existed in non-leachable form (alloy, oxidized, locked).

Note that after 24 h of leaching there was a distinct reduction in bacterial population in the PLS—from 10^9 to 10^1 cells/mL— as shown in Table 2. This phenomenon seemed intuitive due to the inhibitory effect of PCBs on bacterial activity given their heterogenous composition carrying variety of hazardous compounds.

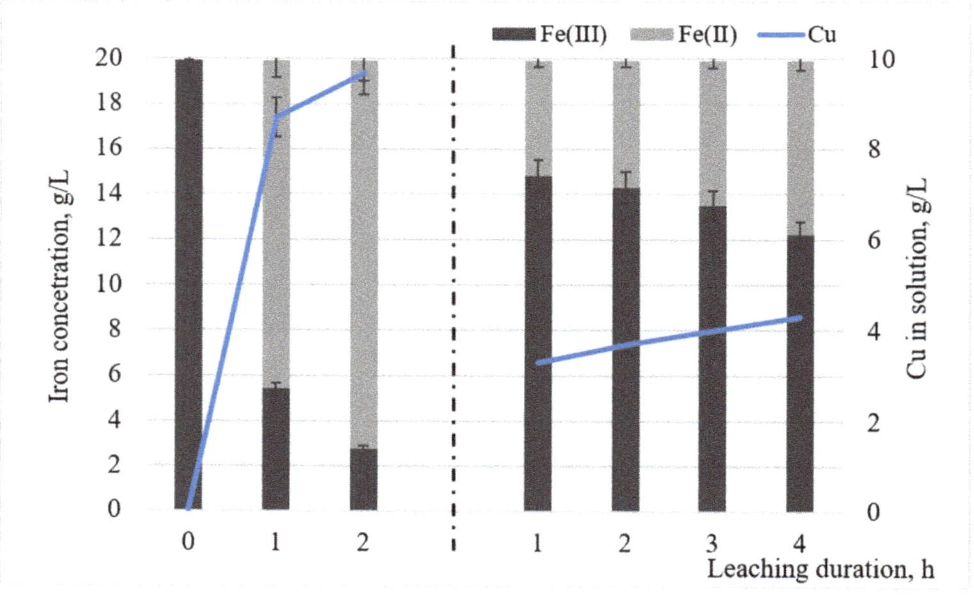

Figure 8. Kinetics of two-stage Cu leaching and iron speciation in the biogenic Fe^{3+} system.

Table 2. Bacteria enumeration during bioleaching (PD 5%, initial pH-2.0, 20 g/L Fe^{3+}, 40 °C, air -1 lpm, 600 rpm, duration-24 h).

Duration	Number of Cells, Cells/mL
Initial (0 min)	0.6×10^9
1 h	0.25×10^7
3 h	0.6×10^5
5 h	0.25×10^5
24 h	0.6×10

The oxidation–reduction potential (ORP) is an important metric that reflects the oxidation and reduction reactions taking place in the medium. Before bioleaching, the ORP of the bacterial solution was as high as 671 mV and 606 mV (Ag/AgCl) at pH 1.1 and 2, respectively. After one hour of leaching, ORP dropped sharply to 423 mV and 428 mV (Ag/AgCl), respectively, due to the reduction of Fe^{3+} ions to Fe^{2+}—Figure 9. After the first hour and until the tests end, ORP declined at a slower pace from 423 and 428 to 400 and 354 mV (Ag/AgCl), respectively.

Analysis of Figure 9 also shows that the initial pH increased over the course of the bioleaching. This increase may have resulted from the alkaline compounds found in the PCBs, which were consumed due to the acidic action of the bacterial culture. Furthermore, the increase in pH may have resulted from the bacterial oxidation of Fe^{2+} to Fe^{3+} (see Equation (2)), which is known as an acid (proton) consumption process. As seen from Figure 9, the initial pH of 1.1 and 2 increased to 1.5 and 2.6, respectively, after 24 h, resulted in precipitation of Fe^{3+} in the form of jarosite. It might be that metallic particles trapped inside the jarosite were reflected in a lower overall leaching efficiency for all the metals.

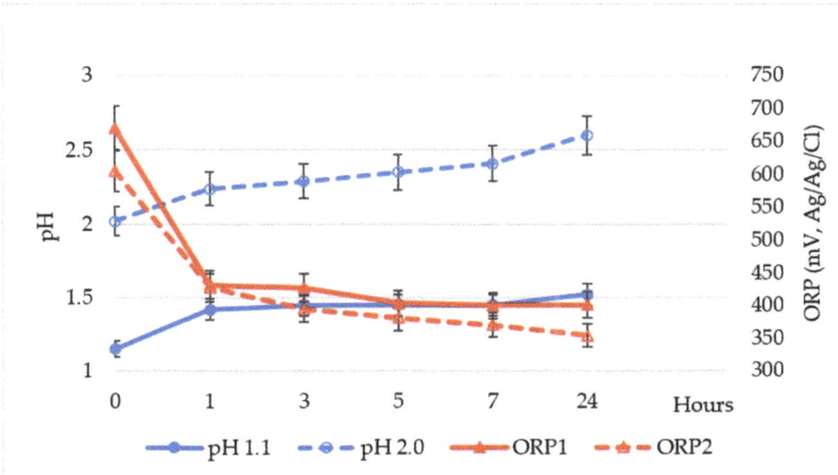

Figure 9. Variation of pH and ORP during bioleaching of PCBs (PD 5%; initial pH 2.0; 20 g/L Fe^{3+}; 40 °C, air 1 lpm; 600 rpm; duration 24 h).

3.2. Mineralogy of the Leached PCBs

The reddish-brown precipitate observed after PCBs bioleaching was collected and sent for mineralogical analysis, being determined as jarosite—Figure 10.

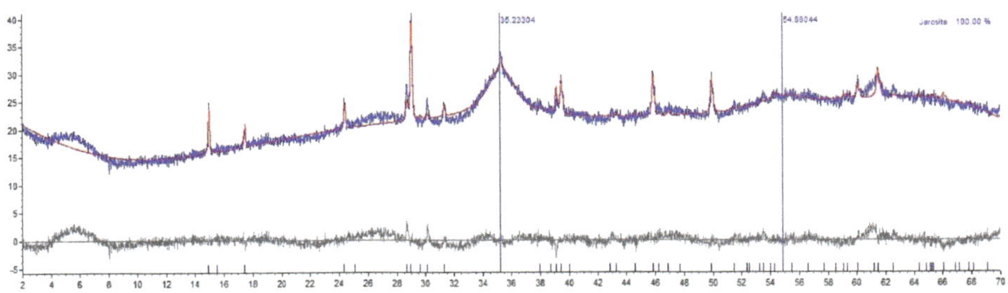

Figure 10. Mineralogy of a residue (jarosite) after bioleaching of PCBs (bio Fe^{3+} 20 g/L; PD 5%; 40 °C; air 1 lpm; 600 rpm, pH 2).

From Figure 10 it can be seen that the occupancy factor for the K site was very high, which may indicate other heavy atoms on that site (Ag, Pb or Ba). The background was also high, indicating the presence of amorphous phases. As noted earlier in the Materials and Methods, given the PCBs' composition, in 100 g of PCB we could expect 3.2 g Pb, 1.98 g Sn and 0.097 g Ag. In view of the elevated content of Pb and the low amount of Ag in our samples, it was likely that the very large electronic density observed in the large site of jarosite was mainly due to Pb presence. This was supported by the known low dissolution degree of Pb in acidic media, which remained in the PCBs, a fact being in accordance with the literature [60]. It has to be noted that a higher amount of jarosite precipitation was observed when pH was kept around 2.

3.3. Visual Appearance of PCBs before and after Leaching

It was important to visually characterize and follow the chemical composition evolution of the remaining PCBs after the leaching through SEM-EDS inspection. Figures 11 and 12 illustrate images from a PCB sample before leaching. The backscattered electrons (BSE) images provided signals from a polished section in grayscale (Figures 11a and 12a) making it possible to distinguish plastics (dark) from metallic particles (bright) thanks to their chemical composition contrast. As atomic numbers of the metal (e.g., Cu, Zn) carriers are close proximity range, the level of brightness could be similar, rendering their distinguishing difficult. Therefore, extra information acquired from energy dispersive spectroscopy (EDS) was considered, enabling us to differentiate the composition of the metallic particles—Figures 11b and 12b.

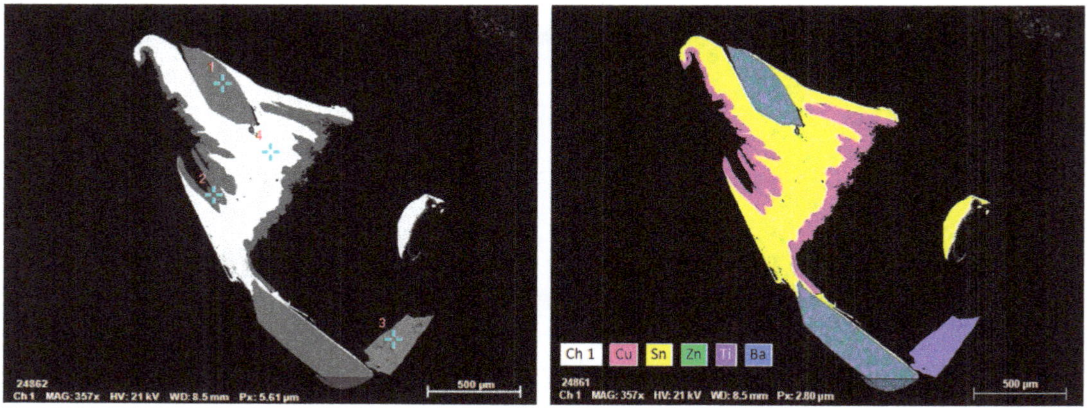

(a) (b)

Figure 11. Micrograph (BSE-SEM) of a feed PCB, (**a**) BSE-SEM image; (**b**) EDS-SEM image.

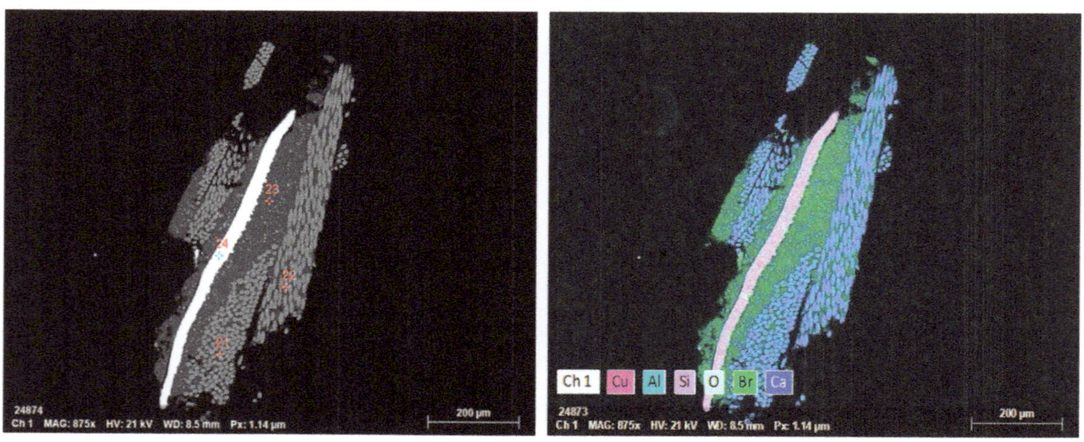

(a) (b)

Figure 12. Micrographs (BSE-SEM) of a feed PCB, (**a**) BSE-SEM image; (**b**) EDS-SEM image.

Analysis of the images in Figure 11a shows flagged composite particles (probably fiberglass or silica laminate) and metals being partly interlocked inside the same PCB piece. Figure 11b shows that tin (Sn) belonging more likely to a soldering material was met in the fragmented PCBs as a single fragment and as joined with copper assemblage.

A situation similar to the one seen in Figure 11 was observed in Figure 12, with non-liberated Cu being wrapped within inert layers, possibly fiberglass.

After leaching completion, the leached residue was recovered as two different granulometric fractions: "coarse" (1–2 mm) and "fine" (below 1 mm). Figure 13 provides view of a "coarse" leached residue indicating presence of copper surrounded by a fiberglass layers.

Two relatively large particles of encapsulated native Cu were visible. The EDS chemistry of points 3–5 mapped in Figure 13 are shown in Table 3. The EDS mapping revealed that the two visible copper fragments were surrounded by Sn with traces of jarosite, and likewise silicon, the latter being part of the PCB composite matrix—EDS points 4 and 5.

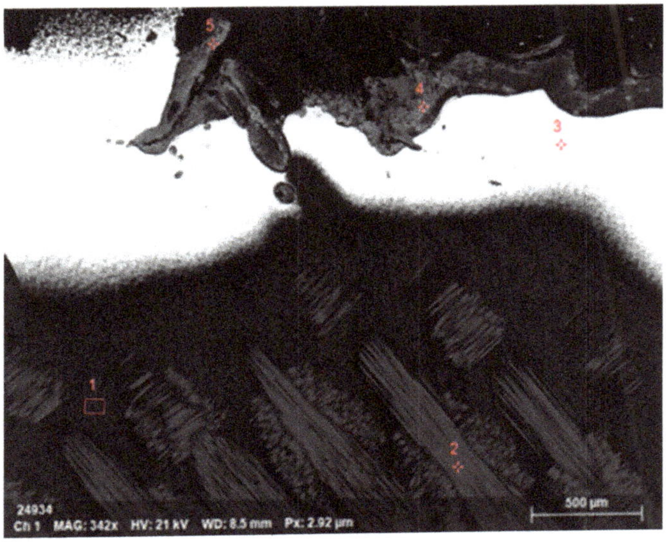

Figure 13. Micrograph (BSE-SEM) of a leached PCB ("coarse" fraction).

Table 3. EDS mapping of leached PCB (%) ("coarse" residue shown in Figure 13).

Spectrum	Oxygen	Aluminum	Silicon	Phosphorus	Sulfur	Calcium	Iron	Copper	Bromine	Tin
1	40.22		20.65			13.11		4.17	21.85	
2	45.45	8.76	21.5			17.39		1.45	5.45	
3								100		
4	36.09		1.84	4.76	0.87	1.25	10.74	2.69	17.3	24.46
5	29.21		0.95	5.91	1.36		17.42	4.01	4.3	36.85

The above observations could explain the impossibility to overcome 87% copper leaching even at a lower PD of 1%. It seems that the liberation degree resulting from the fragmentation of the PCBs was sufficient to release native copper to a certain degree; however, part of it remained trapped inside the PCBs layering.

Figure 14a illustrates a native copper fragment found in the fine residue, seen as spectrum 1 in Table 4, together with two finer Cu-Sn-bearing particles—spectra 3 and 4. This finding would indicate that passivation may have occurred during the leaching or that the hydrodynamic conditions were not optimal with possible creation of dead zones inside the reactor, allowing part of the material to escape from contact with the leachate. The rest of the finer fraction was composed of heterogenous, mixed with all different components present in the PCBs, which were not leachable—Figure 14b.

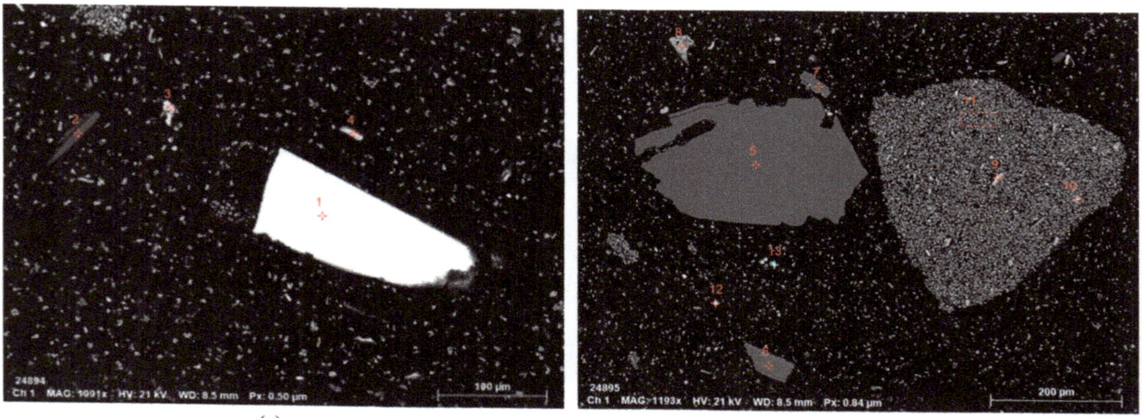

(a) (b)

Figure 14. Micrographs (BSE-SEM) of a leached PCB ("fine" fraction), (**a**) and (**b**) BSE-SEM images of "fine" fraction.

Table 4. EDS mapping of leached PCB (%) ("fine" residue shown in Figure 14a).

Spectrum	Al	Si	P	S	Ca	Fe	Cu	Sn
1							100	
2	8.87	24.86			17.82		5.68	
3		0.57	3.03	0.78		13.07	6.51	53.33
4			2.46	1.2		15.22	8.12	50.13

4. Conclusions

Several leaching experiments were performed on a fragmented end-of-life printed circuit board samples (21.2 wt% Cu, 1 wt% Zn, 0.1 wt% Ni, 2.2 wt% Al) pre-treated in NaOH solution to remove the protective organic coating, while the pulp density, initial pH, Fe^{3+} origin and number of leaching steps were varied.

It was determined that indirect bioleaching outperformed chemical leaching when realized under a two-stage mode. During bioleaching, the relative amount of Fe^{3+} decreased while a proportionate amount of jarosite was generated. The corresponding trend of pH was the opposite of the one for Fe^{3+}. For the highest degree of pulp density (10%), the copper leaching efficiency for both chemical and biological route was nearly similar—62%.

At the lowest pulp density (1%), 87% bioleaching of copper was achieved at 15.5 g/L initial Fe^{3+}concentration. There was an observed difference in the leaching efficiency of nickel as function of the initial concentration of Fe^{3+} and pH. Nearly 100% recovery was observed at 5% pulp density, 18 g/L Fe^{3+} and pH 1.1. As for zinc, the bio-assisted leaching system brought more improvement in leaching efficiency. A double-stage bioleaching with a total duration of 6 h and 10% pulp density led to 61% Cu recovery.

Microscopical inspection of input PCBs and leached residues confirmed that metals were not entirely liberated from their surrounding matrix, which reflected in incomplete leaching. The intrinsic form under which metals are present in the PCBs (alloys, oxides) also plays role in their leachability.

The approach presented in this study enabled us to outline the optimal bioleaching parameters to maximize or achieve the given recovery criteria for non-ferrous metals contained in the PCBs. The results open an avenue for further investigations with the ultimate aim of rendering bioleaching as an enabling technology for urban mining of metals from low-grade WEEE streams that usually escape from the mainstream metallurgical processing routes.

The limitation of the suggested method could be the concentration of Fe^{3+} (lixiviant). To solve this problem, we intent to carry out bioleaching in two stages using an additional amount of Fe^{3+} in the second stage. In order to obtain up-scalable results, tests in larger volume reactors (5 L) that allow for a larger amount of homogenized input material to be treated are underway.

Author Contributions: Conceptualization, A.V. and S.G.; methodology, A.V. and S.G.; validation, A.V., M.A., N.V., P.M. and S.G.; formal analysis, A.V.; investigation, A.V.; resources, N.V. and S.G.; data curation, A.V.; writing—original draft preparation, A.V.; writing—review and editing, M.A., P.M., N.V. and S.G.; visualization, S.G.; supervision, S.G.; funding acquisition, S.G. All authors have read and agreed to the published version of the manuscript.

Funding: This research has been funded by ERAMIN-2 Call 2019, BaCLEM project (Walloon region grant n 2010023).

Data Availability Statement: Not applicable.

Acknowledgments: A.K. Vardanyan is grateful to the Wallonia-Brussels International (WBI) Post-doctoral Excellence Grant and the Science Committee of the Republic of Armenia Research grant no 21T-1F124. The authors are thankful to F. Hattert and H. Bouzahzah for the XRD and SEM-EDS analysis and interpretations.

Conflicts of Interest: The authors declare no conflict of interest.

References

1. Kumari, A.; Jha, M.; Singh, R.P. Recovery of metals from pyrolyzed PCBs by hydrometallurgical techniques. *Hydrometallurgy* **2016**, *165*, 97–105. [CrossRef]
2. Hadi, P.; Xu, M.; Lin, C.S.K.; Hui, C.-W.; Mckay, G. Review: Waste printed circuit board recycling techniques and product utilization. *J. Hazard. Mater.* **2015**, *283*, 234–243. [CrossRef] [PubMed]
3. Sodha, A.B.; Shah, M.B.; Qureshi, S.A.; Tipre, D.R.; Dave, S.R. Decouple and compare the role of abiotic factors and developed iron and sulfur oxidizers for enhanced extraction of metals from television printed circuit boards. *Sep. Sci. Technol.* **2019**, *54*, 4. [CrossRef]
4. Sadia, I.; Srivastava, R.R.; Kima, H.; Nimral, I. Biotechnological recycling of hazardous waste PCBs using *Sulfobacillus thermosulfidooxidans* through pretreatment of toxicant metals: Process optimization and kinetic studies. *Chemosphere* **2022**, *286*, 131978.
5. Islam, A.; Ahmed, T.; Awual, M.R.; Rahman, A.; Sultana, M.; Aziz, A.A.; Monir, M.U.; Teo, S.H.; Hasan, M. Advances in sustainable approaches to recover metals from e-waste—A review. *J. Clean. Prod.* **2020**, *244*, 118815. [CrossRef]
6. Lehner, T. Integrated recycling of non-ferrousmetal at Boliden Ltd. In Proceedings of the 1998 IEEE International Symposium on Electronics and the Environment, Oak Brook, IL, USA, 6 May 1998; pp. 42–47.
7. Bleiwas, D.; Kelly, T. *Obsolete Computers, "Gold Mines", or High-Tech Trach? Resource Recovery from Recycling*; United States Geological Survey 7; USDS: Beaverton, OR, USA, 2001. [CrossRef]
8. He, W.; Li, G.; Ma, X.; Wang, H.; Huang, J.; Xu, M.; Huang, C. WEEE recovery strategies and the WEEE treatment status in China. *J. Hazard. Mater.* **2006**, *136*, 502–512. [CrossRef]
9. Deveci, H.; Yazıcı, E.Y.; Aydın, U.; Yazıcı, R.; Akcil, A. Extraction of copper from scrap TV boards by sulphuric acid leaching under oxidising conditions. In Proceedings of the Going Green-CARE INNOVATION 2010 Conference (2010), Vienna, Austria, 8–11 November 2010; p. 45.
10. Kaya, M. Recovery of metals from electronic waste by physical and chemical recycling processes. *Int. J. Chem. Mol. Nucl. Mater. Metall. Eng.* **2016**, *10*, 232–243. [CrossRef]
11. Moyo, T.; Chirume, B.H.; Petersen, J. Assessing alternative pre-treatment methods to promote metal recovery in the leaching of printed circuit boards. *Resour. Conserv. Recycl.* **2020**, *152*, 104545. [CrossRef]
12. Dave, S.; Sodha, A.; Tipre, D. Microbial technology for metal recovery from e-waste printed circuit boards. *J. Bacteriol. Mycol.* **2018**, *6*, 241–247. [CrossRef]
13. Vasile, C.; Brebu, M.A.; Totolin, M.; Yanik, J.A.L.E.; Karayildirim, T.A.M.E.R.; Darie, H. Feedstock recycling from the printed circuit boards of used computers. *Energy Fuels* **2008**, *22*, 1658–1665. [CrossRef]
14. Birloaga, I.; De Michelis, I.; Ferella, F.; Buzatu, M.; Vegliò, F. Study on the influence of various factors in the hydrometallurgical processing of waste printed circuit boards for copper and gold recovery. *Waste Manag.* **2013**, *33*, 935–941. [CrossRef] [PubMed]
15. Oishi, T.; Koyama, K.; Alam, S.; Tanaka, M.; Lee, J.C. Recovery of high purity copper cathode from printed circuit boards using ammoniacal sulfate or chloride solutions. *Hydrometallurgy* **2007**, *89*, 82–88. [CrossRef]
16. Fogarasi, S.; Imre-Lucaci, F.; Imre-Lucaci, Á.; Ilea, P. Copper recovery and gold enrichment from waste printed circuit boards by mediated electrochemical oxidation. *J. Hazard. Mater.* **2014**, *273*, 215–221. [CrossRef] [PubMed]

17. Hubau, A.; Minter, M.; Changes, A.; Joulian, C.; Silvente, C.; Guezennec, A.G. Recovery of metals in a double-stage continuous bioreactor for acid bioleaching of printed circuit boards (PCBs). *Sep. Purif. Technol.* **2020**, *238*, 116481. [CrossRef]
18. Sodha, A.; Qureshi, S.; Khatri, B.; Tipre, D.; Dave, S. Enhancement in iron oxidation and multi-metal extraction from waste television printed circuit boards by iron oxidizing *Leptospirillum feriphillum* isolated from coal sample. *Waste Biomass Valor.* **2017**, *10*, 671–680. [CrossRef]
19. Jadhav, U.; Hocheng, H. Enzymatic bioleaching of metals from printed circuit board. *Clean Technol. Environ. Policy* **2015**, *17*, 947–956. [CrossRef]
20. Kiddee, P.; Naidu, R.; Wong, M.H. Metals and polybrominated diphenyl ethers leaching from electronic waste in simulated landfills. *J. Hazard. Mater.* **2013**, *252–253*, 243–249. [CrossRef]
21. Mankhand, T.; Singh, K.; Kumar, S.; Gupta, S.; Das, S. Pyrolysis of printed circuit boards. *Int. J. Metall. Mater. Sci. Eng.* **2012**, *6*, 102–107. [CrossRef]
22. Akcil, A.; Erust, C.; Sekhar, G.C.; Ozgun, M.; Sahin, M.; Tuncuk, A. Precious metal recovery from waste printed circuit boards using cyanide and non cyanide lixiviants—A review. *Waste Manag.* **2015**, *45*, 258–271. [CrossRef] [PubMed]
23. Jadhao, P.; Chauhan, G.; Pant, K.K.; Nigam, K.D.P. Greener approach for the extraction of copper metal from electronic waste. *Waste Manag.* **2015**, *57*, 102–112. [CrossRef]
24. Krebs, W.; Brombacher, C.; Bosshard, P.P.; Bachofen, R.; Brandl, H. Microbial recovery of metals from solids. *FEMS Microbiol. Rev.* **1997**, *20*, 605–617. [CrossRef]
25. Vermeulen, F.; Nicolay, X. Sequential bioleaching of copper from brake pads residues using encapsulated bacteria. *Miner. Eng.* **2017**, *106*, 39–45. [CrossRef]
26. Tunchuk, A.; Stazi, V.; Akcil, A.; Yazici, E.Y.; Deveci, H. Aqueous metal recovery techniques ftrom e-scrap: Hydrometallurgy in recycling. *Miner. Eng.* **2012**, *25*, 28–37. [CrossRef]
27. Mrazikova, A.; Kadukova, J.; Marcincakova, R.; Velgosova, O.; Willner, J.; Fornalczyk, A.; Saturnus, M. The effect of specific conditions on Cu, Ni, Zn and Al recovery from PCBs waste using acidophilic bacterial strains. *Arch. Metall. Mater.* **2016**, *61*, 261–264. [CrossRef]
28. Tapia, J.; Duenas, A.; Cheje, N.; Soclle, G.; Patino, N.; Ancalla, W.; Tenorio, S.; Denos, J.; Taco, H.; Cao, W.; et al. Bioleaching of heavy metals from printed circuit boards with an Acidophilic iron-oxidizing microbial consortium in stirred tank reactor. *Bioengineering* **2022**, *9*, 79. [CrossRef] [PubMed]
29. Modin, O.; Fuad, N.; Rauch, S. Microbial electrochemical recovery of zinc. *Electrochim. Acta.* **2017**, *248*, 58–63. [CrossRef]
30. Hong, Y.; Valix, M. Bioleaching of electronic waste using acidophilic sulfur oxidizing bacteria. *J. Clean. Prod.* **2014**, *65*, 465–472. [CrossRef]
31. Brandl, H.; Bosshard, R.; Wegmann, M. Computer-munching microbes: Metal leaching from electronic scrap by bacteria and fungi. *Hydrometallurgy* **2001**, *59*, 319–326. [CrossRef]
32. Yang, C.; Zhu, N.; Shen, W.; Zhang, T.; Wu, P. Bioleaching of copper from metal concentrates of waste printed circuit boards by a newly isolated *Acidithiobacillus ferrooxidans* strain Z1. *J. Mater. Cycles Waste Manag.* **2017**, *19*, 247–255. [CrossRef]
33. Zhu, N.; Xiang, Y.; Zhang, T.; Wu, P.; Danga, Z.; Li, P.; Wu, J. Bioleaching of metal concentrates of waste printed circuit boards by mixed culture of acidophilic bacteria. *J. Hazard. Mater.* **2011**, *192*, 614–619. [CrossRef] [PubMed]
34. Xia, M.C.; Wang, Y.P.; Peng, T.J.; Shen, L.; Yu, R.L.; Liu, Y.D.; Chen, M.; Li, J.K.; Wu, X.L.; Zeng, W.M. Recycling of metals from pretreated waste printed circuit boards effectively. *J. Biosci. Bioeng.* **2017**, *123*, 714–721. [CrossRef] [PubMed]
35. Mahmoud, A.; Cezac, P.; Hoadley, A.F.; Contamine, F.; D'hugues, P. A review of sulfide minerals microbially assisted leaching in stirred tank reactors. *Int. Biodeterior. Biodegrad.* **2017**, *119*, 118–146.
36. Murugesan, M.P.; Kannan, K.; Selvaganapathy, T. Bioleaching recovery of copper from printed circuit boards and optimization of various parameters using response surface methodology (RSM). *Mater. Today Proc.* **2020**, *26 Pt 2*, 2720–2728. [CrossRef]
37. Ilyas, S.; Anwar, M.A.; Niazi, S.; Ghauri, M.A. Bioleaching of metals from electronic scrap by moderately thermophilic acidophilic bacteria. *Hydrometallurgy* **2007**, *88*, 180–188. [CrossRef]
38. Shah, M.; Tipre, D.; Purohit, M.; Dave, S. Development of two-step process for enhanced biorecovery of Cu-Zn-Ni from computer printed circuit boards. *J. Biosci. Bioeng.* **2015**, *120*, 167–173. [CrossRef]
39. Marques, A.C.; Cabrera, J.M.; Malfatti, C.F. Printed circuit boards: A review on the perspective of sustainability. *J. Environ. Manag.* **2013**, *131*, 298–306. [CrossRef] [PubMed]
40. Liang, G.; Tang, J.; Liu, W.; Zhou, Q. Optimizing mixed culture of two acidophiles to improve copper recovery from printed circuit boards (PCBs). *J. Hazard. Mater.* **2013**, *250–251*, 238–245. [CrossRef] [PubMed]
41. Jagannath, A.A.; Shetty, V.K.; Saidutta, M.B. Bioleaching of copper from electronic waste using *Acinetobacter* sp. Cr B2 in a pulsed plate column operated in batch and sequential batch mode. *J. Environ. Chem. Eng.* **2017**, *5*, 1599–1607. [CrossRef]
42. Anshu, P.; Subrata, H. Feasibility of bioleaching of selected metals from electronic waste by *Acidiphilium acidophilum*. *Waste Biomass Valorization* **2017**, *9*, 871–877. [CrossRef]
43. Amiri, F.; Mousavi, S.M.; Yaghmaei, S.; Barati, M. Bioleaching kinetics of a spent refinery catalyst using *Aspergillus niger* at optimal conditions. *Biochem. Eng. J.* **2012**, *67*, 208–217. [CrossRef]
44. Santhiya, D.; Ting, Y.P. Use of adapted Aspergillus niger in the bioleaching of spent refinery processing catalyst. *J. Biotechnol.* **2006**, *121*, 62–74. [CrossRef] [PubMed]

45. Falguni, P.; Lakshmi, B. Bioleaching of copper and nickel from mobile phone printed circuit board using Aspergillus fumigatus A2DS. *Environ. Chem. Lett.* **2019**, *17*, 1873–1879.
46. Pradhan, J.; Kumar, S. Metals bioleaching from electronic waste by *Chromobacterium violaceum* and *Pseudomonads* sp. *Waste Manag. Res.* **2012**, *30*, 1151–1159. [CrossRef] [PubMed]
47. Tay, S.B.; Natarajan, G.; Abdul Rahim, M.N.; Tan, H.T.; Chung, M.C.; Ting, Y.P.; Yew, W.S. Enhancing gold recovery from electronic waste via lixiviant metabolic engineering in *Chromobacterium violaceum*. *Sci. Rep.* **2016**, *3*, 2236. [CrossRef] [PubMed]
48. Shin, D.; Jeong, J.; Lee, S.; Pandey, B.; Lee, J. Evaluation of bioleaching factors on gold recovery from ore by cyanide-producing bacteria. *Miner. Eng.* **2013**, *48*, 20–24. [CrossRef]
49. Hubau, A.; Minter, M.; Changes, A.; Joulian, C.; Perez, C.; Guezennec, A.G. Continuous production of a biogenic ferric iron lixiviant for the bioleaching of printed circuit boards (PCBs). *Hydrometallurgy* **2018**, *180*, 180–191. [CrossRef]
50. Bas, A.D.; Deveci, H.; Yazici, E.Y. Bioleaching of copper from low grade scrap TV circuit boards using mesophilic bacteria. *Hydrometallurgy* **2013**, *138*, 65–70. [CrossRef]
51. Guezennec, A.G.; Joulian, C.; Delort, C.; Bodenan, F.; Hedrich, S.; d'Hugues, P. CO_2 mass gtransfer in bioleaching reactors: CO_2 enrichment applied to a complex copper concetrate. *Hydrometallurgy* **2018**, *180*, 277–286. [CrossRef]
52. Ilyas, S.; Ruan, C.; Bhatti, H.N.; Ghauri, M.A.; Anwar, M.A. Column bioleaching of metals from electronic scrap. *Hydrometallurgy* **2010**, *101*, 135–140. [CrossRef]
53. Sodha, A.B.; Tipre, D.R.; Dave, S.R. Optimisation of biohydrometallurgical batch reactor process for copper extraction and recovery from non-pulverized waste printed circuit boards. *Hydrometallurgy* **2020**, *191*, 105170. [CrossRef]
54. Boxall, N.J.; Cheng, K.Y.; Bruckard, W.; Kaksonen, A.H. Application of indirect non-contact bioleaching for extracting metals from waste lithium-ion batteries. *J. Hazard. Mater.* **2018**, *360*, 504–511. [CrossRef]
55. Yken, J.V.; Cheng, K.Y.; Boxall, N.J.; Nikoloski, A.N.; Moheimani, N.; Valix, M.; Sahajwalla, V.; Kaksonen, A.H. Potential of metals leaching from printed circuit boards with biological and chemical lixiviants. *Hydrometallurgy* **2020**, *196*, 105433. [CrossRef]
56. Erkmen, O. Practice 4—Most probable number technique. In *Microbiological Analysis of Foods and Food Processing Environments*; Elsevier: Amsterdam, The Netherlands, 2022; pp. 31–37.
57. Lucchesi, C.A.; Hirn, C.F. EDTA Titration of total Iron in Iron(II) and Iron(III) mixtures. Application to Iron driers. *Anal. Chem.* **1960**, *32*, 1191–1193. [CrossRef]
58. Bouzahzah, H.; Benzaazoua, M.; Mermillod-Blondin, R.; Pirard, E. A novel procedure for polished section preparation for automated mineralogy avoiding internal particle settlement. In Proceedings of the 12th International Congress for Applied Mineralogy (ICAM), Istanbul, Turkey, 10–12 August 2015.
59. Vardanyan, A.; Vardanyan, N.; Abrahamyan, N.; Aatach, M.; Gaydardzhiev, S. Sequential biologically assisted extraction of Cu and Zn from printed circuit boards (PCB). *Int. J. Environ. Stud.* **2022**. [CrossRef]
60. Guezennec, A.G.; Bru, K.; Jacob, J.; d'Hugues, P. Co-processing of sulfidic mining wastes and metal-rich post-consumer wastes by biohydrometallurgy. *Miner. Eng.* **2015**, *75*, 45–53. [CrossRef]

metals

MDPI

Article

Extraction and Recovery of Metals from Spent HDS Catalysts: Lab- and Pilot-Scale Results of the Overall Process

Nertil Xhaferaj [1,2] and Francesco Ferella [3,*]

[1] Department of Food Technology, Faculty of Food and Biotechnology, Agricultural University of Tirana, KodërKamëz, SH1, 1000 Tirana, Albania
[2] School of Pharmacy, University of Camerino, Via S. Agostino 1, 62032 Camerino, Italy
[3] Department of Industrial and Computer Engineering and Economics, University of L'Aquila, P.le E. Pontieri 1, Monteluco di Roio, 67100 L'Aquila, Italy
* Correspondence: francesco.ferella@univaq.it; Tel.: +39-0862-434238

Abstract: The present study proposes an overall recycling process for spent hydrodesulfurization (HDS) catalysts. The process put together stages already known in the technical literature, tested again with samples coming from the roasting stage in a pilot kiln, which is the most limiting stage of metal recovery from spent catalysts. These catalysts contain valuable metals such as cobalt (Co), molybdenum (Mo), nickel (Ni), and vanadium (V). In particular, a dry Co-Mo catalyst was used and treated in order to optimize the roasting step (time, soda ash, and temperature) at a pilot scale and thus maximize the extraction yield of molybdenum (Mo) and vanadium (V). In particular, a dry Co-Mo catalyst was used. After roasting at 700 °C for 2.5 h, the best conditions, the catalysts underwent water leaching, separating Mo and V from Co and the alumina carrier, which remained in the solid residue. The pregnant solution was treated to remove arsenic (As) and phosphorus (P), representing the main impurities for producing steel alloys. V was precipitated with NH_4Cl, and further calcined to obtain commercial-grade V_2O_5, whereas Mo was recovered as molybdic acid by further precipitation at a pH of around one. Thus, molybdic acid was calcined and converted into commercial-grade MoO_3 by calcination. The hydrometallurgical section was tested on a lab scale. The total recovery yield was nearly 61% for Mo and 68% for V, respectively, compared with their initial concentration in the spent Co-Mo catalysts.

Keywords: catalyst; molybdenum; vanadium; cobalt; hydrodesulfurization; HDS; recycling; circular economy

Citation: Xhaferaj, N.; Ferella, F. Extraction and Recovery of Metals from Spent HDS Catalysts: Lab- and Pilot-Scale Results of the Overall Process. *Metals* **2022**, *12*, 2162. https://doi.org/10.3390/met12122162

Academic Editor: Daniel Assumpcao Bertuol

Received: 31 October 2022
Accepted: 13 December 2022
Published: 15 December 2022
Corrected: 10 July 2023

1. Introduction

Since 2009, the maximum sulfur content in petrol and diesel fuels has been limited to 10 ppm in the European Union [1]. Based on this strict regulation, oil companies must perform thorough hydrodesulfurization (HDS) consisting of removing sulfur from petrol; this process is carried out in the presence of a catalyst [2,3]. Catalysts are essential in petroleum refining, especially in hydrotreating; they are used to produce distillates like gasoline, naphtha, jet fuel, kerosene, and gas oils. Such catalysts commonly consist of molybdenum (Mo) supported on an alumina or silica carrier together with promoters such as cobalt (Co) or nickel (Ni). Hydrodesulfurization (HDS), hydrodenitrogenation (HDN), and hydrodemetallization (HDM) are typical reactions in which various impurities such as sulfur, nitrogen, and metals, for instance, vanadium (V) are removed [4,5]. The remarkable flexibility of the catalytic chemistry allows the refiners to have rapid production to respond to market needs and product specifications. The volume of spent catalysts discharged from different process units depends on several factors, such as the volume of fresh catalysts used, their activity, and the poisoning deposits formed, and thus on their deactivation and regeneration cycles [6]. In most refineries, a considerable amount of the spent catalyst wastes come from the hydrodesulfurization stages [4]. After several cycles, the activity of the catalyst decreases below the acceptable level, so it is usually regenerated and reused. After some thousand working hours, the catalyst activity may decrease to deficient levels after several cycles. Further regeneration may not be economically feasible: the spent catalysts are discarded as solid wastes [7]. Among secondary industrial resources, spent catalysts are undoubtedly most important for their economic value and environmental concerns when discharged, as they have been classified as hazardous wastes [8]. Furthermore, Mo and V are widely used for preparing different types of steel, whereas Co, besides the steelmaking industry, is also extensively used for manufacturing Li-ion accumulators [9,10].

Various methods for recovery of HDS catalysts have been proposed in the literature: acid leaching followed by solvent extraction [11], primary leaching of spent catalyst, and then separation of metals through selective precipitation [12] recovery from biotechnological routes [13,14], carbon adsorption [15], polyelectrolyte extraction [16] and solvent extraction [17–19]. To all methods mentioned above, the recovery of metals was at different rates; however, the extraction of metals can achieve around 90% [11–13]. Presently, vanadium recovery is always greater than 90% concerning the concentration in spent catalysts [8,20]. Considering that these catalysts contain sulfur, coke, and sometimes hydrocarbons like naphtha, a thermal pre-treatment is required to convert these metals into oxides and enhance the leaching yield. The hydrometallurgical way is still the most used over the pyrometallurgical one [21]. Spent HDS catalysts are classified as hazardous materials: their disposal causes environmental threats because of the potential release of heavy metals into the soil and the groundwater table [20–22].

The present work reports the results of a one-step industrial process that includes roasting catalysts in the air at 700 °C in the presence of soda ash (Na_2CO_3) in a pilot-scale rotary kiln furnace, i.e., under operating conditions much close to those of industrial applications. The rest of the process was tested in the laboratory, whose hydrometallurgical trials are much easier to scale up on pilot and thus full scale. The aim is the conversion of molybdenum and vanadium oxides into soluble sodium molybdate and vanadate, easily leached by hot water [23–25]. Although sodium hydroxide, sodium bicarbonate, and sodium sulfate have been used in direct salt roasting, sodium carbonate is the most used due to its lower cost [26,27]. This method allows the separation of Mo and V from Co, Ni, and Al directly in the leaching stage, facilitating the downstream recovery. Other research groups tested the acid leaching, but in this case, all metals are leached, i.e., Co, Mo, and Ni (and V if present), together with a consistent amount of aluminum contained in the carrier: this makes the further separation and recovery of each metal difficult, often with a low grade of the recovered oxides/salts [8,28,29].

2. Materials and Methods

2.1. Characterization of Samples

The shape of Co-Mo catalysts was cylindrical with an approximate size of L = 4 mm and D = 2 mm. Prior to the characterization, the catalyst was dried overnight in an oven at 100 °C. The sample was grounded in a mortar and sieved at 100 µm. Afterward, it was characterized by X-ray fluorescence (XRF) (Spectro Xepos) and X-ray diffraction (XRD) (X-Pert, Philips). Furthermore, 1 g of catalyst was immersed in 10 mL of aqua regia ($HCl:HNO_3$ = 1:3) at 80 °C for 2 h: the suspension was filtered, and the solution was transferred into a 10 mL calibrated flask. The concentration of metals, i.e., Mo, V, Al, Co, As, and P, was determined by Inductively Coupled Plasma-Mass Spectrometry (ICP-MS, Agilent 7900 spectrometer). The content of metals was quantified from Equation (1):

$$X = \frac{C \cdot V}{1000 \cdot p} \cdot 100 \quad (1)$$

where X is the concentration of metal, wt%; C is the concentration of metal in solution after dissolution in aqua regia, mg L^{-1}; V is the final volume of solution (10 mL), L, and p is the weight of the solid sample (unroasted catalyst), g. Sulfur, carbon, and hydrogen were quantified with a Fisons Instruments 1108 CHNS elemental analyzer.

2.2. Roasting of Samples

Unground samples were roasted at various temperatures in different time lapses to remove carbon, sulfur, and other impurities or organic compounds using a pilot rotary kiln (length 200 cm, inner diameter 18 cm). The kiln was used in a batch operating mode, filling around 15% of the cross-section to let the air flows throughout the catalysts bed and foster the oxidation. A 1 kW electric motor with a speed regulator put the kiln in rotation at around 4 rpm. The kiln was made up of AISI 316L stainless steel: the cylinder was surrounded by an alumina covering, housing the 15 kW resistances for heating the kiln to 650–700 °C, which was the optimal temperature range determined by tests in a muffle oven. The temperature was monitored and regulated by a temperature indicator and controller (TIC) that enabled or disabled the electrical resistances. One of the ends was covered by one mall hood that sucked the flue gas and sent it to an alkaline scrubber before the release into the atmosphere via the chimney. Air was continuously insufflated by a small fan installed on the other edge. When the thermal process ended, the kiln was inclined to empty it. The roasted catalyst was recovered in a steel vessel after cooling in air. Some catalysts cylinder were manually broken to check that the inner part was not black anymore, i.e., the oxidation was complete even in the inner part where air takes more time to penetrate.

Soda ash (Na_2CO_3) was added to the catalyst to form sodium vanadate and sodium molybdate, water-soluble salts. They can be easily separated from other metals like Al and Co during the leaching stage.

The roasting temperature ranged from 400 to 750 °C. The amount of soda ash varied from 10 to 45%wt, referring to the catalyst's weight. The roasting time was tested from 40 to 180 min to optimize the thermal process, which is also one of the most energy-consuming stages of the recycling process.

Roasting time, soda ash concentration, and temperature were investigated to determine the best route of roasting catalyst to get as much metal extraction as possible in the leaching stage. The soda ash used during the roasting process was in excess concerning the stoichiometric amount required by Mo and V. This choice was made as most of it is consumed by the side reaction that captures SO_2, converting into Na_2SO_4.

2.3. Leaching Tests

Leaching tests with water were performed in 250 mL flasks at room temperature. Noteworthy, various temperatures (25–80 °C) were tried for optimization of leaching, but it didn't affect the leaching, so for further investigation, the room temperature was selected as the best option. Water was used as a solvent, and the mass was mechanically stirred at 250 rpm. The solid-to-liquid ratio (S/L) was 20% w/v for all experiments. Samples were taken after the end of the tests to monitor the extraction of the metals. The concentrations of Mo, V, Co, and Al were determined by ICP-MS, and the extraction yield was calculated by comparing the mass of the metals extracted with the initial mass contained in the catalysts and calculated by Equation (1).

2.4. Recovery Tests

The purification tests were carried out to remove As and P from the pregnant solution. These two elements, together with sulfur, lower the quality of steel and alloys. Their accumulation in steel is detrimental since they decrease toughness, increase temper brittleness, reduce corrosion resistance, and impairs welding.

Purification tests were performed in a 250 mL flask at 40 °C for 1 h under mechanical stirring at 200 rpm. Arsenic and phosphorus were precipitated via a chemical reaction with a stoichiometric amount of MgO and NH_4Cl (Equations (4) and (5)) [30–32]. A few drops of diluted HCl were added to adjust the pH in the range of around nine.

Mo and V were recovered by sequential precipitations. The chemical reaction was conducted in a 250 mL Erlenmeyer flask at 2000 rpm. The temperature of the precipitation ranged from 25 to 100 °C. The amount of NH_4Cl and HCl was calculated by Equations (6)–(8) to know the stoichiometric concentration of the precipitating agent. The pH varied from 1–9, and the reaction time from 20–60 min. Precipitation of V occurred at pH around 9 and a temperature of 40 °C, whereas the precipitation of Mo was carried out at pH1 and a temperature of 90 °C. Mo and V were recovered with high-grade as oxides (MoO_3 and V_2O_5) (Equations (9) and (10)), respectively, by calcination of their salts (NH_4VO_3 and H_2MoO_4) in a muffle furnace for 1 h at 500 °C.

3. Results

3.1. Characterization of Samples

The concentration of the main metals, determined by ICP-MS and XRF and contained in the Co-Mo sample, is reported in Table 1.

Table 1. The concentration of metals in the spent unroasted catalyst (%wt, average values).

Sample	Mo	V	Co	Ni	Al	P *	As *
Co-Mo	8.7 ± 0.9	0.25 ± 0.04	2.6 ± 0.4	0.10 ± 0.02	19.2 ± 3.3	1.5 ± 0.3	0.5 ± 0.1

* XRF analysis.

Aluminum was found to have the highest concentration, around 19%, of the spent Co-Mo mass. Instead, the vanadium concentration was rather low. Molybdenum and cobalt are the most valuable metals in the sample. The vanadium content is certainly due to its presence in the oil feedstock treated in the hydrodesulfurization unit. Concentrations of carbon, hydrogen and sulfur in dried Co-Mo are listed in Table 2.

Table 2. The concentration of C, H, and S in the spent unroasted catalysts (%wt).

Constituents	C	H	S
Co-Mo	11.24 ± 1.4	2.18 ± 0.6	9.24 ± 1.1

Regarding the XRD analysis, some phases were identified, although the sample had low crystallinity and no well-defined peaks. The main probable crystalline phases were Al_2O_3, $Co_{0.8}V_{1.6}Mo_{0.4}$, $CoMoO_4$, MoS_2 and $CoMoS_2$.

3.2. Roasting and Leaching Stages

Regarding molybdenum, based on XRD peaks, a hypothetical reaction that took place is described below in Equation (2). Concerning vanadium, we found it hard to identify any clear peaks from XRD. However, from the literature, it can be inferred that V2S3 is often present, together with other minor phases. The problem lies in the present sample, where the low vanadium concentration makes it difficult to find its main phase. Nevertheless, for our purposes, the reaction that usually takes place during roasting is given by Equation (3) [25]:

$$2\,MoS_2 + 6\,Na_2CO_3 + 9\,O_2 \rightarrow 2\,Na_2MoO_4 + 4\,Na_2SO_4 + 6\,SO_2 \quad (2)$$

$$V_2S_3 + 4\,Na_2CO_3 + 5\,O_2 \rightarrow 2\,NaVO_3 + Na_2SO_4 + 2\,SO_2 + 4\,CO_2 \quad (3)$$

In the presence of Na_2CO_3 and O_2, vanadium and molybdenum sulfide were converted into soluble salts, e.g., Na_2MoO_4 and $NaVO_3$. Note of worth, gaseous SO_2 and CO_2 resulting from the reactions above were mainly captured in the kiln and converted into sodium salts (Na_2CO_3 and Na_2SO_4) before being scrubbed in a NaOH aqueous solution. During the leaching stage, additional Na_2CO_3 (5%wt) and H_2O_2 (5%wt) were added to the solution, as explained in Section 4.1. The best extraction yields obtained in the tests are listed in Table 3 (ICP analysis).

Co, Ni, and the alumina carrier were not leached by hot water, so they remained in the solid residue after filtration. Co and Ni could be recovered with further leaching with acids.

However, aluminum is leached contemporaneously, making the subsequent purification difficult and the final grade of the recovered metals. For this reason, such residue is often sent to metallurgical treatment by electric arc furnaces.

Table 3. Leaching of metals (%).

Sample	Mo	V	Co	Ni	Al	P	As
Co-Mo	80 ± 2.5	70 ± 3.7	-	-	-	<0.1	<0.05

3.2.1. Effect of Roasting Temperature

The percentage of molybdenum and vanadium extracted as a function of temperature is shown in Figure 1. The maximum recovery is observed around 700 °C, where the molybdenum leaching is 68%. On the other side, the leaching of vanadium is slightly higher, reaching 80% at 650 °C. A further increase in temperature decreases molybdenum's recovery, which may be attributed to the sublimation of MoO_3. At variance, the increase in temperature did not affect vanadium leaching. Hence, an optimum temperature of 700 °C was chosen for further investigation.

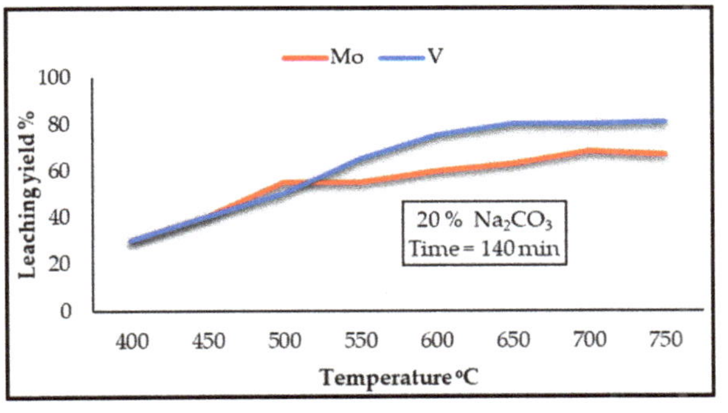

Figure 1. Leaching of molybdenum and vanadium vs. temperature.

3.2.2. Effect of Soda Ash

The effect of soda ash addition on molybdenum and vanadium leaching from the spent catalyst was investigated for 140 min at 700 °C. The results are shown in Figure 2. As can be inferred from Figure 2, at 20 wt.% of soda ash content in the mass mixture, molybdenum leaching reaches 70% and vanadium 80%. Further increase in soda ash content did not improve the molybdenum and vanadium leaching yield. In light of this, 20 wt.% soda ash addition was considered to be the optimum value.

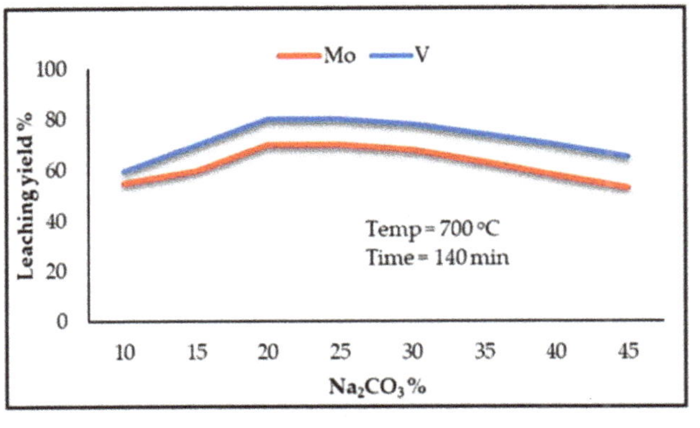

Figure 2. Leaching of molybdenum and vanadium vs. soda ash amount.

3.2.3. Effect of Roasting Time

The trials shown here refer to the soda ash content of 20%wt added in the feed mixture with a constant reaction temperature of 700 °C. The experimental results are given in Figure 3. It is possible to recognize that a roasting time of 140 min seems sufficient to leach about 70% of molybdenum and 80% of vanadium from the spent catalyst. The entire roasting was not completed within 140 min even though a further increase in this time does not affect the molybdenum leaching.

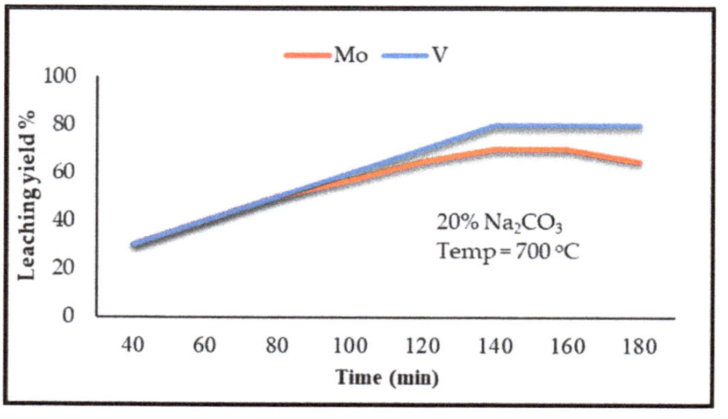

Figure 3. Leaching of molybdenum and vanadium vs. time (T: 700 °C).

The above parameters optimization reveals that, on a pilot scale, it is possible to extract around 70% of molybdenum and 80% of vanadium at 700 °C after 140 min roasting with 20%wt of soda ash. From the elemental analysis, it was clear that some fractions of sulfur and carbon were still present in the roasted catalyst, explaining the percentage of Mo and V extraction in water during the leaching step. Accordingly, around 20%wt of vanadium and 30%wt of molybdenum were still in the leaching solid, as the sulfides did not oxidize in the abovementioned conditions. Moreover, molybdenum and vanadium were found in the solid residue after leaching, after a chemical acid digestion was carried out to close the material balance.

After roasting, the carbon, hydrogen, and sulfur concentrations were 1.60%wt, 0.71%wt, and 1.40%wt, respectively. This was confirmed by a visual inspection of the roasted material. Some cylinders were randomly collected and manually broken to check the oxidation rate achieved in the inner core. All the cylinders showed a blue color on the surface, but some of those were still partially black inside. The roasting was thus not efficient at 100%.

3.3. *Purification of Leach Liquor*

Equations (4) and (5) are the reactions that take place in the purification stage. Taking into account the pH of the leach solution, which is alkaline, the addition of a stoichiometric quantity of MgO, and performing a reaction for 1 h under gently warming, almost all of As and P were removed since such elements were not found in the solution after the precipitation. Additionally, the loss of Mo and V that might have been co-precipitated altogether with As and P is very small, i.e., 0.5% of Mo and 0.3% of V, compared to the amount contained in the pregnant solution.

$$NH_4Cl + HCl + MgO + Na_2HAsO_4 \rightarrow MgNH_4AsO_4\downarrow + 2\,NaCl + H_2O \quad (4)$$

$$NH_4Cl + HCl + MgO + Na_2HPO_4 \rightarrow MgNH_4PO_4\downarrow + 2\,NaCl + H_2O \quad (5)$$

3.4. Precipitation of Metals and Their Separation

Vanadium is precipitated as NH_4VO_3 by mixing NH_4Cl into the hot water-leached solution from the Na_2CO_3-roasted waste catalysts. The precipitation occurs as per the reaction in Equation (6), as explained in [33]:

$$NaVO_3 + NH_4Cl \rightarrow NH_4VO_3 + NaCl \quad (6)$$

After precipitation, around 0.4%wt of the total vanadium mass contained in the solution after As and P purification remained dissolved; afterward, molybdenum was precipitated by adding HCl to very acidic pH and high temperature, i.e., pH around 1 and 90 °C [34,35]. The reactions proposed are the following:

$$Na_2MoO_4 + 2\,NH_4Cl \rightarrow (NH_4)_2MoO_4\downarrow + 2\,NaCl \quad (7)$$

$$(NH_4)_2MoO_4 + 2\,HCl \rightarrow H_2MoO_4\downarrow + 2\,NH_4Cl \quad (8)$$

3.4.1. Effect of Time on Vanadium and Molybdenum Precipitation

Figure 4 shows the impact of reaction time on the precipitation of vanadium and molybdenum from the leaching solutions under specific conditions (precipitation of V at pH 9 and 40 °C; precipitation of Mo was precipitated at pH 1 and approximately 90 °C). Vanadium precipitates quickly, reaching 60% after 20 min, but after that, it climbs steadily, achieving 95% after 60 min with no positive impact on response time. In these conditions, molybdenum remained dissolved in solution. However, when the pH was lowered to 1 in a boiling solution, molybdenum precipitation occurred at the same rate as vanadium.

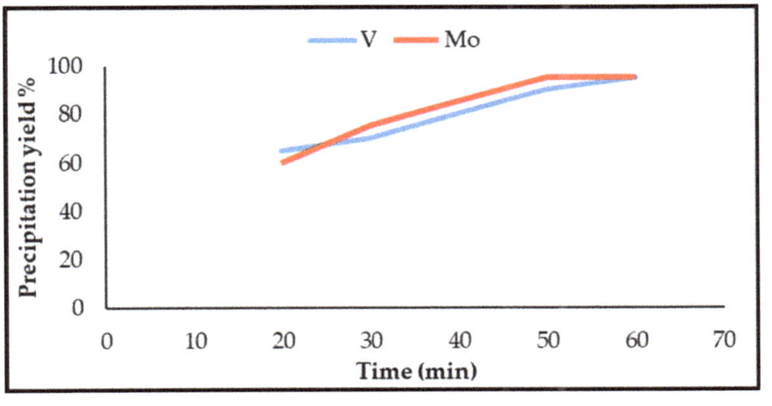

Figure 4. Precipitation yield of molybdenum and vanadium vs. time (V: pH 9, 40 °C; Mo: pH 1, 90 °C).

3.4.2. Effect of Temperature on Vanadium and Molybdenum Precipitation

As seen in Figure 5, the influence of temperature on the precipitation of vanadium and molybdenum was studied in the range of 20–100 °C. The effects of temperature on molybdenum precipitation are significant (pH = 1 and reaction time of 1 h at 90 °C with over 95% of Mo precipitate). Under these conditions, precipitation of vanadium did not take place. However, at 40 °C and pH = 1, 95% of the V precipitation was unaffected by temperature change. A higher temperature did not impact the amount of vanadium that precipitated, and it has the opposite effect at very high temperatures close to the boiling point.

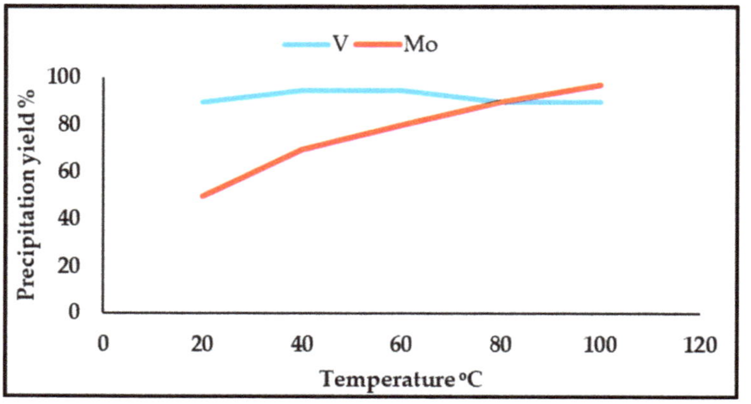

Figure 5. Precipitation yield of molybdenum and vanadium vs. temperature (V: pH 9, time 60 min.; Mo: pH 1, time 60 min.).

3.4.3. Effect of pH on Vanadium and Molybdenum Precipitation

The effect of pH on the precipitation of V and Mo is reported in Figure 6. As shown, pH plays a fundamental role in precipitating metals from the leach solution. At pH 9 (reaction time of one hour and 40 °C as explained above), with over 95% vanadium precipitate. On the other hand, molybdenum requires a highly acidic environment and higher temperatures. At pH 1 (reaction time of one hour and 90 °C), almost all molybdenum in the solution precipitated at a high rate.

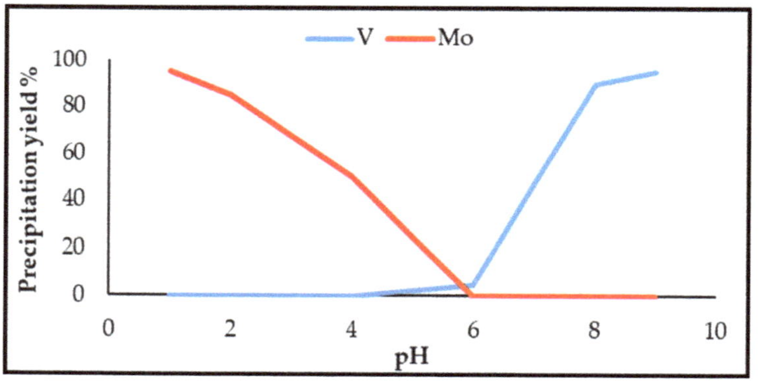

Figure 6. Precipitation yield of molybdenum and vanadium vs. pH (V: 40 °C, time 60 min; Mo: 90 °C, time 60 min).

3.5. *Production of MoO_3 and V_2O_5*

Equations (9) and (10) describe the decomposition of molybdate and vanadate. Ammonia and water were released because of the decomposition of these compounds, i.e., NH_4VO_3 and $(NH_4)_2MoO_4$ [36]. Different temperatures and times were tried to get the best possible parameters on these recovery steps, and the best option was calcination at 500 °C for 1 h.

$$NH_4VO_3 \rightarrow V_2O_5 + H_2O\uparrow + NH_3\uparrow \tag{9}$$

$$H_2MoO_4 \rightarrow MoO_3 + H_2O\uparrow \tag{10}$$

$$(NH_4)_2MoO_4 \rightarrow MoO_3 + 2\,NH_3\uparrow + H_2O\uparrow \tag{11}$$

There is also some ammonium molybdate that can precipitate together with molybdic acid: this does not represent a problem as that salt is also decomposed during the calcination, according to Equation (11).

Once roasted, both Mo and V salts were cooled down and manually ground; the two products were analyzed by XRD, which showed that the products were V_2O_5 and MoO_3, as expected (XRD patterns not reported). 0.2 g of each sample were dissolved by aqua regia to measure Mo and V quantitatively in the two oxides recovered according to the procedure reported in Paragraph 2.1: the results showed the grade of the V_2O_5 and MoO_3 was 98.2%wt and 98.4%wt, respectively. Besides that, a further qualitative ICP scan was carried out to detect the highest absorbance relevant to the other elements. The most concentrated impurities were aluminum, sodium, and magnesium, probably in the form of oxides and sulfur (as sulfates).

4. Discussion

4.1. Roasting and Leaching Process

Among various recovery processes already published in the literature [5], we wanted to find an economically and environmentally feasible way. In this context, roasting spent catalysts in a rotary kiln furnace in the presence of Na_2CO_3, turning Mo and V into soluble salt, and later leaching with water might be an alternative approach. Based on the results, leaching did not achieve high yields for both metals even though different parameters for the roasting were investigated. Elemental analysis of the roasted catalyst revealed the presence of carbon and sulfur, suggesting an unroasted fraction of the catalyst. Moreover, pH is a critical parameter during the leaching and further recovery of metals. Maintaining a pH above eight-nine during the leaching is fundamental to avoiding the partial dissolution of other metals like Co and Ni. In light of this data, powdered Na_2CO_3 (5% wt) was added to keep the pH alkaline, together with an oxidizing agent such as H_2O_2 (5% wt), to try to break the remaining metal-S bond [23]. As a consequence of this new strategy, leaching was improved by 80% for V and 70% for Mo. A further improvement was not possible, and a significant fraction of V and Mo remained in the residue.

4.2. Purification and Selective Precipitation of V and Mo

It is conceivable for some alumina, As, and P to be leached during the leaching process. As and P may then react with molybdenum and vanadium in solution to produce a variety of different compounds that strongly influence the subsequent recovery process, despite the amounts of the components above being very low [30]. ICP data from leached solution revealed small quantities of As and P in the solution. The following route applied elsewhere [30,31] was taken into consideration. After optimization of all reaction parameters, it was highlighted that As and P were no longer present in the solution.

Regarding the selective precipitation of Mo and V, the experiments tested were carried out under specific conditions (see Paragraphs 3.4 and 3.5). Before adjusting the molybdenum parameters, a small amount of vanadium remains in the solution. Nevertheless, vanadium did not precipitate with molybdenum. Otherwise, V should have been reduced from V^{5+} to V^{4+} to avoid precipitation. Molybdenum precipitates with a very high yield at a strongly acidic pH under boiling. Calcination of precipitated salt produces good grade and purified V_2O_5 and MoO_3.

The overall recovery process is shown below in Figure 7. Table 4 shows the material balance of the whole process based on data from a pilot- and lab-scale test conducted under carefully chosen processing conditions. Molybdenum and vanadium were recovered at 67.9% and 60.8%, according to the initial concentration of metals in the HDS catalyst. The suggested roasting, leaching, and separation methods offer an affordable option for handling the water-leaching waste of used catalysts and may be used in the thorough processing of spent catalysts generated as hazardous waste [37].

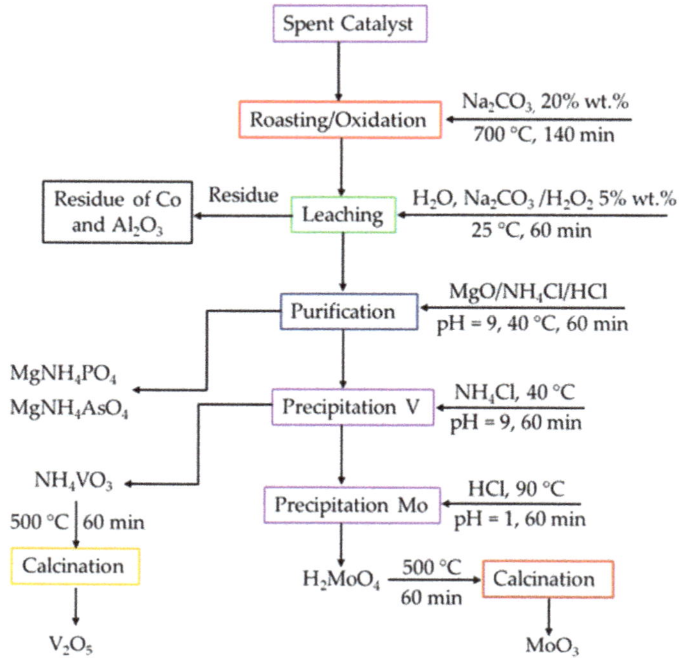

Figure 7. Flowsheet of the overall process.

Regarding the upscaling of the process, it would be necessary to test the hydrometallurgical section on a pilot scale. This way shall be pursued to have more reliable data for robust profitability analysis of the overall recycling route.

The most challenging stage, i.e., roasting, was already tested with a pilot rotary kiln, as described in the present paper. Roasting results are usually very different from simple trials carried out with a laboratory muffle oven, where a small mass is roasted. Thus air can diffuse efficiently through the heap.

Compared to the overall extraction and Mo reported in the most recent literature, higher yields were obtained using oxidative acid leaching with H_2O_2 without roasting as pre-treatment, i.e., 90% of Mo and 83% Co [38]. Park et al. [34] tested alkaline oxidative leaching with Na_2CO_3 and H_2O_2, recovering about 85% of Mo by carbon adsorption. An alkaline leaching of Co-Mo catalyst with a NaOH solution was tested by Ruiz et al. [39] after roasting at 800 °C. An overall Mo extraction yield of 90% was achieved. Other authors conducted a preliminary roasting with different time and temperature conditions, followed by alkaline leaching with NH_3, Na_2CO_3, and NaOH: the extraction yield of Mo—roasting plus leaching—was in the range of 90–98%, compared to the initial concentration in the spent catalyst [40–46].

In most of the studies handling the treatment of Co-Mo catalysts and reviewed in the scientific literature, vanadium extraction was seldom considered.

Vanadium was recovered with Mo from spent Ni-Mo catalysts by Chen et al. [47], after NaOH roasting, by adding barium hydroxide and barium aluminate to the leaching solution: a final recovery of nearly 95% and 93% for V and Mo was achieved, respectively.

A similar process using Na_2CO_3 roasting and water leaching was reported by Chen et al. [48], but Mo and V were extracted and purified by solvent extraction. 88.2% molybdenum and 87.1% vanadium were recovered from such catalysts.

Other researchers used solvent extraction, even after the preliminary roasting obtaining Mo and V extractions up to 98% [49–55].

Acid leaching is another recycling route tested after roasting: the main drawback is that aluminum, cobalt, or nickel are extracted in addition to vanadium and molybdenum: this makes the downstream recovery and separation operations more complicated and expensive [56].

These works were all performed on a laboratory scale, including the most critical stage, i.e., roasting. All the methods tested in the literature show some critical issues compared to our process, in the perspective of upgrading to a bigger scale: acid leaching dissolves Al, Co, and Ni together with Mo and V, so the purification stages are complicated and expensive; moreover, solvent extractions need expensive reagents that have to be reused, producing several spent solutions to treat. After roasting, alkaline leaching consumes more Na_2CO_3 or NaOH, the latter being more expensive.

The thermal process by the rotary kiln is undoubtedly more efficient in removing carbon, sulfur, and hydrocarbons from a big amount of catalyst mass. Nevertheless, a further second stage will be required to improve the leaching stage. In the first roasting stage, C, S, and HC are burnt off quantitatively, whereas, in the second one, soda ash is added: this choice will allow an enhanced reaction between Na and Mo/V ions, leading to the formation of more Na_2MoO_4 and $NaVO_3$ [21]. After the optimization, the recycling route proposed in this paper could be entirely updated to a pilot scale. It shows clear technical advantages over other recycling methods, as hot water is used as a lixiviant, and high-grade Mo and V are recovered by selective precipitation.

Table 4. Yields of the process stages.

Stage (Yield, %)	V	Mo	Al	Si	P	As	Co	Ni
Leaching	80.0	70.2	ND	ND	<0.1	<0.05	ND	ND
After purification	79.5	69.7	-	-	ND	ND	-	-
Precipitation of V	95.2	-	-	-	-	-	-	-
Precipitation of Mo	-	97.3	-	-	-	-	-	-
Calcination of V (V_2O_5)	89.5	-	-	-	-	-	-	-
Calcination of Mo (MoO_3)	-	90.1	-	-	-	-	-	-
Recovery of V and Mo in the whole process %	67.9	60.8	-	-	-	-	-	-

5. Conclusions

In the present paper, an overall process for the recovery of Mo and V from spent HDS catalysts was reported. The thermal pre-treatment stage was tested by a pilot rotary kiln, whereas the laboratory carried out hydrometallurgical trials.

Spent catalysts generated from refining and petrochemical plants can be a secondary source of molybdenum, vanadium, cobalt, and aluminum. Roasting with soda ash is efficient and cheap for industrial applications and allows the extraction of molybdenum and vanadium after the leaching of the roasted catalyst in water. Under selected experimental conditions, i.e., roasting at 700 °C and adding 20 wt.% soda ash for 140 min, 70% molybdenum, and 80% vanadium were extracted from a Co-Mo catalyst.

After leaching, the pregnant solution was purified by the precipitation of As and P, which represent harmful impurities in steel alloys, as Mo and V are mainly used in the steelmaking industry. Cobalt and nickel remain in the leaching residue, together with the alumina carrier: this material needs further treatment for the extraction of Co and Ni.

V is recovered by selective precipitation at 40 °C and pH 9 as ammonium vanadate, which is thus calcined to obtain commercial-grade V_2O_5. In the last stage, Mo is mainly recovered as H_2MoO_4 after precipitation at 90 °C and pH 1: molybdic acid is thus calcined at 500 °C for 1 h to get commercial grade MoO_3. The total recovery yields were around 68% and 61% for V and Mo, respectively.

Regarding future work, a two-stage calcination treatment on the pilot scale is foreseen: the goal is to remove all carbon, coke, sulfur, and residual hydrocarbons before the second treatment with soda ash and enhance the conversion into sodium molybdate and vanadate.

The present work demonstrated that the limiting step in achieving high Mo and V recovery yields, commercially required for the development to the full scale, is the roasting stage.

This double stage will result in an enhancement of the leaching yield and, thus, of the overall process. Furthermore, metals have to be concentrated more in the pregnant solutions, as the precipitation yield certainly improves; for this reason, a counter-current leaching stage must also be tested in future work.

Author Contributions: Conceptualization, N.X.; methodology, F.F.; data curation, N.X.; writing—original draft preparation, N.X.; writing—review and editing, F.F.; supervision, F.F. All authors have read and agreed to the published version of the manuscript

Funding: This research received no external funding.

Data Availability Statement: Not applicable.

Conflicts of Interest: The authors declare no conflict of interest.

References

1. European Parliament, Directive 2003/17/EC. Available online: https://eur-lex.europa.eu/LexUriServ/LexUriServ.do?uri=OJ:L:2003:076:0010:0019:EN:PDF (accessed on 13 October 2022).
2. Kim, H.I.; Park, K.H.; Mishra, D. Influence of sulfuric acid baking on leaching of spent Ni-Mo/Al_2O_3 hydro-processing catalyst. *Hydrometallurgy* **2009**, *98*, 192–195. [CrossRef]
3. Wang, J.Z.; Du, H.; Olayiwola, A.; Liu, B.; Gao, F.; Jia, M.L.; Wang, M.H.; Gao, M.L.; Wang, X.D.; Wang, S.N. Recent advances in the recovery of transition metals from spent hydrodesulfurization catalysts. *Tungsten* **2021**, *3*, 305–328. [CrossRef]
4. Marafi, M.; Stanislaus, A. Spent catalyst waste management: A review. Part I. Developments in hydro processing catalyst waste reduction and use. *Resour. Conserv. Recy.* **2008**, *52*, 859–873. [CrossRef]
5. Marafi, M.; Stanislaus, A. Spent hydro processing catalyst management: A review. Part II. Advances in metal recovery and safe disposal methods. *Resour. Conserv. Recy.* **2008**, *53*, 1–26. [CrossRef]
6. Akcil, A.; Vegliò, F.; Ferella, F.; Okudan, D.; Tunkun, A. A Review of metal recovery from spent petroleum catalyst sandash. *Waste Manag.* **2015**, *45*, 420–433. [CrossRef]
7. Furimsky, E. Spent refinery catalysts: Environment, safety and utilization. *Catal. Today* **1996**, *30*, 223–286. [CrossRef]
8. Ferella, F.; Ognyanova, A.; DeMichelis, I.; Taglieri, G.; Vegliò, F. Extraction of metals from spent hydro treating catalysts: Physico-mechanical pre-treatment sand leaching stage. *J. Hazard. Mater.* **2011**, *192*, 176–185.
9. Wang, M.; Wang, X.; Liu, W. A novel technology of molybdenum extraction from low grade Ni-Moore. *Hydrometallurgy* **2009**, *97*, 126–130. [CrossRef]
10. Feng, J.; Yan, S.; Zhang, R.; Gu, S.; Qu, X.; Bi, J. Recycling and reuse performance of cobalt catalyst for coal hydrogasification. *Fuel* **2023**, *335*, 126939. [CrossRef]
11. Banda, R.; Nguyen, T.H.; Sohn, S.H.; Lee, M.S. Recovery of valuable metals and regeneration of acid from the leaching solution of spent HDS catalysts by solvent extraction. *Hydrometallurgy* **2013**, *133*, 161–167. [CrossRef]
12. Pinto, I.S.S.; Soares, H.M.V.M. Recovery of molybdates from an alkaline leachate of spent hydro desulphurization catalyst proposal of an early-closed process. *J. Clean. Prod.* **2013**, *52*, 481–487. [CrossRef]
13. Beolchini, F.; Fonti, V.; Ferella, F.; Veglio, F. Metal recovery by means of biotechnological strategies. *J. Hazard. Mater.* **2010**, *178*, 529–534. [CrossRef] [PubMed]
14. Rocchetti, L.; Fonti, V.; Vegliò, F.; Beolchini, F. An environmentally friendly process for the recovery of valuable metals from spent refinery catalysts. *Waste Manag. Res.* **2013**, *31*, 568–576. [CrossRef] [PubMed]
15. Pagnanelli, F.; Ferella, F.; DeMichelis, I.; Vegliò, F. Adsorption on to activated carbon for molybdenum recovery from leach liquors of exhausted hydro treating catalysts. *Hydrometallurgy* **2011**, *110*, 67–72. [CrossRef]
16. Shalchian, H.; Ferella, F.; Birloaga, I.; DeMichelis, I.; Vegliò, F. Recovery of molybdenum from leach solution using polyelectrolyte extraction. *Hydrometallurgy* **2019**, *190*, 105167. [CrossRef]
17. Park, K.; Kim, H.I.; Parhi, P.; Mishra, D.; Nam, C.; Park, J.; Kim, D. Extraction of metals from Mo–Ni/Al_2O_3 spent catalyst using H_2SO_4 baking-leaching-solvent extraction technique. *J. Ind. Eng. Chem.* **2012**, *18*, 2036–2045. [CrossRef]
18. Wu, H.; Duan, S.; Liu, D.; Guo, X.; Yi, A.; Li, H. Recovery of nickel and molybdate from ammoniacal leach liquor of spent hydro desulfurization catalyst using LIX84 extraction. *Sep. Purif. Technol.* **2021**, *269*, 118750. [CrossRef]
19. Kim, H.I.; Moon, G.; Choi, I.; Lee, J.Y.; Jyothi, R.K. Hydro metallurgical process development for the extraction, separation and recovery of vanadium from spent desulfurization catalyst bio-leach liquors. *J. Clean. Prod.* **2018**, *187*, 449–458. [CrossRef]
20. Pradhan, D.; Kim, D.J.; Ahn, J.G.; Chaudhury, J.R.; Lee, S.W. Kinetics and statistical behavior of metals dissolution from spent petroleum catalyst using acido philiciron oxidizing bacteria. *J. Ind. Eng. Chem.* **2010**, *16*, 866–871. [CrossRef]
21. Krinan, S.; Zulkapli, N.S.; Kamyab, H.; Taib, S.M.; Din, M.F.B.M.; AbdMajid, Z.S.; Chaiprapat, S.; Kenzo, I.; Ichikawa, Y.; Nasrullah, M. Current technologies for recovery of metals from industrial wastes: An overview. *Environ. Technol. Innov.* **2021**, *22*, 101525. [CrossRef]

22. Le, M.N.; Lee, M.S. A Review on Hydrometallurgical Processes for the Recovery of Valuable Metals from Spent Catalysts and Life Cycle Analysis Perspective. *Min. Proc. Ext. Met. Rev.* **2021**, *42*, 335–354.
23. Ruiz, V.; Meux, E.; Diliberto, S.; Schneider, M. Hydrometallurgical Treatment for Valuable Metals Recovery from Spent CoMo/Al_2O_3 Catalyst. Improvement of Soda Leaching of an Industrially Roasted Catalyst. *Ind. Eng. Chem. Res.* **2011**, *50*, 5295–5306. [CrossRef]
24. Huang, S.; Liu, J.; Zhang, C.; Hu, B.; Wang, X.; Wang, M.; Wang, X. Extraction of Molybdenum from Spent HDS Catalyst by Two-Stage Roasting Followed by Water Leaching. *JOM* **2019**, *71*, 4681–4686. [CrossRef]
25. *Ullmann's Encyclopedia of Industrial Chemistry*, 5th ed.; WILEY-VCH: Weinheim, Germany, 1996.
26. Zeng, L.; Cheng, C.Y. A literature review of the recovery of molybdenum and vanadium from spent hydro desulphurization catalysts. Part I: Metallurgical processes. *Hydrometallurgy* **2009**, *98*, 1–9. [CrossRef]
27. Zeng, L.; Cheng, C.Y. A literature review of the recovery of molybdenum and vanadium from spent hydro desulphurization catalysts. Part II: Separation and purification. *Hydrometallurgy* **2009**, *98*, 10–20. [CrossRef]
28. Barik, S.; Park, K.H.; Parhi, P.; Park, J.; Nam, C. Extraction of metal values from waste spent petroleum catalyst using acidic solutions. *Sep. Purif. Technol.* **2012**, *101*, 85–90. [CrossRef]
29. Barik, S.; Park, K.H.; Parhi, P.; Park, J. Direct leaching of molybdenum and cobalt from spent hydro desulphurization catalyst with sulphuric acid. *Hydrometallurgy* **2012**, *111*, 46–51. [CrossRef]
30. Enbo, W.; Jingyang, N.; Lin, X. Study on thermal property of hetero polyacids with kegging structure. *Acta Chimica Sinica* **1995**, *53*, 757–764.
31. Stratful, M.D.; Scrimshaw, L.; Lester, J.N. Conditions in fluencing the precipitation of magnesium ammonium phosphate. *Water Res.* **2001**, *35*, 4191–4199. [CrossRef]
32. Battistoni, P.; Fava, G.; Pavan, P.; Musacco, A.; Cecchi, F. Phosphate removal in an aerobic liquors by struvite crystallization without addition of chemicals: Preliminary results. *Water Res.* **1997**, *31*, 2925–2929. [CrossRef]
33. Liu, G.Z.; Sui, Z.T. The study of extracting vanadium and molybdenum from HDS spent catalyst. *Compr. Util. Miner.* **2002**, *4*, 39–41.
34. Park, K.H.; Mohapatra, D.; Reddy, B.R. Selective recovery of molybdenum from spent HDS catalyst using oxidative soda ashleach/carbon adsorption method. *J. Hazard. Mater.* **2006**, *1*, 311–316. [CrossRef] [PubMed]
35. Park, K.H.; Reddy, B.R.; Mohapatra, D.; Nam, C.W. Hydro metallurgical processing and recovery of molybdenum trioxide from spent catalyst. *Int. J. Miner. Process.* **2006**, *80*, 261–265. [CrossRef]
36. Shao, Y.; Feng, Q.; Chen, Y.; Ou, L.; Zhang, G.; Lu, Y. Studies on recovery of vanadium from desilication residue obtained from processing of a spent catalyst. *Hydrometallurgy* **2009**, *96*, 166–170. [CrossRef]
37. Xhaferaj, N. Development of Technologies of Hydrometallurgy in Organic for the Separation of Metal Sand/or their Salt Exploit Able on the Market, EurekaProject. Ph.D. Thesis, University of Camerino, Camerino (MC), Italy, 2016.
38. Ruiz, V.; Meux, E.; Schneider, M.; Georgeaud, V. Hydro metallurgical Treatment for Valuable Metals Recovery from SpentCoMo/Al_2O_3 Catalyst. 2. Oxidative Leaching of an Unroasted Catalyst Using H_2O_2. *Ind. Eng. Chem. Res.* **2011**, *50*, 5307–5315. [CrossRef]
39. Angelidis, T.N.; Tourasanidis, E.; Marinou, E.; Stalidis, G.A. Selective dissolution of critical metals from diesel and naptha spent hydro desulphurization catalysts. *Resour. Conserv. Recycl.* **1995**, *13*, 269–282. [CrossRef]
40. Sun, D.D.; JooHwa, T.; HeeKiat, C.; Leung, D.L.K.; Qian, G. Recovery of heavy metals and stabilization of spent hydro treating catalyst using a glass-ceramic matrix. *J. Hazard. Mater.* **2001**, *87*, 213–223. [CrossRef]
41. Mohapatra, D.; Park, K.H. Selective recovery of Mo, Co and Al from spent Co/Mo/gamma-Al_2O_3 catalyst: Effect of calcination temperature. *J. Environ. Sci. Health. Part A Toxic Hazard. Subst. Environ. Eng.* **2007**, *42*, 507–515. [CrossRef]
42. Kar, B.B.; Datta, P.; Misra, V.N. Spent catalyst: Secondary source for molybdenum recovery. *Hydrometallurgy* **2004**, *72*, 87–92. [CrossRef]
43. Wang, J.; Wang, S.; Olayiwola, A.; Yang, N.; Liu, B.; Weigand, J.J.; Wnzel, M.; Du, H. Recovering valuable metals from spent hydro desulfurization catalyst via blank roasting and alkaline leaching. *J. Hazard. Mater.* **2021**, *416*, 125849. [CrossRef]
44. Huang, S.; Zhao, Z.; Chen, X.; Li, F. Alkali extraction of valuable metals from spent Mo–Ni/Al_2O_3 catalyst. *Int. J. Refract. Met. Hard Mater.* **2014**, *46*, 109–116. [CrossRef]
45. Yang, Y.; Xu, S.; Li, Z.; Wang, J.; Zhao, Z.; Xu, Z. OilremovalofspenthydrotreatingcatalystCoMo/Al_2O_3 via a facile method with enhanced metal recovery. *J. Hazard. Mater.* **2016**, *318*, 723–731. [CrossRef] [PubMed]
46. Chen, Y.; Feng, Q.; Shao, Y.; Zhang, G.; Ou, L.; Lu, Y. Research on the recycling of valuable metals in spent Al_2O_3-based catalyst. *Miner. Eng.* **2006**, *19*, 94–97. [CrossRef]
47. Chen, Y.; Feng, Q.; Shao, Y.; Zhang, G.; Ou, L.; Lu, Y. Investigations on the extraction of molybdenum and vanadium from ammonia leaching residue of spent catalyst. *Int. J. Miner. Process.* **2006**, *79*, 48. [CrossRef]
48. Imam, D.M.; El-Nadi, Y.A. Recovery of molybdenum from alkaline leach solution of spent hydro treating catalyst by solvent extraction using methyl tricapryl ammonium hydroxide. *Hydrometallurgy* **2018**, *180*, 172–179. [CrossRef]
49. Cai, Y.; Ma, L.; Xi, X.; Nie, Z.; Yang, Z. Comprehensive recovery of metals in spent Ni–Mo/γ–Al_2O_3 hydro fining catalyst. *Hydrometallurgy* **2022**, *208*, 105800. [CrossRef]
50. Sahu, K.K.; Agrawal, A.; Mishra, D. Hazardous waste to materials: Recovery of molybdenum and vanadium from acidic leach liquor of spent hydro processing catalyst using alamine 308. *J. Environ. Manag.* **2013**, *125*, 68–73. [CrossRef] [PubMed]
51. Parhi, P.K.; Misra, P.K. Environmental friendly approach for selective extraction and recovery of molybdenum (Mo) from a sulphate mediated spent Ni–Mo/Al_2O_3 catalyst baked leach liquor. *J. Environ. Manag.* **2022**, *306*, 114474. [CrossRef] [PubMed]

52. Zhang, D.; Liu, Y.; Hu, Q.; Ke, X.; Yuan, S.; Liu, S.; Ji, X.; Hu, J. Sustainable recovery of nickel, molybdenum, and vanadium from spent hydro processing catalysts by an integrated selective route. *J. Clean. Prod.* **2020**, *252*, 119763. [CrossRef]
53. Li, H.; Feng, Y.; Wang, H.; Li, H.; Wu, H. Separation of V(V) and Mo(VI) in roasting-water leaching solution of spent hydro desulfurization catalyst by co-extraction using P507—N235 extractant. *Sep. Purif. Technol.* **2020**, *248*, 117135. [CrossRef]
54. Gao, B.; Jiang, H.; Zeng, M.; Peng, M.; Hu, L.; Zhang, W.; Mao, L. High-efficiency recycling method for Mo and Ni from spent catalyst via soda roasting and solvent extraction. *J. Clean. Prod.* **2022**, *367*, 132976. [CrossRef]
55. Feng, C.; Zhang, C.; Yuan, S.; Liu, M.; Chen, R.; Hu, H.; Hu, J. Sustainable recovery of surface-deposited oils and valuable metals from uncrushed spent hydro processing catalysts. *J. Clean. Prod.* **2022**, *338*, 130564. [CrossRef]
56. Huang, Y.; Shi, K.; Su, S.; Liu, B.; Sun, H.; Han, G. Selectively stepwise separation and recovery of molybdenum and vanadium from simulated leaching solution of spent hydrodesulfurization catalysts. *J. Environ. Chem. Eng.* **2022**, *10*, 108462. [CrossRef]

metals

MDPI

Article

Economic and Environmental Sustainability of an Innovative Cryo-Mechano-Hydrometallurgical Process Validated at Pilot Scale for the Recycling of Li Batteries

Francesca Pagnanelli [1,*], Pier Giorgio Schiavi [1], Pietro Altimari [1], Francesca Beolchini [2], Alessia Amato [2], Jacopo Coletta [3], Flavia Forte [3], Emanuela Moscardini [3] and Luigi Toro [3]

[1] Department of Chemistry, La Sapienza University of Rome, Piazzale Aldo Moro 5, 00185 Rome, Italy
[2] Department of Life and Environmental Sciences, Università Politecnica delle Marche, Piazza Roma 22, 60121 Ancona, Italy
[3] Eco Recycling Srl, Viale Palmiro Togliatti 1639, 00155 Rome, Italy
* Correspondence: francesca.pagnanelli@uniroma1.it

Abstract: An innovative cryo-mechano-hydrometallurgical process (named LIBAT) was demonstrated at pilot scale for the treatment of EOL lithium primary batteries with chemistry Li(0)-MnO_2. The process allowed the recycling of steel scraps from external cases after cryomechanical dismantling, and the recovery of Mn and Li products after hydrometallurgical processing. During demonstration activities, about nine tons of batteries were treated in the cryomechanical section, and one ton of black mass was treated in the hydrometallurgical section for the recovery of Mn hydroxides and Li_2CO_3. The environmental impacts of the process were evaluated in comparison with an innovative pyrometallurgical approach allowing Li recovery, confirming the benefits of the proposed process due to a reduction in energy consumption. Process simulations were performed considering different mixture feeds (only Li primary, or mixture with Li-ion batteries) and process options (only the cryomechanical section of the integral process) to assess the lower limits of potentiality that would ensure economic sustainability.

Keywords: Li primary batteries; recycling; hydrometallurgy; life cycle analysis; process analysis

Citation: Pagnanelli, F.; Schiavi, P.G.; Altimari, P.; Beolchini, F.; Amato, A.; Coletta, J.; Forte, F.; Moscardini, E.; Toro, L. Economic and Environmental Sustainability of an Innovative Cryo-Mechano-Hydrometallurgical Process Validated at Pilot Scale for the Recycling of Li Batteries. *Metals* **2023**, *13*, 497. https://doi.org/10.3390/met13030497

Academic Editors: Alexandre Chagnes and Denise Crocce Romano Espinosa

Received: 12 January 2023
Revised: 16 February 2023
Accepted: 23 February 2023
Published: 1 March 2023

1. Introduction

End-of-life (EOL) primary batteries of Li(0)-MnO_2 are an environmental problem due to their low value content and to the dangerous materials they contain (Li(0) and the flammable solvents of the electrolytes) [1].

These aspects have depressed research on Li(0) recycling, with all the attention being dedicated to Li-ion batteries, which also contain the same flammable solvents but are more widely diffused on the market and contain valuable metals such as Co, Ni and Cu [2].

Li(0) batteries are currently a cost for battery collectors, who after manual sorting of the different battery types, can obtain revenues from some types (Li-ion batteries), but must pay smelters for the disposal of others (Li(0) type).

The research gap addressed in this paper is the lack of a process able to treat the Li(0)-MnO_2 battery type and ensuring effective material recovery.

In fact, the most commonly adopted treatment for EOL batteries in the EU is the pyro-metallurgical one in which whole batteries, or some battery fractions, are smelted at high temperature (1500 °C) with other solids [3]. This leads to the downcycling of the main battery components of Li-MnO_2 batteries, such as lithium, manganese and iron, which are lost in the slag [4].

Only a few works can be found in the literature addressing Li(0) treatment, with Li-ion being the most diffuse and targeted technology both in manufacturing and recycling. These few works use thermal processes for battery deactivation [5–9], or for direct dismantling [10,11], followed by hydrometallurgical operations.

All these works have operated at lab scale without evaluation of heterogeneity effects and impurity effects on final product composition [5,6,8–11], or in pilot scale [7], but without reporting analyses of environmental and economic impacts.

A patent survey was also performed evidencing that Toxco company developed a combined hydro-pyro-metallurgical process for Li recycling from Li(0)-containing wastes including primary batteries [12].Nevertheless, to the best of the authors' knowledge, no process or plant is now operating in the EU for dedicated hydrometallurgical treatment of Li(0) primary batteries, and no economic analysis or LCA is available regarding these batteries except a previous preliminary process analysis reported from one group [13].

In this context, this work presents an innovative process for treating Li(0) batteries and recovering all their valuable components. In particular, the demonstration of the process at pilot scale is reported, evidencing its technical feasibility in terms of the safety of the cryomechanical opening of batteries, and also for the assessment of metal impurities in the final products.

The proposed process is able to recycle the different components of Li(0) batteries (steel scraps, Mn and Li products) in a sequence of physical and chemical operations. The technical solution for the treatment of Li(0) batteries includes a cryomechanical treatment that avoids fire and explosions due to residual Li(0) (present in EOL primary batteries because of incomplete discharge) and the flammable solvents used to dissolve electrolytes in all Li batteries. In this way, the safe liberation of battery components is achieved and, after dry separation, fractions are neutralized in water for safe Li(0) oxidation and dissolution. Electrode materials rich in Li and Mn can then be fed to the hydrometallurgical section, where products of Mn and Li can be recovered for the manufacturing of new electrode materials. The specific combination of cryomechanical and hydrometallurgical operations here reported can be used also for the treatment of other types of batteries, such as Li-ion batteries, which present similar chemical composition and safety issues. Li-ion batteries contain also other valuable materials (copper, aluminum, cobalt, nickel, graphite) which can be recovered according to the same sequence of operations used for Li(0)-MnO_2 batteries [14,15]

Specific contributions of this work include an environmental analysis performed by LCA, comparing this process with landfilling and pyrometallurgical treatment, and economic analyses performed on the basis of experimental results elaborated as mass and energy balances and considering different process conditions in terms of sequence of operations (only cryomechanical, or the complete process), feed (only Li(0), or mixed Li(0)-Li-ion), and potentiality.

This work is a continuation of two previous works addressing upcycling of $LiMnO_2$ for the resynthesis of $LiMnPO_4$ electrode nanomaterials [16], and reporting a preliminary process simulation of the same process before demonstration campaigns were concluded [13]. In particular, in [13], preliminary data from the cryomechanical section were used along with lab-scale data for the hydrometallurgical section because the pilot-scale prototype was not yet started up. Preliminary process analysis was performed considering only Li(0) feed with different potentialities (up to 900 ton/y) showing how economic feasibility changes with these conditions [13].

In the present paper, final mass and energy balances from demonstration campaigns were adopted, economic analysis was performed for a mixed feed of Li(0) and Li-ion batteries, and environmental impact assessment by LCA was reported for the first time both for the LIBAT process and for an innovative pyrometallurgical process aiming at Li recovery.

2. Materials and Methods

2.1. Materials

EOL Li(0) batteries were manually sorted by S.E.Val. staff and then further classified in order to separate Li(0)-MnO_2 chemistry from thionyl chloride and iron disulfide types, which were not specifically addressed in this study. Li(0)-MnO_2 types were further sorted as coin type and button type by sieving. Demonstration activities were then performed using 1:1 mixtures of coin and button cells of Li(0)-MnO_2 type. During all the demonstration activities, a total amount of 8900 kg of Li was used.

2.2. Process Demonstration and Product Characterization

The proposed process included two main sections specifically designed and constructed by Eco Recycling staff: a cryomechanical section and a hydrometallurgical section, as detailed in the block diagram in Figure 1, in which also the material flow and energy consumption are reported, based on a feed of 100 kg batteries per batch.

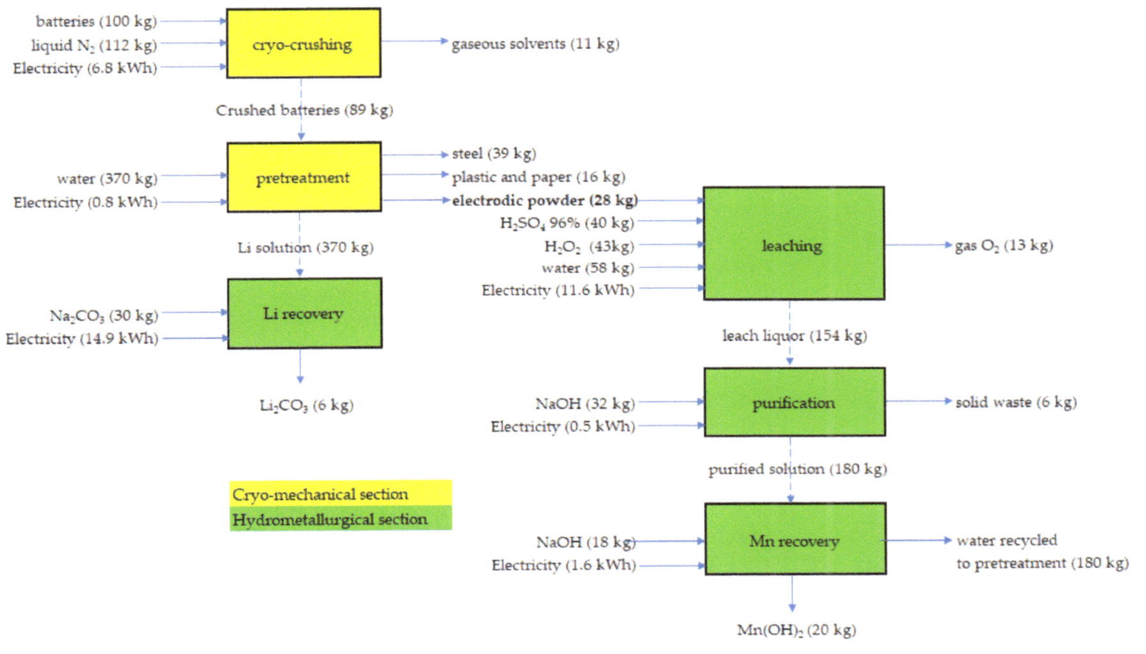

Figure 1. Process flow diagram for the cryomechanical and hydrometallurgical sections of the process for Li(0) recycling.

The cryomechanical section (400 kg/day potentiality) included a cryogenic pretreatment performed on the batteries in order to prevent explosions and control flame formation during crushing. Thermally stabilized batteries (−80 °C for 45 min using a N_2 liquid shower in a cabinet) were then crushed in a hammer mill with 10 mm undersieve. Crushed material was sieved (1 mm) and the oversieve was separated by magnetic separation using a magnetic overbelt placed above the sieving belt. After this treatment, three fractions were collected: a fine powder made up of electrode materials, a coarse magnetic fraction made up of external steel cases, and a non-magnetic coarse fraction made of separators of plastic and papers. All these fractions were periodically sampled and characterized for their chemical composition by digestion and analytical determination of metal in solution by atomic absorption spectrophotometer (ContrAA 300—Analytik Jena AG, Jena, Germany)

All fractions were then neutralized in water to dissolve residual Li(0), giving three Li-bearing streams joined together and used for Li recovery in the hydrometallurgical section.

The hydrometallurgical section included:

- Lithium recovery from neutralization waters by precipitation as Li_2CO_3.

- Manganese extraction from neutralized electrode powder.
- Manganese-bearing solution purification.
- Manganese recovery by precipitation as manganese hydroxides.

The lithium-bearing solutions obtained after neutralization constituted the feedstock for the lithium carbonate precipitation reactor. Here, after sodium carbonate addition, lithium carbonate precipitated due to its low solubility when the mother solution was heated. After precipitation, the final Li product was recovered by filtration.

The neutralized electrode powder was fed to the leaching reactor for Mn extraction using H_2SO_4 and H_2O_2. Extraction of Mn was linked with solubilization of Fe coming from external case steel impurity ending in the electrode powder. Fe was removed by selective precipitation tuning leachate pH. Iron hydroxide and the insoluble residual carbon fraction were separated by filtration of the leachate. Manganese hydroxide was precipitated by increasing the pH of the Mn-bearing solution. After precipitation, the final Mn product was recovered by filtration.

Li and Mn products were characterized by X-ray diffraction (Rigaku, D-Max Ultima – Rigaku Corporation, Tokyo, Japan) and chemical composition by digestion and atomic absorption spectrophotometry.

2.3. *Life Cycle Assessment*

On the basis of mass and energy balances obtained during demonstration activities, the environmental impact of the developed process was assessed by life cycle impact assessment (LCIA), which includes both classification and characterization steps and normalization and weighting phases. The study was carried out using the Gabi Software System and Database for Life Cycle Engineering (version 10.6.2.9, Sphera Solutions GmbH, Leinfelden-Echterdingen, Germany) integrated with Professional Database v. 2022.2, used for the production processes of energy and raw materials and the quantification of the environmental impact of treatments, following the recommendation of the ISO 14040:2006 norm. The impact categories and the related characterization methods were selected in agreement with the Product Environmental Footprint (PEF) guidelines (environmental categories, recommended models at midpoint, together with their indicators, characterization factors, units and sources) and in accordance with the nomenclature and properties of the International Reference Life Cycle Data System ILCD recommendation. In this regard, the ILCD Handbook represents the basis of the EF method. The analysis considered the impacts on the categories of acidification terrestrial and freshwater, climate change, ecotoxicity freshwater, eutrophication (freshwater, marine, terrestrial), ionizing radiation (human health), land use, cancer/non cancer human health effects, ozone depletion, photochemical ozone formation (human health), resource use, energy carriers, resource use minerals and metals, respiratory inorganics and water scarcity. The normalization and weighting step allow the determination of the relevance of the different environmental impact categories. Although they are not mandatory, normalization ("calculating the magnitude of category indicator results relative to reference information") and weighting ("converting and possibly aggregating indicator results across impact categories using numerical factors based on value-choices") are strongly recommended. They allow a global assessment to both overcome the possible heterogeneity of classification and characterization results and compare the criticality of the different environmental aspects.

Environmental impacts of the proposed process were compared with the other two scenarios, which considered landfilling and an innovative pyrometallurgical approach developed by Accurec at pilot scale [8]. In this innovative pyro-metallurgical process, exhausted batteries are fed to a vacuum pyrolysis heat treatment for their deactivation, an alternative to the N_2 shower in the LIBAT process. The gaseous compounds produced in this step are re-condensed to produce crake oil. The deactivated batteries are fed into two consecutive grinding steps and then to size separation for the recovery of the steel coarse fraction from the fine electrode powder. This powder is then pelletized with bentonite binder and fed to an oven in a nitrogen atmosphere in which the lithium is reduced, evaporated (1342 °C) and then recovered by condensation [8]. In the impact evaluation, the solids leaving the furnace are fed to the hydrometallurgical treatment according to the same process as in the LIBAT process. In Table A1, mass and energy balances for the pyro-metallurgical process used in this evaluation are reported as elaborated according to information extracted from the literature [8] and thermodynamic simulations. For the LCA of the LIBAT process, also thermal valorization for the non-magnetic coarse fraction made of plastic and paper was considered.

2.4. Process Simulations

The process was simulated in batch mode using SuperPro Designer 12 (Intelligen, Inc., Koch Plains, NJ, USA). The multi-parametric analysis was performed by simulating the complete process (cryomechanical followed by hydrometallurgical process for Mn and Li recovery), or a part of it (cryomechanical section and Li recovery), and varying the following inputs:

- Different plant potentialities in the range of 12–550 t/yr.
- With and without the "consortium fee", the fee paid by consortia and local authorities to private recycling companies performing the end-of-life treatment of batteries.
- Different "treatment fees" paid for end-of-life battery treatment by private customers, i.e., battery collectors not having dedicated infrastructures for the treatment of Li(0) batteries.
- Different feed compositions (only Li(0), and mixtures of Li(0) and Li-ion with a 1:10 weight ratio).

Mass and energy balances were obtained for cryomechanical treatment and hydrometallurgical treatment after the demonstration at pilot scale, as reported in Figure 1.

Capital investment was estimated for the different potentialities, using a power scaling law with 0.6 exponent starting from EUR 320,000, which was the cost of the pilot plant built.

Operating costs of chemicals, waste and wastewater treatment, and electric energy were set on the basis of costs sustained during the demonstration, while prices of the products were estimated on the basis of market analysis (as reported in Table A2).

The economic feasibility of the process was assessed using various measures of economic profitability such as gross margin and payback time (PBT).

The gross margin is a measure of profit that tells what portion of the revenues becomes gross profit. It is calculated as the ratio between gross profit and revenues, as in (1):

$$\text{Gross Margin (\%)} = (\text{Gross profit})/\text{Revenues} \times 100 \quad (1)$$

which, for the definition of gross profit as revenues minus annual operating costs (AOC), becomes:

$$\text{Gross Margin (\%)} = (\text{Revenues} - \text{AOC})/\text{Revenues} \times 100 \quad (2)$$

The PBT refers to the amount of time needed to recover the cost of the investment.

$$\text{PBT} = (\text{Average annual cash flow})/(\text{Cost of investment}) \quad (3)$$

The cash-flow analysis was conducted under the assumption of an inflation rate of 1%, an income tax of 40% and a depreciation time of 7 years. Such a depreciation time was set based on an overall plant depreciation coefficient of 15%. Such a coefficient value was adopted based on current Italian legislation (Ministerial Decree 31 December 1988) that sets as 10% the depreciation coefficient for large mechanical machines, 12% for electricity generation plants, engines and pumps, and 17.5% for chemical equipment within the metallurgy industry.

More detailed information on process economics can be found in the relevant literature [17].

3. Results

3.1. Process Demonstration and Product Characterization

The analysis of the batteries manually sorted at S.E.Val. as Li(0)-MnO_2 type denoted the presence of two main formats: coin cells (33% as weight) and cylindrical cells (67%). These formats have different sizes and were easily separated by sieving, and dedicated tests using single type feed denoted different contents of Li ranging from 0.7 to 2.4% in coin and cylindrical type, respectively. Due to the simplicity of format separation by sieving, and for security reasons, the shredding activity recombining battery formats in a 1:1 ratio was performed in order to have standard composition and then controlled conditions during cryo-milling and Li-neutralization.

In these conditions, the batches of batteries were fed to the cryomechanical section and separated by sieving and magnetic belt, giving the three fractions with % weight distribution and composition reported in Table 1.

Table 1. Weight distribution and chemical composition of the three fractions obtained after cryomechanical treatment and dry separation.

Fractions	[%]	Li [mg/g]	Mn [mg/g]
Electrodic Powder	32 ± 9	40 ± 2	364 ± 6
Magnetic Fraction	39 ± 7	6 ± 7	6 ± 1
Non-Magnetic Fraction	16 ± 2	29 ± 7	288 ± 76

Chemical characterization of samples from these fractions evidenced a high content of Li and Mn in the electrodic powder, as expected, but also in the non-magnetic coarse fraction, due to residues of electrodic powder still adhering to plastic and paper layers. About 64% of the weight of the non-magnetic fraction was made of this residual electrodic powder, which was then separated and recovered after neutralization in the bottom of the reactor.

During the demonstration activities, a total of 8900 kg of batteries were treated, producing the following solid fractions: 3450 kg of steel scraps, 2800 kg of electrode powder and 1400 kg of non-metal fraction; this last fraction included still adhering electrode powder, which was separated from the plastic and paper in a neutralization bath.

The distribution of processed batteries among mechanical section outlet streams was then: 32% fine fraction (electrode powder), 39% steel and 16% non-metals. A total of 64% of this last fraction was indeed electrode powder recovered after neutralization, so another +10% should be considered for the fine fraction, which then would have reached 42%. Therefore, the % distribution of the different fractions emerging from mechanical treatment was 42% fine fraction, 39% iron and 6% non-metals. A weight loss of 14% was registered mainly related to organic solvents evaporated and entrapped by the gas treatment unit operating in the industrial site.

As for the hydrometallurgical section, Li_2CO_3 was recovered from washing waters, obtaining 2.7 kg of Li_2CO_3 per 100 kg of Li(0)-MnO_2 treated batteries (56% of the total Li in batteries) with a precipitation yield of 70%. Li_2CO_3 purity could be increased from 97% to 99.8% by removing sodium impurities in filter press washings at the expense of the Li recycling rate, which was 39%.

The electrode powder was also fed to the hydrometallurgical section for Mn hydroxides recovery. Under optimized conditions, the recovery of 20 kg of Mn hydroxides per 100 kg of Li(0)-MnO_2-treated batteries was achieved, corresponding to a recycling rate of 94% for manganese.

As with the solid composition, Mn hydroxide purity could be increased up to 98.7% by washing in filter press, but without a recycling rate decrease.

During the demonstration, lithium was recovered by neutralization baths; as reported above, Li dissolved in this phase was about 56% of the total Li in batteries, thus limiting the Li recycling rate. In order to increase the Li recycling rate, it was also recovered after Mn recovery from the Li-bearing effluents, using these effluents instead of water in the neutralization reactors. Consequently, a further increase in the Li recycling rate was obtained (+29%), reaching a 68% recycling rate and without changes in the final purity of the carbonate. Reusing Li-bearing effluents in the neutralization section also allowed a reduction in water consumption.

Mn and Li products were also characterized by X-ray diffraction, evidencing relevant peaks of Mn_3O_4 and Li_2CO_3 phase, respectively (Figure 2); then, in the case of Mn hydroxides (generally amorphous products), the presence of both Mn(II) and Mn(III) could be assumed due to partial oxidation of Mn during precipitation.

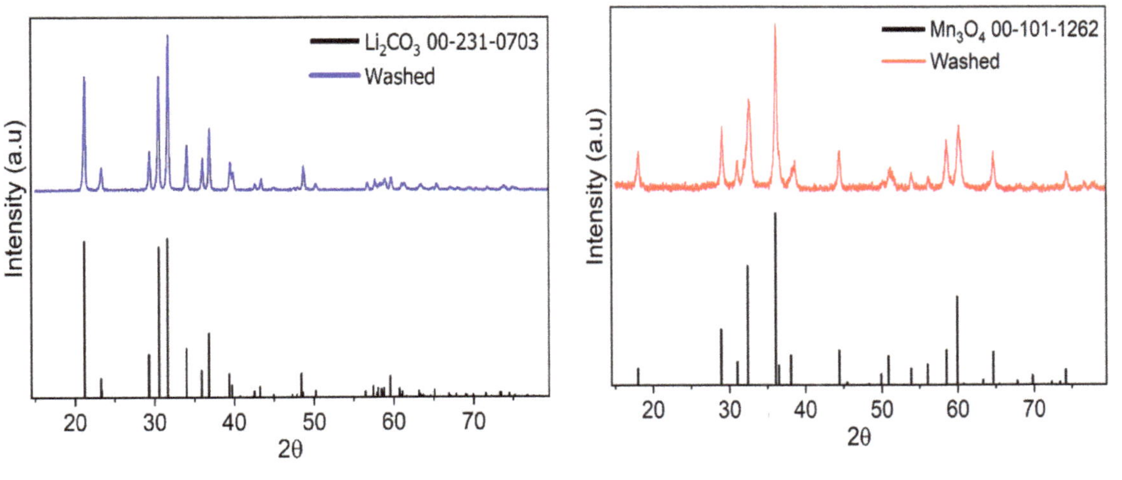

(a) (b)

Figure 2. X-ray diffraction spectra of washed products: Li_2CO_3 (**a**) and Mn hydroxides (**b**) after washing in filter press.

The mass balances denoted that the overall recycling rate of the process (given by the recycling of steel scraps, Mn and Li products) was 65%.

3.2. *Life Cycle Assessment*

In order to compare the environmental impacts of the proposed process with possible alternatives, a dedicated analysis was performed using LCA for two other scenarios: landfilling and a pyrometallurgical approach.

This comparison was necessary because, in the literature, no available environmental analysis has been found for a Li(0) recycling process, as attention to Li-containing batteries is mainly focused on Li-ion batteries. The LCA results for Li-ion batteries cannot be compared with Li(0) batteries because these types of batteries present different chemical composition. Literature analysis evidenced the effect of chemical composition even inside the class of Li-ion batteries due to the heterogeneity of chemistry of cathodic materials which affected the impact categories in different ways [18,19].

Therefore, in order to have a suitable comparison with the proposed process, LCA was performed also for an advanced pyrometallurgical process developed by Accurec at pilot scale [8]. This choice was motivated by the fact that conventional pyrometallurgical processes currently used for Li-ion batteries are not suitable for treating Li(0)-MnO_2 primary batteries. In fact, Li(0)-MnO_2 primary batteries consist of Fe, Li and Mn, and feeding the whole batteries into the furnace (as in the case of Li-ion batteries) would mean losing Fe, Li and Mn in the slag, and then all the components are lost in slag without any recycling. In fact, re-dissolving such slag for metal recovery is not convenient and the metal-containing slag is generally used as concrete additive for building materials or road pavements (downcycling). The alternative pyrometallurgical process used as a comparison for the LIBAT process includes high temperature deactivation of the batteries for dismantling and component liberation, and then high temperature pyrometallurgical treatment for Li reduction, volatilization and condensation. In this way, Li can be recovered and the residue treated according to the same hydrometallurgical treatment proposed in LIBAT. Therefore, the advanced pyro- process here considered as a comparison allows the recycling of the same metals as the LIBAT process.

The LIBAT process impact showed negative values (credits towards the environment) in many categories: acidification, cancer human health effect, eutrophication (both marine and terrestrial), land use, ozone depletion, photochemical ozone formation, resource use and respiratory inorganics. This result was mainly due to the recycling of lithium, manganese and steel scraps. The avoided impacts resulting from material recycling overcome the emissions of the process. Furthermore, there was also a positive effect for the environment due to the plastic exploitation for energetic scopes.

Considering the loads of the LIBAT process, the chemicals used in leaching had the highest environmental loads in most categories. This was due to the use of hydrogen peroxide, which accounted for an 80% contribution to the load of the leaching process. In fact, even if H_2O_2 does not produce any toxic residue in redox processes, its production is energy intensive and requires the use of catalysts, which are critical raw materials. Its use in the process thus heavily weighs down the load on the environment. One possible alternative reducing agent could be carbohydrates or organic acids, for instance from agro-industrial wastes, as green and sustainable reducing agents allowing Mn dissolution [20]. In addition, sodium hydroxide causes a relevant impact (with an average contribution in the whole process around 40%) in the categories of climate change, eutrophication, ionizing radiation, land use and photochemical ozone formation. Also in this case, an alternative precipitating agent could be used such as carbonate or oxalate to recover Mn from solution. As concerns the environmental assessment of the pyro- process, the main issue was related to energy consumption in the step of lithium recovery, and energy consumption affected almost all the impact categories. The energy consumption due to initial thermal deactivation and battery smelting (estimated by thermodynamic data) was dramatically higher with respect to cryo-crushing. LIBAT energy consumption was 13% of that used in the advanced pyro- process. Of course, this difference generated a cascade of negative loads for the pyro-process in the different impact categories.

As reported above, LCA allowed the estimation of process impacts in different impact categories; considering the traditional category of climate change, the LIBAT process allowed a 91% reduction of CO_2 emissions with 7.2×10^2 kg_{CO2eq}/ton batteries versus 8.4×10^3 kg_{CO2eq}/ton batteries in the advanced pyro-process.

The normalization and weighting steps offered an overview of the environmental load of the innovative treatment (Figure 3). The assessment produced a negative value, around −2.5, which confirmed the environmental benefits of the process, thanks to the recovered fractions. A relevant environmental gain is highlighted in the category of cancer human health effects, mainly connected to the possibility of steel recovery. The most affected category identified was that of ionizing radiation, with leading contributions by both purification and leaching, which had 48% (mainly due to NaOH use) and 24% (due to both H_2O_2 and electricity consumption), respectively, of the total emissions in this category.

In order to have a comparison for all impact categories, an environmental performance index was evaluated for landfilling, the LIBAT process and the pyro-metallurgical process, normalizing and weighting the different impact categories. This index showed that landfilling had a load of 0.7, the pyro-metallurgical process had 2.9 (mainly due to energy consumption) and LIFE-LIBAT resulted in a credit of −2.5 towards the environment (Figure 4), confirming the benefits of the proposed process and the fact that using a pyro-metallurgical approach has high impact due to high energy consumption which cannot be balanced by the environmental credit of recovered fractions. Furthermore, a sensitivity analysis was performed to estimate the necessary energy decrease needed to make the pyrometallurgical approach sustainable from an environmental point of view. The assessment (considering possible energy savings between 10 and 50%) showed that the balance between impacts and recovery benefit could be reached only by an energy reduction of 50%, which resulted in an index of −0.5. This hypothesis, almost impossible to achieve at a real scale, confirmed the relevant advantage of the LIBAT option.

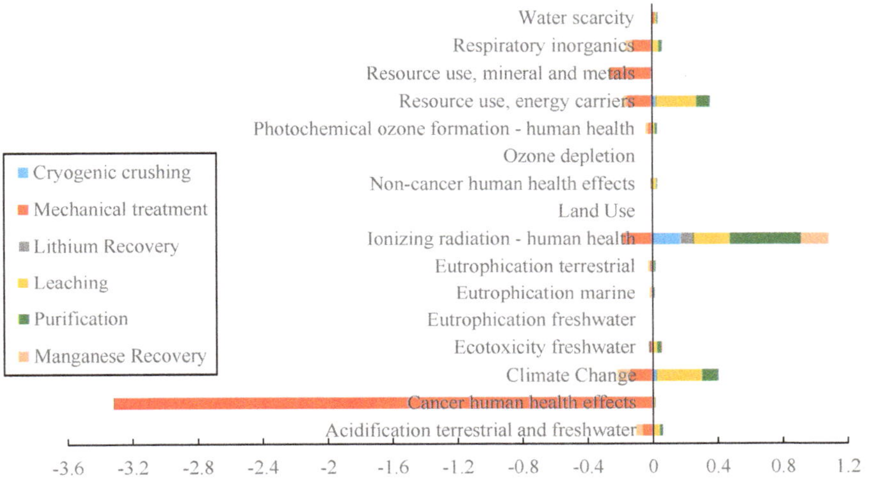

Figure 3. Results of normalization and weighting steps of LCA related to LiBat process (functional unit: 100 kg of treated waste).

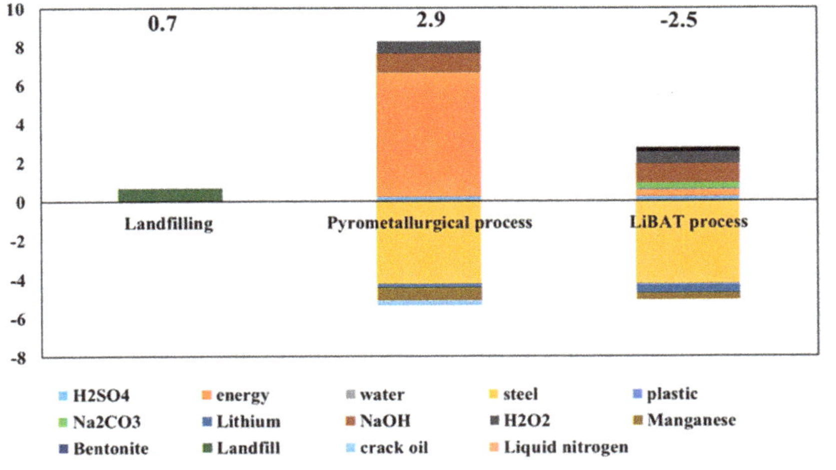

Figure 4. Environmental performance index comparison between landfilling, the LIFE-LIBAT process and the advanced pyro-metallurgical process, including Li reduction, evaporation and condensation.

These results were supported by the very good inventory data quality, confirmed by all the impact categories considered; by their temporal representativeness (data not older than 4 years); and by their technological and geographical representativeness (energy and mass balances were based on real experiments performed at pilot scale and on European territory).

3.3. Process Simulations

In process simulations, the current EU situation in EOL battery management was considered. Dedicated consortia and local authorities collect EOL battery mixtures and then pay EOL treatment performers; this is what is called a "consortium fee" and it is estimated around 0.4 EUR/kg according to the current quotation offered by S.E.Val. The following simulations were performed considering the consortium fee, but also the lack of this incentive that is a negative cost for battery recyclers.

The stocks of batteries taken by recyclers contain all types of batteries and require, as a first operation, a sorting according to the different chemistries, each one requiring a different EOL treatment. After this sorting, not all recyclers can treat all types of sorted batteries. Some kinds of batteries can be sold by a recycling company to others that are able to treat them. Li(0) batteries are one of the most difficult types to treat due to Li(0) and explosive solvents, but also due to their low intrinsic value compared with Li-ion batteries containing Co and Ni. Therefore, after sorting, battery recyclers typically pay specialized recyclers for treating Li(0) batteries; this is the "treatment fee" which, according to S.E.Val information, is currently 1.850 EUR/kg. In the following simulations, different treatment fees were considered in a competitive range according to the current price.

The products obtained by the developed process are steel scraps, Li_2CO_3 and Mn hydroxides. As for steel scraps, the recycling of ferrous metals is already a consolidated route with a market price of about 180 EUR/ton for these scraps [21].

As for lithium carbonate, the main application is in the manufacturing of Li-ion batteries, while minor applications are found in glass, ceramics and grease. As for its use in Li-ion batteries manufacturing, the production of materials is concentrated in few large enterprises (BASF, Umicore, Johnson Matthey). The Li carbonate obtained by the LIBAT process was suitable in terms of purity (>99%) and a commercial price of 59 EUR/kg was assumed [22].

As for manganese, it is mainly used in steel production starting from ores, while high purity Mn products (oxides and sulfates) are used in battery manufacturing (mainly alkaline and NMC Li-ion batteries, respectively) and fertilizers. As for NMC battery manufacturing (the Li-ion type used in electric vehicles and industrial batteries for energy storage from renewables), high purity Mn hydroxides obtained by the LIBAT process can be used as raw material for the production of high purity sulfate; the market price for Mn hydroxide can be estimated as 80% of that of the high-purity $MnSO_4$ used for LIB production (0.85 EUR/kg) [22].

Process simulations were first performed using the complete LIBAT scheme with potentialities ranging from 12 ton/y to 500 ton/y, and with negative costs for batteries ranging from 400 EUR/ton (the cost paid by consortia to battery recyclers for taking a battery mix) to 1500 EUR/ton (applying a competitive fee with respect to the current price asked by smelters for EOL Li(0) treatment, which is 1850 EUR/ton); these analyses showed that the process is economically sustainable for this range of potentialities, taking into consideration a cost of treatment (which will be a revenue for the battery recycler) lower than that currently paid for Li(0) treatment in pyrometallurgical processes. Nevertheless, the gross margin remains low, evidencing the low remunerability of the process.

Larger potentialities were not tested at this stage, even though the total amount of Li(0) in the EU can reach about 2000 ton/y, because it is not realistic that a new player can capture the whole market.

Other simulations were performed considering only cryomechanical treatment, neutralization and Li recovery, thus excluding Mn extraction and recovery and assuming disposal costs for the electrodic powder. In this case, the operating costs of the LIBAT process at prototype scale are competitive (lower) than those required by smelters for Li(0) treatment. Process simulations performed for different potentialities of the prototype (50 and 100 ton/y, corresponding to 200 and 400 kg/d) showed that operating costs diminished by revenues were lower than smelter prices (410 and 820 EUR/ton, respectively, versus 1850 EUR/ton for pyro- treatment). However, for these potentialities, with a negative cost for treatment of 400 EUR/ton, the process is not profitable yet. Nevertheless, by increasing the potentialities (to 300 and 500 ton/y) and considering a price for Li(0) treatment being paid to recyclers, the process becomes economically feasible for 300 ton/y and a treatment fee of 1500 EUR/ton (still lower than the smelter price) and the payback time is 2.85 years, while for 500 ton/y and a treatment fee of 1000 EUR/ton, the PBT is 4.03. These figures demonstrate the competitiveness of the LIBAT process against pyro-metallurgical treatment. Even in this case, gross margin values are comparable with the complete LIBAT process at the same potentialities.

Another set of simulations was performed considering the possibility of treating other types of wastes (Li-ion batteries) with the same process. In this case, simulations were performed considering typical weight ratios observed in the collected battery mix (Li(0):Li-ion = 1:10), mass balances from laboratory tests using Li-ion batteries cryo-crushed in the LIBAT prototype [23], and additional revenues from the Cu scraps (2.6 EUR/kg) [21], Al scraps (0.7 EUR/kg) [21], and graphite (7 EUR/kg) [22] present in Li-ion batteries. The final product of the hydrometallurgical section in this case was a mixed hydroxide containing nickel, manganese and cobalt (22.8 EUR/kg estimated market price, calculated as 80% of the commercial mixed oxide price, 28.5 EUR/kg [22]), which is the precursor for the production of cathodic material in NMC Li-ion batteries. In this case, considering only the negative cost received by recyclers for battery mix treatment (400 EUR/ton), the complete process was economically feasible with a payback time of 2.78 years for the 550 ton/y potentiality. This potentiality could be easily reached considering that about 3000 ton/y of Li-ion batteries are currently collected only in Italy. In this case, a significantly larger gross margin was estimated, evidencing the advantages in recovering high value products such as Li-ion battery precursors.

In Table 2, the main economic figures for the different simulations are reported.

Table 2. List of adopted conditions for process simulations performed.

Process	Feed	Potentiality (t/y)	Consortium Fee (EUR/kg)	Treatment Fee (EUR/kg)	CAPEX (kEUR)	OPEX (kEUR)	Gross Margin (%)	PBT (y)
LIBAT	Li(0)	12	0.4	0	320	58	−60.90	N/A
LIBAT	Li(0)	300	0	1	619	572	20.40	3.72
LIBAT	Li(0)	500	0	1	840	859	24.51	2.53
Cryo-mech	Li(0)	50	0.4	0	200	83	−33.23	N/A
Cryo-mech	Li(0)	100	0.4	0	200	125	−14.01	N/A
Cryo-mech	Li(0)	300	0	1	387	395	9.26	8.65
Cryo-mech	Li(0)	300	0	1.5	387	395	23.62	2.85
Cryo-mech	Li(0)	500	0	1	525	565	16.22	4.03
Cryo-mech	Li(0)	500	0	1.5	525	756	29.47	1.87
LIBAT	Li-ion + Li(0)	550	0.4	0	2.792	1.400	42.03	2.78

It should be noted that in these simulations, the potentiality was fixed on the basis of available information and estimates regarding the waste flux of Li(0) batteries in the EU. This amount is quite limited, thus also limiting the possible advantages in scale law for decreasing products' costs. Possible future improvements in waste collection schemes could lead to an increase in waste flows, and then also the construction of large-scale plants with a reduction in production costs. On the other hand, the outputs of the economic analysis were based on a preliminary optimization at pilot scale with equipment not fully automated or optimized as generally is the case in large scale. This would lead to a decrease in operating costs for manpower, but perhaps also an increase in plant costs related to automatization and control. Finally, the economic figures were strongly affected by the revenues due to selling products and the battery market during the last year, which saw a significant increase in battery materials' costs; as an example, Li_2CO_3 increased from about 13 up to 67 EUR/kg during 2022. If these trends continue, considering the global investments in electric vehicles, the remunerability of Li-battery recycling processes will increase, allowing even small-scale plants to be distributed in the different countries, and thus not requiring high initial investment and simultaneously reducing the transport of hazardous wastes.

4. Discussion

The analysis reported in this work is a milestone in the development of hydrometallurgical processes for the treatment of Li(0) batteries. However, the process development for such complex wastes requires technical, economic and environmental validation in order to have a chance of being adopted in the future.

The implemented procedure for battery sorting and the optimization of the operating conditions in pilot scale evidenced that cryo-stabilization allowed safe operations for Li(0) batteries dismantling. This was the first step in the battery recycling process, allowing products to be obtained at pilot scale and thus assessing the effect of waste heterogeneity and impurities on final product composition. Even the effects of the complex initial composition of this waste flux was addressed by the treatment of about nine tons of Li(0) batteries, allowing a wide proven effectiveness for both process sections. Performing a demonstration campaign with significant amounts of wastes then allowed evaluation of not only the technical feasibility of the process, but also the capacity of the process for producing standardized commercial grade materials despite the heterogeneity of the input materials. Therefore, demonstration of the process confirmed the robustness towards waste variability. In addition, automatic dismantling determined the formation of metal impurity fragments (Fe, Al, Cu) in the electrodic powders. Then, by performing automated opening of batteries and further component separation, the capacity of the process for eliminating such impurities in the final products was also validated.

The quantification of energy consumption in the different sections allowed establishment of a clear distinction between cryo-stabilization and hydrometallurgical product recovery (Libat process), and thermal stabilization and pyro-metallurgical product recovery for Li (Accurec process). In particular, these two options presented almost one order of magnitude difference in their energy consumption, thus determining the consequent environmental impacts.

The economic feasibility of the overall process evidenced the importance of the battery collection system having large amounts of wastes to treat. Alternatively, the integral recycling of batteries can be economically feasible if more types of batteries are treated in the same plant. The demonstration activities evidenced the ability of the cryomechanical section to safely open Li-ion batteries. This is an important added value because, as explained above, the Li-ion battery market and range of applications are far larger than those for Li(0), and they are also expected to grow significantly in the future. Fortunately, lithium-ion batteries are an alternative type which can be treated by the LIBAT process route with only minor adjustments regarding the hydrometallurgical recipes [23]. This would allow faster development of the proposed cryo-mechano-hydrometallurgical route and benefit from the economy of scale law for reducing process costs. In fact, the European market for Li-ion batteries, mainly driven by electric vehicles [24], could grow from around 450 million EUR/year (2017) to 3–14 billion EUR/year (2025). In the long term, the global market could exceed EUR 200 billion EUR/year. To tap into it, Europe would need to further expand its manufacturing base and support the competitiveness of its industry, first of all by ensuring the internal supply of raw materials now completely dependent on non-EU countries. Battery recycling could then represent a contribution to the supply of materials, especially if, not only the more noble metals such as Co, Ni and Cu (obtained as alloys in the pyro-metallurgical process), but also Li and graphite (which are critical raw materials downcycled in the pyro-metallurgical process), become specific targets of recycling processes.

Even though this process was tested with both Li(0)-MnO_2 and Li-ion batteries, one issue that remains to be further addressed is the possibility of feeding other types of Li primary batteries into the cryomechanical and hydrometallurgical sections. In particular, the thionyl chloride and iron disulfide types were not addressed yet because of complications associated with toxic vapor release and the generation of iron impurities in the leach liquor.

5. Conclusions

In this work, quantitative analysis was performed for an innovative recycling process for Li(0) primary batteries including dismantling, fraction separation, and hydrometallurgical product recovery. This analysis evidenced the technical feasibility of the proposed process in terms of technological readiness and the high quality of the recycled products. Life cycle analysis also quantitatively assessed the environmental advantages of the proposed process in comparison with another innovative process which is based on the pyro-metallurgical route. Economic analysis provided the limits for process profitability, evidencing the importance of improving the collection scheme and using flexible hydrometallurgical processes for implementing at scale and the upcycling of EOL batteries.

Future activities are now in course for a demonstration at pilot scale of the production of Li-ion battery precursors as mixed oxides of nickel, manganese and cobalt (NMC) according to results previously obtained at lab scale [23]. In this case, the cryomechanical section of the LIBAT process will be used for battery opening and component separation, as already done, while the hydrometallurgical section will be specifically dedicated to leaching, purification and NMC recovery within the activities of the LIFE DRONE project [25].

Author Contributions: Conceptualization, F.P.; methodology, F.P.; software, F.B., A.A. and E.M.; validation, J.C., F.F. and P.G.S.; formal analysis, F.F., P.A. and E.M.; data curation, F.F. and P.G.S.; writing—original draft preparation, F.P.; writing—review and editing, F.P.; project administration, E.M. and P.A.; funding acquisition, F.P, E.M., P.A. and L.T. All authors have read and agreed to the published version of the manuscript.

Funding: This work was co-funded by the EU LIFE+ program within the project "Recycling of primary LIthium BATtery by mechanical and hydrometallurgical operations" Life LiBat—LIFE16 ENV/IT/000389.

Conflicts of Interest: The authors declare no conflict of interest. The funders had no role in the design of the study; in the collection, analyses, or interpretation of data; in the writing of the manuscript; or in the decision to publish the results.

Appendix A

Table A1. Mass and energy balances for the innovative pyro-metallurgical process developed by Accurec.

Input	Output
Step 1: Thermal pretreatment	
Batteries: 100 kg Electricity: 64.3 kWh	Crack oil: 26.8
Step 2: Mechanical pretreatment	
Pyrolyzed batteries: 73.17 kg Electricity: 41.6 kWh	Steel: 35 kg

Table A1. *Cont.*

Input	Output
Step 3: Lithium thermal recovery	
Black mass: 37.9 kg Bentonite: 0.6 kg Methane: 258.5 Electricity: 158.7 kWh	Gaseous emission: 8.9 kg Lithium carbonate cake: 4 kg
Step 4. Leaching	
Electrodic powder: 28.8 kg Sulfuric acid (96%): 41 kg Water: 60 kg Electricity: 11.6 kWh Hydrogen peroxide: 45 kg	Leaching solution: 155 kg
Step 5. Purification	
Leaching solution: 155 kg Sodium hydroxide: 39 kg Electricity: 0.5 kWh	Solid waste: 7 kg
Step 6. Manganese recovery	
Purified solution: 187 kg Sodium hydroxide: 16 kg Electricity: 1.6 kWh	Wastewater: 185 kg Manganese hydroxide cake: 18 kg

Table A2. Costs of raw materials (EUR/kg), utilities (EUR/kWh), and waste treatment (EUR/kg), and the selling prices (EUR/kg) of the products [21,22].

Production Cost		
Raw materials	Sodium carbonate	0.22
	Hydrogen peroxide	1.37
	Sodium hydroxide	0.25
	Sulfuric acid	0.18
	Water	0.0005
	Liquid nitrogen	0.34
Waste treatment	Solid waste treatment	0.64
	Wastewater	0.14
Utility	Electricity	0.27
Selling prices of the products		
Products	Steel scraps	0.18
	Mn hydroxide	0.68
	NMC precursor	22.8
	Graphite	7
	Sodium sulphate	0.1
	Copper scraps	2.6
	Aluminium scraps	0.7
	Lithium carbonate	59

References

1. Lisbona, D.; Snee, T. A review of hazards associated with primary lithium and lithium-ion batteries. *Process Saf. Environ. Prot.* **2011**, *89*, 434–442. [CrossRef]
2. Wang, Y.; An, N.; Wen, L.; Wang, L.; Jiang, X.; Hou, F.; Yin, Y.; Liang, J. Recent progress on the recycling technology of Li-ion batteries. *J. Energy Chem.* **2021**, *55*, 391–419. [CrossRef]
3. Hagelüken, C. Recycling of Electronic Scrap at Umicore's Integrated Metals Smelter and Refinery. *Erzmetall* **2006**, *59*, 152–161.
4. Assefi, M.; Maroufi, S.; Yamauchi, Y.; Sahajwalla, V. Pyrometallurgical recycling of Li-ion, Ni–Cd and Ni–MH batteries: A minireview. *Curr. Opin. Green Sustain. Chem.* **2020**, *24*, 26–31. [CrossRef]
5. Pavón, S.; Kahl, M.; Hippmann, S.; Bertau, M. Lithium recovery from production waste by thermal pre-treatment. *Sustain. Chem. Pharm.* **2022**, *28*, 100725. [CrossRef]
6. Paulino, J.F.; Busnardo, N.G.; Afonso, J.C. Recovery of valuable elements from spent Li-batteries. *J. Hazard. Mater.* **2008**, *150*, 843–849. [CrossRef] [PubMed]
7. Kondás, J.; Jandová, J.; Nemeckova, M. Processing of spent Li/MnO_2 batteries to obtain Li_2CO_3. *Hydrometallurgy* **2006**, *84*, 247–249. [CrossRef]
8. Friedrich, B.; Weyhe, R.; Träger, T. Recovery Concept of Value Metals from Automotive Lithium-Ion Batteries. *Chem. Ing. Tech.* **2015**, *87*, 1550–1557.
9. Pindar, S.; Dhawan, N. Microwave processing of spent coin cells for recycling of metallic values. *J. Clean. Prod.* **2021**, *280*, 124144. [CrossRef]
10. Pindar, S.; Dhawan, N. Recycling of discarded coin cells for recovery of metal values. *Miner. Eng.* **2020**, *159*, 106650. [CrossRef]
11. Contestabile, M.; Panero, S.; Scrosati, B. A laboratory-scale lithium battery recycling process. *J. Power Sources* **1999**, *83*, 75–78. [CrossRef]
12. McLaughlin, W.; Adams, T.S. Li Reclamation Process. U.S. Patent US5888463A, 30 March 1999.
13. Schiavi, P.G.; Baldassari, L.; Altimari, P.; Moscardini, E.; Toro, L.; Pagnanelli, F. Process Simulation for Li-MnO_2 Primary Battery Recycling: Cryo-Mechanical and Hydrometallurgical Treatments at Pilot Scale. *Energies* **2020**, *13*, 4546. [CrossRef]
14. Schiavi, P.G.; Zanoni, R.; Branchi, M.; Marcucci, C.; Zamparelli, C.; Altimari, P.; Navarra, M.A.; Pagnanelli, F. Upcycling Real Waste Mixed Lithium-Ion Batteries by Simultaneous Production of rGO and Lithium-Manganese-Rich Cathode Material. *ACS Sustain. Chem. Eng.* **2021**, *9*, 13303–13311. [CrossRef] [PubMed]
15. Abo Atia, T.; Elia, G.; Hahn, R.; Altimari, P.; Pagnanelli, F. Closed-loop hydrometallurgical treatment of end-of-life lithium ion batteries: Towards zero-waste process and metal recycling in advanced batteries. *J. Energy Chem.* **2019**, *35*, 220–227. [CrossRef]
16. Schiavi, P.G.; dos Santos Martins Padoan, F.C.; Altimari, P.; Pagnanelli, F. Cryo-Mechanical Treatment and Hydrometallurgical Process for Recycling Li-MnO_2 Primary Batteries with the Direct Production of $LiMnPO_4$ Nanoparticles. *Energies* **2020**, *13*, 4004. [CrossRef]
17. Peters, M.; Timmerhaus, K. *Plant Design and Economics for Chemical Engineers*, 4th ed.; McGraw-Hill, Inc.: New York, NY, USA, 1991.
18. Mohr, M.; Peters, J.F.; Baumann, M.; Weil, M. Toward a cell-chemistry specific life cycle assessment of lithium-ion battery recycling processes. *J. Ind. Ecol.* **2020**, *24*, 1310–1322. [CrossRef]
19. Xiong, S.; Ji, J.; Ma, X. Environmental and economic evaluation of remanufacturing lithium-ion batteries from electric vehicles. *Waste Manag.* **2020**, *102*, 579–586. [CrossRef] [PubMed]
20. Furlani, G.; Moscardini, E.; Pagnanelli, F.; Ferella, F.; Vegliò, F.; Toro, L. Recovery of manganese from zinc alkaline batteries by reductive acid leaching using carbohydrates as reductant. *Hydrometallurgy* **2009**, *99*, 115–118. [CrossRef]
21. Available online: https://www.scrapmetalbuyers.com (accessed on 12 January 2023).
22. Available online: https://www.metal.com (accessed on 12 January 2023).
23. Schiavi, P.G.; Branchi, M.; Casalese, E.; Altimari, P.; Navarra, M.A.; Pagnanelli, F. Resynthesis of NMC111 Cathodic Material from Real Waste Lithium Ion Batteries. *Chem. Eng. Trans.* **2021**, *86*, 463–468.
24. Martins, L.S.; Guimarães, L.F.; Botelho Junior, A.B.; Tenório, J.A.S.; Espinosa, D.C.R. Electric car battery: An overview on global demand, recycling and future approaches towards sustainability. *J. Environ. Manag.* **2021**, *295*, 113091. [CrossRef] [PubMed]
25. Available online: https://www.lifedrone.eu/ (accessed on 12 January 2023).

metals

MDPI

Article

Study of Metal Leaching from Printed Circuit Boards by Improved Electrochemical Hydrochlorination Technique Using Alternating Current

Vera Serga [1], Aleksej Zarkov [2,3], Andrei Shishkin [4,5,*], Edgars Elsts [2], Maksims Melnichuks [1], Mikhail Maiorov [6], Ervins Blumbergs [4,6,7] and Vladimir Pankratov [2]

1 Institute of Materials and Surface Engineering, Faculty of Materials Science and Applied Chemistry, Riga Technical University, 3/7 P. Valdena Street, LV-1048 Riga, Latvia
2 Institute of Solid State Physics, University of Latvia, 8 Kengaraga Street., LV-1063 Riga, Latvia
3 Institute of Chemistry, Vilnius University, 24 Naugarduko Street, LT-03225 Vilnius, Lithuania
4 ZTF Aerkom SIA, 32 Miera Steet., Salaspils nov., LV-2169 Salaspils, Latvia
5 Rudolfs Cimdins Riga Biomaterials Innovations and Development Centre of RTU, Institute of General Chemical Engineering, Faculty of Materials Science and Applied Chemistry, Riga Technical University, 3 Pulka Street, LV-1007 Riga, Latvia
6 Institute of Physics, University of Latvia, 32 Miera Street, LV-2169 Salaspils, Latvia
7 Faculty of Civil Engineering, Riga Technical University, 21/1 Azenes Street, LV-1048 Riga, Latvia
* Correspondence: andrejs.siskins@rtu.lv

Abstract: This paper presents the results of the leaching of metals from computer PCBs by electrochemical hydrochlorination using alternating current (AC) with an industrial frequency (50 Hz). Leaching was carried out with a disintegrator-crushed computer motherboard with a particle size (d) of <90 μm. In the course of the research, the leaching efficiency of metals including Fe, Sn, Mn, Al, Cu, Zn, Pb, Ni, Ti, Sb, Cr, Co and V was evaluated depending on process parameters, such as AC density, experiment duration, hydrochloric acid concentration in the electrolyte solution, solid/liquid ratio, electrolyte temperature, and the loading option of raw material (loading option 1 involving loading into the electrolyte solution, and loading option 2 involving loading into the filter containers attached to electrodes). The research results showed that AC superimposition significantly intensifies the leaching of metals. It was established that the complete leaching of metals including Al, Mn, Sn, Ti and Zn, under experimental conditions (loading option 2, $C_{HCl} = 6$ mol·L^{-1}, i = 0.80 A·cm^{-2}, S/L = 8.6 g·L^{-1}), is reached after 1.5 h, and that of Cu and Ni is reached after 2 h from the beginning of the experiment. At the same time, the degree of leaching of other metals after 2 h is Co-78.8%, Cr-84.4%, Sb-91.7%, Fe-98.9%, V-98.1% and Pb-5.1%. The paper also reports the results on the leaching of all abovementioned metals, as well as Ag and Pd, with disintegrator-crushed mixed computer PCBs with d < 90 μm and loading option 1.

Keywords: printed circuit boards; electrochemical hydrochlorination; leaching; alternating current; base metals; silver; palladium

Citation: Serga, V.; Zarkov, A.; Shishkin, A.; Elsts, E.; Melnichuks, M.; Maiorov, M.; Blumbergs, E.; Pankratov, V. Study of Metal Leaching from Printed Circuit Boards by Improved Electrochemical Hydrochlorination Technique Using Alternating Current. *Metals* **2023**, *13*, 662. https://doi.org/10.3390/met13040662

Academic Editor: Stefano Ubaldini

Received: 5 March 2023
Revised: 21 March 2023
Accepted: 23 March 2023
Published: 27 March 2023

1. Introduction

Currently, active scientific research on the recycling of waste PCBs is ongoing within the framework of hydrometallurgical technologies. Moreover, a significant part of these studies is related to extracting the most valuable components of PCBs—noble metals (mainly gold) and copper [1–6].

Leaching with cyanide solutions is the main method of the hydrometallurgical recycling of noble metal ores [7]. However, the exceptional toxicity of cyanides has led to the long-term search for alternative leaching agents that can be used to recover gold from both different ores [8,9] and secondary raw materials, i.e., e-waste [10–12]. Thiosulfate, thiourea and various halides, including chlorine/chloride leach systems, are considered the most promising replacement for cyanide [13].

Halide leaching is used to leach platinum metals from such secondary raw materials as spent automotive catalysts. Chloride leaching of palladium from spent catalysts has been studied using various oxidants (Fe^{3+} or Cu^{2+} ions) [14,15]. The research results showed [14] that in the absence of oxidant at 60 °C and with an experiment duration of 120 min, the degree of Pd leaching (R_{Pd}) increases from 18.0% to 41.2% with an increase in HCl concentration from 2.0 mol·L^{-1} to 6.0 mol·L^{-1}. The results also showed that, using a chloride solution with the composition 2.0 mol·L^{-1} HCl, 4.0 mol·L^{-1} NaCl and 0.67 mol·L^{-1} Fe^{3+} at 80 °C for 90 min, R_{Pd} reached 99.5%. Pd leaching using Fe^{3+} as an oxidant is represented as a reaction:

$$2Fe^{3+} + Pd + 4Cl^{-} \rightarrow [PdCl_4]^{2-} + 2Fe^{2+} \quad (1)$$

A significant increase in the rate of dissolution of palladium from spent catalysts in a 2 mol·L^{-1} HCl solution was also observed in the presence of Cu^{2+} ions [15].

A process for cuprous chloride synthesis and the simultaneous leaching of Ag and Pd from PCBs was developed [16]. In the chloride system, cupric ions also played the role of an oxidant. Metal leaching efficiency was 93.9%, 95.3% and 98.5% for Ag, Pd and Cu, respectively, in the two stages of the process. The dissolution of Ag and Pd is presented by the authors as the following reaction:

$$\frac{1}{2}\,Pd\,(Ag) + Cu^{2+} + 4Cl^{-} = [CuCl_2]^{-} + \frac{1}{2}\,[PdCl_4]^{2-}\,([AgCl_2]^{-}) \quad (2)$$

A combination of sodium chloride and calcium hypochlorite as the lixiviant was proposed for silver leaching from a magnetic fraction of PCBs [17]. Under experimental conditions, the degree of silver leaching was about 95%. The dissolution of silver is represented by the equation:

$$HClO + H^{+} + 7Cl^{-} + 2Ag = 2[AgCl_4]^{3-} + H_2O \quad (3)$$

Authors [18] have also studied the dissolution of metallic copper with molecular chlorine, generated in situ as a result of the chemical reaction between sodium hypochlorite and hydrochloric acid:

$$NaClO + 2HCl \rightarrow NaCl + Cl_2 + H_2O \quad (4)$$

The presented results showed that the rate of metal dissolution is mainly determined by the concentration of chlorine.

Chlorine, bromine and iodine are well-known lixiviants for leaching gold [19,20]. The solubility of chlorine increased in relatively concentrated HCl solutions due to the formation of tri-chloride ions as follows [21]:

$$Cl_2 + Cl^{-} = Cl_3{}^{-} \quad (5)$$

Moreover, in a solution of hydrochloric acid at pH < 2, the dominant species of chlorine are $Cl_{2(aq)}$ and $Cl_3{}^{-}$, but at pH > 2, HClO is the dominant species. The distribution of chlorine species in a solution depends on HCl concentration and temperature [22].

To avoid the release of chlorine into the atmosphere during the hydrochlorination of both ore and secondary raw materials, it is preferable to perform a process with low consumption of chlorine to increase the environmental friendliness of the process. Therefore, in connection with the possibility of the complex recycling of raw materials with a low content of valuable components, new directions in hydrochlorination have been outlined in recent years. A separate leaching reactor connected to the anode chamber of a chlorine gas generator was used when studying the leaching process of both copper plates [23] and the non-magnetic fraction of PCBs [24]. The degree of metal leaching from PCBs into a 2 $mol \cdot L^{-1}$ HCl solution at the temperature of 50 °C in 4 h was 98, 96, 96 and 71% for Zn, Sn, Pb and Cu, respectively. Gold leaching from mobile phone PCBs was also investigated in a separate leaching reactor [25]. The results showed that copper and gold could be separated in two leaching stages by adjusting the solution's acidity and chlorine concentration. In the 1st stage of leaching in a 2 $mol \cdot L^{-1}$ HCl solution, the degree of copper leaching was 97%, while the gold leaching was below 5%. High leaching of gold (93%) was obtained along with residual copper (770 $mg \cdot L^{-1}$) in the second stage of leaching with a diluted acid solution (0.1 $mol \cdot L^{-1}$ HCl). The metal leaching process from scrap PCBs with chlorine in the hydrochloric acid medium in a leaching reactor coupled to an electrochemical reactor was also studied in [26]. As a result, a multi-component solution containing nearly all of the available metals in the scrap was obtained. The reaction of metals dissolution in this process is described in general form:

$$Me_{scrap} + (n - z)\, Cl^{-} + z/2Cl_2 \rightarrow [MeCl_n]^{(n - z)-} \quad (6)$$

Here, the composition of chloride complexes of metals varies depending on the concentration of chloride ions in the solution [27]. In a recent study [28], electro-generated chlorine produced in a separate electrochemical cell was also used as an oxidising agent to leach Cu, Sn, Pb, Zn and Al from PCBs into an HCl solution (3 $mol \cdot L^{-1}$). To increase the efficiency of chlorine utilisation, it is proposed to use a series of sequentially connected leaching cells. As a result, in 100 min of the experiment, the maximum degree of Cu leaching, 87.22%, was achieved, while the degree of Zn, Sn, Al and Pb leaching was more than 90%.

The hydrochlorination method was used for gold leaching from the solder-removed PCBs fraction with a particle size 4–5 mm pretreated in 4.0 $mol \cdot L^{-1}$ of an HCl solution with the addition of 5 vol% H_2O_2 [29]. Chlorine, as in the studies mentioned above [23–26,28], was produced by the electrolysis of an HCl solution in a separate electrochemical cell. The research results showed that in 2.0 $mol \cdot L^{-1}$ of NaCl solution (pH = 1.0) at a chlorine supply rate of 0.62 $mmol \cdot min^{-1}$, a solution temperature of 50 °C and an experiment duration of 75 min, more than 99% of Au was leached into the solution. Under these experimental conditions, the leaching degree of copper and nickel was 71.6% and 59.6%, respectively.

The authors of [30] proposed a scheme for leaching critical and valuable metals from scrap mobile electronics. This scheme consisted of two stages: the first stage was the dissolution of Cu, Sn, Ag and magnetic materials, using Fe^{3+} formed in an acid sulfate solution; the second stage was the dissolution of Pd and Au, using Cl_2 electro-generated in an HCl solution in a separate electrochemical cell.

When a direct electric current (DC) is passed through an electrochemical system with hydrochloric acid electrolyte, metal ionisation occurs at the anode: the anode material is electrically dissolved, chlorine and oxygen are released, and hydrogen is released at the cathode [31]. Feasible reactions of the metal oxidation process on alternating current (AC) differ from DC processes in that the electrodes change their polarity with a given frequency (50 Hz). Both oxidation and reduction processes take place on the same electrode. The AC of various shapes contributes to the intensification of the processes of anodic dissolution/dispersion of metals and the elimination of the passivation of the metal electrode. In addition, the use of AC makes it possible to eliminate the diffusion limitations that arise during electrolysis [32,33].

A study of the electrochemical behavior of Pt in hydrochloric acid solutions under AC polarisation conditions, using titanium as an auxiliary electrode, showed [34] that a significant increase in the current efficiency of platinum dissolution is observed in HCl concentrations ranging from 5 $mol \cdot L^{-1}$ to 7 $mol \cdot L^{-1}$. When using two iron electrodes in the anode half-cycle of the AC, active dissolution of the electrode material is observed with the formation of Fe^{2+} ions in a solution of hydrochloric acid electrolyte; in the cathode half-cycle, the hydrogen ion reduction reaction occurs on the electrode. The process is characterised by a high iron dissolution rate of 88 $g \cdot m^{-2} \cdot h^{-1}$ at C_{HCl} = 80 $g \cdot L^{-1}$ [35]. The stability of pairs of graphite–graphite electrodes during AC polarisation makes it possible to carry out processes at high currents and potentials on electrodes that cannot be used with DC [32].

In a previous work [36], the possibility of producing an oxidising agent—chlorine and leaching metals (with the examples of Au and Cu) from PCBs powders in one electrochemical cell by passing an AC of industrial frequency (50 Hz)—was demonstrated for the first time. The main advantages of the proposed scheme for leaching metals from PCBs are simplicity (the absence of any thermal or chemical pretreatment) and minimal environmental risk: chlorine is produced and consumed directly in the process (in situ), meaning there is no need for its production, transportation and storage. In this scheme, disintegrator-crushed PCBs powders with a non-conductive matrix content ~50 wt% were loaded into electrolyte and a suspension was created using a stirrer (loading option 1). With this loading option of raw materials, the oxidation of metals occurred in the volume of the hydrochloric acid electrolyte by, predominantly, forms of dissolved chlorine such as $Cl_{2(aq)}$ and Cl_3^-.

In this paper, to implement the process of electrochemical hydrochlorination, another loading option of raw material powders is proposed: a metal-rich fraction of PCBs (the content of non-conductive matrix ~5 wt%) is loaded into filter containers attached to the electrodes (loading option 2). With this loading option, the metal particles present in the PCB powder are in contact with the graphite electrodes. Therefore, in the anodic half-cycles of AC, both the electrooxidation (electrodissolution) and the electrochlorination of metals during the oxidation of chloride ions on their surface will occur. In this case, the efficiency of metal oxidation will not depend on the concentration of free chlorine in the volume of electrolyte, which is not the case with loading option 1. This technique (loading option 2) could increase the completeness of the leaching of both base and noble metals from PCB powders.

The results of studies on the influence of process parameters, such as the AC density, experiment duration, concentration of hydrochloric acid in an electrolyte solution, and solid/liquid ratio, on the leaching efficiency of metals, including Al, Co, Cr, Cu, Mn, Ni, Sb, Sn, Ti, Zn, Fe, Pb and V, are presented. For comparison, for these metals, and additionally for Pd and Ag, the results of leaching under the experimental conditions described in our previous paper [36] are also presented.

2. Materials and Methods

Computer motherboards and, for comparison, mixed computer PCBs were used as the source materials. Preliminary preparation of raw materials included dismantling the PCBs, double crushing in a hammer mill (for computer motherboards) and single crushing (for mixed computer PCBs) using the high-energy semi-industrial disintegration-milling system (disintegrator DSL-350, Tallinn University of Technology, Tallinn, Estonia) and subsequent sieving using FRITSCH ANALYSETTE 3 PRO Vibratory Sieve Shaker (FRITSCH GmbH, Weimar, Germany). A detailed description of the new method of disintegration-crushing of PCBs is given in [37]. For research, powders of the finest fraction of raw material powders of disintegrator-crushed computer motherboards (raw material 1) and mixed computer PCBs (raw material 2) with a particle size (d) of <90 μm were used.

Experimental setup and experimental technique for electrochemical hydrochlorination of PCBs powders using AC of industrial frequency (50 Hz) are presented in previous work [36]. Within the framework of the proposed experimental technique, the raw material powder under study was loaded into the electrolyte solution (loading option 1) (Figure 1a). Another loading option was also used in this study: the raw material powder was loaded into filter containers attached to graphite electrodes (loading option 2) (Figure 1b). The containers were made of polyethene terephthalate (PET) cloth (average mesh size 135 × 65 μm). This material has good chemical resistance and sufficient mechanical strength. Equal weights of the raw material were loaded into the containers. The experiments used two types of electrochemical cells: a square-shaped diaphragmless polypropylene cell (cell 1) and a water-jacketed glass cell (cell 2). An AC with a density (i) of 0.40, 0.72 and 0.80 A·cm^{-2} (raw material 1 powder, loading option 2) and 0.88 A·cm^{-2} (raw material 2 powder, loading option 1) on the visible surface of graphite electrodes were used. The duration of the experiments was from 0.5 to 5 h.

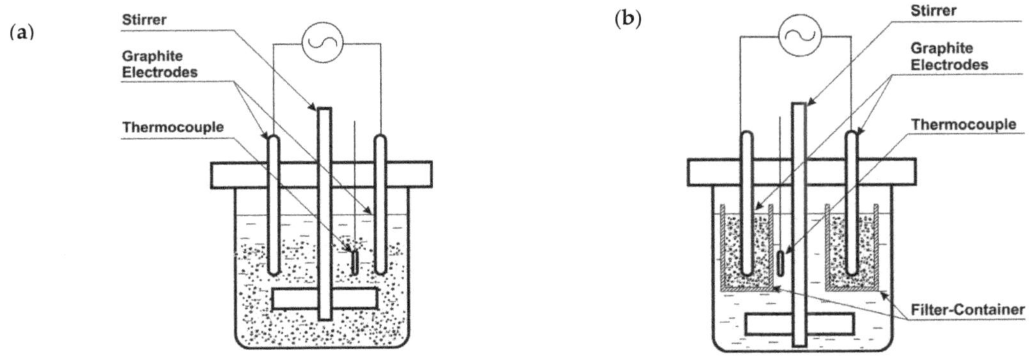

Figure 1. Schematic representation of an electrochemical cell with different loading options of the raw material: (**a**) loading option 1; (**b**) loading option 2.

During the process, the electrolyte temperature (T_{el}) was controlled using a thermocouple (chromel–alumel). Solutions of hydrochloric acid in distilled water with concentrations of 4, 6 and 8 mol·L^{-1} were used as electrolytes. The ratio of solid (PCB powder) and liquid (electrolyte) phases (S/L) in the experiments was 5.7, 8.6, 11.4 and 14 g·L^{-1}.

The chemical treatment of a representative sample was carried out to establish the metal content in raw material powders. In preparing a representative sample, the rolling method was used to mix the powders, and the quartering method was used to reduce the sample. Chemical leaching of the sample included sequentially boiling in a solution of HCl (1:1) and two portions of aqua regia until a wet residue was formed. During the boiling process in aqua regia, HNO_3 excess was removed by adding several portions of concentrated HCl. The resulting wet residue was transferred to the filter with a 3 mol·L^{-1} HCl solution and washed accurately. Filtrate was diluted to a volume of 50 mL with 3 mol·L^{-1} HCl solution.

The quantitative determination of metals in solutions resulting from the chemical and electrochemical leaching of raw materials was carried out by inductively coupled plasma optical emission spectrometry (ICP-OES, Optima 7000 DV ICP-OES, Perkin Elmer Inc., Waltham, MA, USA).

Solid residues of PCBs after both chemical and electrochemical leaching were washed on a filter with distilled water until the pH of the filtrate was ~5–6, dried at 105 °C for an hour, weighed and ground in an agate mortar. X-ray diffraction (XRD, diffractometer Rigaku—MiniFlex 600 with CuKα radiation with λ = 0.15418 nm, Rigaku Corp., Tokyo, Japan) and X-ray fluorescence (XRF, analyser EAGLE III XPL, EDAX Inc., Mahwah, NJ, USA) analyses were used to characterise solid materials (raw material, solid residue after leaching and precipitate from electrolyte). A microscope (KEYENCE VHX-2000, Keyence Corporation, Osaka, Japan) was used for optical imaging of the disintegrator-crushed powder of raw material 1.

The efficiency of electrochemical leaching of metals from raw materials was evaluated by the degree of metal leaching (R), which was defined as the ratio of the quantity of metal transferred into the electrolyte solution (M_{el}) to the quantity of metal contained in the raw material (M_{PCBs}), as demonstrated in Equation (7):

$$R\ (\%) = (M_{el}/M_{PCBs}) \times 100\% \quad (7)$$

The content of the metals in the raw material (M_{PCBs}) was calculated from the results of the ICP-OES analysis of the leachates obtained from the chemical leaching of a representative sample of the raw materials.

3. Results and Discussion

3.1. Raw Material Characterisation

A fine fraction of raw material 1 with d < 90 μm is the main object of this research. Therefore, the powder obtained after crushing and subsequent sieving is relatively homogeneous and contains only a small amount of plastic and fibreglass fragments (Figure 2b).

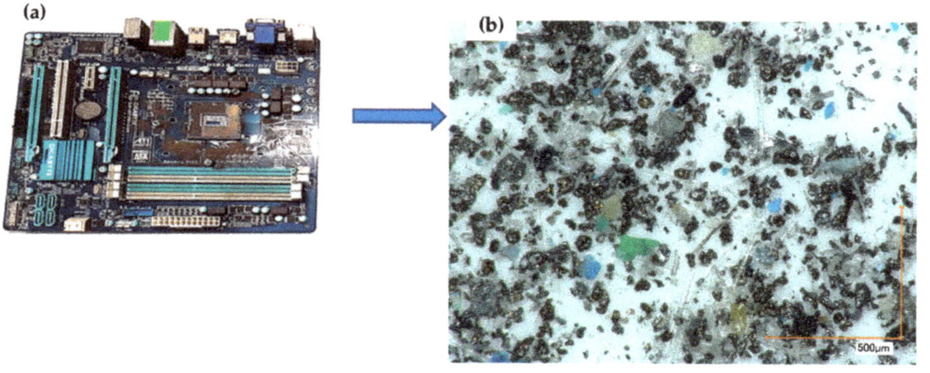

Figure 2. Image of computer motherboard (**a**) and optical image at magnification ×200 of disintegrator-crushed fraction of raw material 1 with d < 90 μm (**b**).

Table 1 shows the results of the ICP-OES analysis of leachates, obtained from the chemical treatment of raw material 1 and raw material 2, which characterised in detail in our previous study [36]. Based on the presented data, it can be concluded that iron is the most significant component of the raw material 1 sample, while the content of each of the other metals does not exceed one per cent. The presence of precious metals (Au, Pd and Ag) in the leachate is not detected. Element content in the raw material 2 sample is significantly different and contains a broader range of elements. Predominantly present are Fe, Pb and Sn. Moreover, the content of iron is ~10 times smaller than in the raw material 1 sample. In addition, the presence of noble metals (Au, Pd, and Ag) and, in trace amounts, Mo, is established.

Table 1 also shows the content of solid residue after the chemical treatment of raw materials and the overall scope of elements (other) that were not determined and/or could not be determined in solutions with the ICP-OES method.

Based on the XRD analysis results of the solid residue obtained from the chemical treatment of raw material 1, such crystalline phases as SiO_2 (PDF 00-046-1045), Si (PDF 00-027-1402) and Al_2O_3 (PDF 00-046-1212) (Figure 3, curve 2) were identified in the sample. The presence of the crystalline phase of α-Fe (PDF 00-006-0696), the dominant crystalline phase in raw material 1 (Figure 3, curve 1), was not established. It should be noted that the phase composition of the solid residue obtained under similar chemical leaching conditions of raw material 2 is identical [36]. However, iron is present in different forms in the raw materials: the crystalline phase of α-Fe, in raw material 1; and the crystalline phase of magnetite Fe_3O_4, in raw material 2.

Table 1. Content of metals in raw materials according to the results of ICP-OES analysis of the solutions prepared by chemical leaching.

Component, Content	Raw Material 1	Raw Material 2 [36]
Fe, wt.%	81.4	7.56
Sn, wt.%	0.887	4.76
Mn, wt.%	0.665	0.170
Al, wt.%	0.488	1.31
Cu, wt.%	0.386	1.40
Zn, wt.%	0.220	2.00
Pb, wt.%	0.208	7.50
Ni, wt.%	0.206	1.03
Ti, wt.%	0.208	1.74
Sb, wt. ppm	129	1365
Cr, wt. ppm	102	188
Co, wt. ppm	19	224
V, wt. ppm	18	29
Mo, wt. ppm	BDL *	17
Ag, wt. ppm	BDL *	8450
Pd, wt. ppm	BDL *	1550
Au, wt. ppm	BDL *	824
Solid residue, wt.%	4.84	45.8
Other, wt.%	10.4	26.5

* BDL—below the detectable limit.

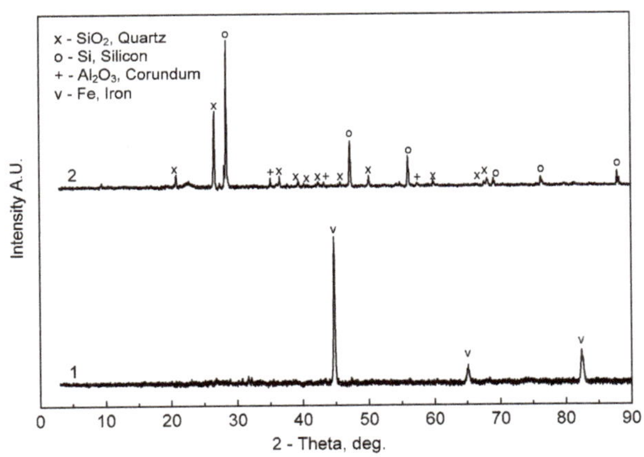

Figure 3. XRD patterns: 1: raw material 1; 2: solid residue after chemical treatment of a raw material 1 sample.

3.2. Influence of Process Parameters on the Efficiency of Metal Leaching from Raw Materials

When raw materials are treated with hydrochloric acid solutions, active metals are dissolved, even in the absence of an oxidising agent. The low dimensionality (d < 90 μm) of the raw materials chosen for the research contributes to an increased dissolution rate. Chlorine, being a strong oxidising agent, interacts with metals and almost all metal oxides to form chlorides and hydrocarbons [38]; therefore, the transition of base metals into solutions within the framework of the proposed method is reasonably expected. The leaching research of all identified metals in raw material 1 and raw material 2 (Table 1) are shown below, both with and without an oxidising agent.

3.2.1. Raw Material 1

Impact of Loading Option on the Degree of Chemical Leaching of Metals

Figure 4 demonstrates the results of metal leaching from raw material 1 in the absence of an electric current with two different loading options of the raw material. With loading option 1, the effective leaching of Mn (R_{Mn} = 100%) and Sn (R_{Sn} = 98.9%) is observed. Moreover, the change from loading option 1 to loading option 2 has a less significant effect on the leaching efficiency of Mn in comparison with Sn. For elements such as Co, Cr, Cu, Ni, Sb, Ti, Zn and Fe, the change in the loading option leads to a significant decrease in the leaching efficiency. The loading option practically does not affect the degree of V leaching into the hydrochloric acid solution. Similarly, the effect is also not very significant for aluminium. Regardless of the loading option used, the transition of lead into a hydrochloric acid solution has not been established. The decrease in the efficiency of metal leaching due to the change from loading option 1 to loading option 2 is associated with diffusion restrictions arising from the presence of a separating diaphragm (container material) between the pulp and the electrolyte volume.

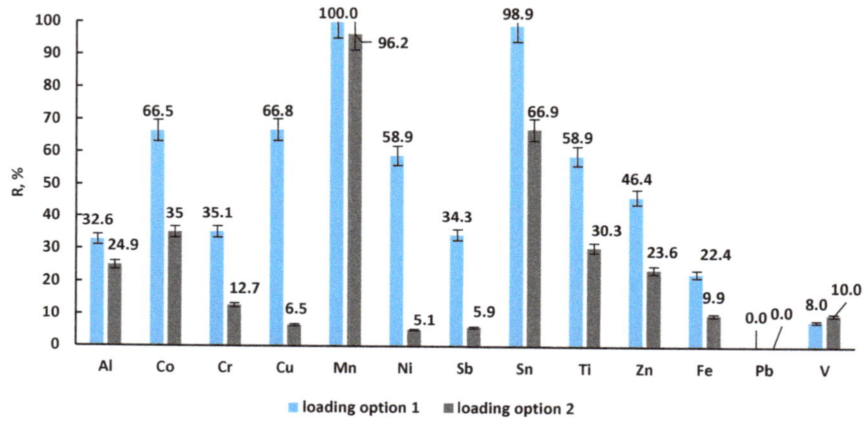

Figure 4. Leaching degree of metals depending on the loading option of the raw material. Experimental conditions: raw material 1; electrochemical cell 1; C_{HCl} = 6 mol·L^{-1}; i = 0; S/L = 8.6 g·L^{-1}; t_{ex} = 2 h; ambient temperature.

Impact of Solid/Liquid Ratio on Metal Electrochemical Leaching Degree

Because of the low content of the non-conductive matrix and the high content of the highly conductive material—metallic iron (Table 1, Figure 3, curve 1)—loading option 2 was used to load this material into the electrochemical cell. In this case, when an AC passes through an electrochemical system with hydrochloric acid electrolyte, during the anode half-cycles, basic reactions such as the ionization of metals that have electrical contact with graphite electrodes, the release of chlorine (both on the surface of graphite electrodes and metal particles of the raw material) and the oxygen evolution reaction will occur. It is known [18] that molecular chlorine at the moment of release upon direct contact with the chlorinated substance is more active than gaseous chlorine supplied from outside. With loading option 2 in the anodic half-cycles of AC, the electrooxidation of Cl^- ions with the formation of atomic chlorine occurs directly on the surface of the electrically conductive particles of raw material 1, and this provides a high metal dissolution rate in the electrochemical process [39]. It should be noted that the true current density, with this loading option, will be less than the calculated one.

To test this loading option, Au powder (99.99%, Premion, Germany) was used as the raw material, and cell 1 was used as the electrolyser. During the tests, 0.250 g of gold powder was placed in each filter container and attached to graphite electrodes, and the cell was filled with 6 mol·L^{-1} of HCl. The process of electrochemical leaching was carried out at i = 0.88 A·cm^{-2} for 2 h. According to the results of the ICP-OES analysis, after the completion of the experiment, the gold concentration in the electrolyte solution reached 212.1 mg·L^{-1}, corresponding to a metal recovery of 14.9%. The process was accompanied by electrolyte heating from 20 °C to 67 °C.

To select the optimal S/L ratio for research on the electrochemical leaching of metals from raw material 1, a series of experiments were carried out with different S/L ratios. The results showed (Figure 5a) that with an increase in the S/L ratio, a gradual decrease in the leaching efficiency of Cr, Co and V is observed. In the case of Sb, an increase in the S/L ratio from 8.6 g·L^{-1} to 11.4 g·L^{-1} leads to a more than twofold decrease in the degree of leaching. Lead leaching occurs with very low efficiency, but is characterised by a gradual increase in the degree of leaching in the studied range of the S/L ratio. It should be noted that metals such as Al, Mn and Ti are leached completely regardless of the S/L ratio. The leaching of Cu, Fe, Ni and Zn deteriorates at an S/L ratio > 8.6 g·L^{-1}, and for Sn, there is only a slight decrease in the degree of leaching, with an S/L ratio = 14.3 g·L^{-1} (Figure 5b).

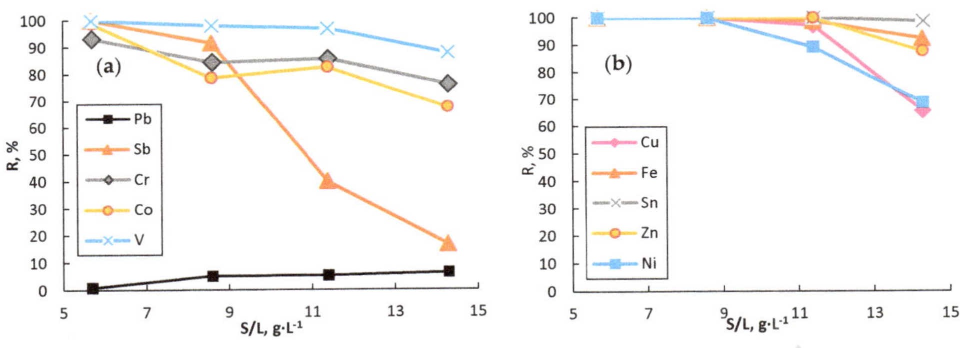

Figure 5. Impact of S/L ratio on the degree of metal leaching. Experimental conditions: electrochemical cell 1; loading option 2; raw material 1; C_{HCl} = 6 mol L^{-1}; I = 0.80 A·cm^{-1};, t_{ex} = 2 h. The electrolyte temperature during the processes increased from 20 °C to 79 °C.

Considering the obtained results and the application of loading option 2 of the raw material, the S/L ratio of 8.6 g·L^{-1} was chosen for the study of the electrochemical leaching processes of the metals.

Impact of Current Density on Metal Leaching Degree

A comparison of the results of the electrochemical (Figure 6) and chemical leaching (Figure 4), obtained under the same experimental conditions, shows that the superimposition of an AC (i = 0.40 A·cm^{-2}) on the studied system significantly intensifies the transition of almost all metals into the hydrochloric acid solution. A twofold increase in AC density has the same effect, except for Sb and Pb. Compared to chemical leaching using raw material loading option 1, the impact of AC is more noticeable at i = 0.80 A·cm^{-2}.

As a result of the current passage, the electrolyte's heating is observed. The final temperature of the electrolyte solution (reaction mixture) depends both on the duration of the experiment at a fixed current density and on the current density on the graphite electrodes (Figure 7).

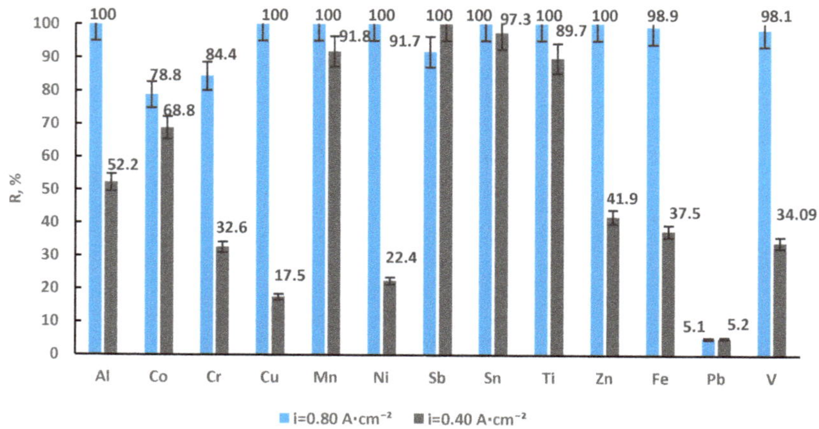

Figure 6. Impact of current density on the degree of metal leaching. Experimental conditions: raw material 1; electrochemical cell 1; loading option 2; C_{HCl} = 6 mol·L^{-1}; S/L = 8.6 g·L^{-1}; t_{ex} = 2 h.

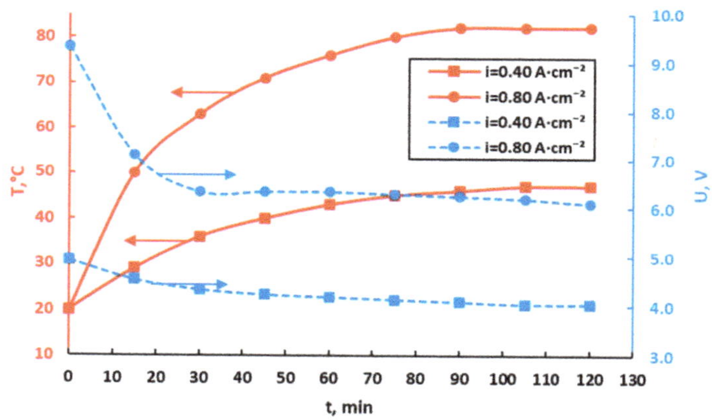

Figure 7. Electrolyte temperature and cell voltage versus the experiment duration. Experimental conditions: raw material 1; electrochemical cell 1; loading option 2; C_{HCl} = 6 mol ·L^{-1}; S/L = 8.6 g·L^{-1}.

Impact of Experiment Duration on Metal Electrochemical Leaching Degree

According to the results of the ICP-OES analysis of the solution obtained from the chemical treatment of raw material 1 (Table 1), the metals, listed by the amount that their content decreased in the raw material, are arranged as follows: Fe > Sn > Mn > Al > Cu > Zn > Pb > Ti > Sb > Cr > Co > V. Based on this series, metals are divided into three groups and the kinetic curves of their electrochemical leaching are shown in Figure 8. The presented data show (Figure 8a) that the complete leaching of Mn is achieved with an experiment duration of 1 h and that the complete leaching of Sn and Al is achieved at 1.5 h. Copper, the least active element of this group, completely dissolves in the electrolyte solution in 2 h. Ti, Zn and Ni are completely leached in 1, 1.5 and 2 h, respectively (Figure 8b). The behaviour of lead is significantly different: the slow leaching of the metal into the electrolyte solution begins only an hour after the beginning of the experiment when the electrolyte temperature reaches 76 °C (Figures 7 and 8b). At the end of the investigation, the degree of lead leaching is only 5.1%. This behaviour is probably because the lead, in this study, is in the form of a mixed oxide, $PbO{\cdot}TiO_2$,, which is non-conductive. The transition of metals (Co, Cr, Sb and V) present in the raw material in microquantities into the solution is characterised by high values of leaching degree, but it is not possible to completely leach them into the electrolyte solution within 2 h of the experiment (Figure 8c). On the one hand, the incomplete leaching of metals may be related to the chemical form in which the metals are present in the raw material. On the other hand, due to a feature of loading option 2, during the leaching process, the amount of the metal component of the sample decreases and the non-conductive matrix content increases, ultimately leading to the loss of electrical contact between the remainder of the conductive material and the electrode. Thus, after completion of the 2 h leaching, the following content of metals remains in the solid residue of raw material 1: Pb—94.9; Co—21.2; Cr—15.6; Sb—8.3; Fe—1.1; and V—1.9%. At the same time, the following content of metals ($mg \cdot L^{-1}$) is established in the electrolyte solution: Al—41.8; Cu—33.0; Mn—56.9; Ni—17.6; Ti—17.8; Zn—18.8; Co—0.135; Cr—0.752; Fe—6903; Pb—0.919; Sb—1.01; V—0.148; and Sn—76.0. It should be noted that in comparison with chemical leaching (Figure 4), an increase in the efficiency of all metal leaching is observed under the conditions of electrochemical leaching.

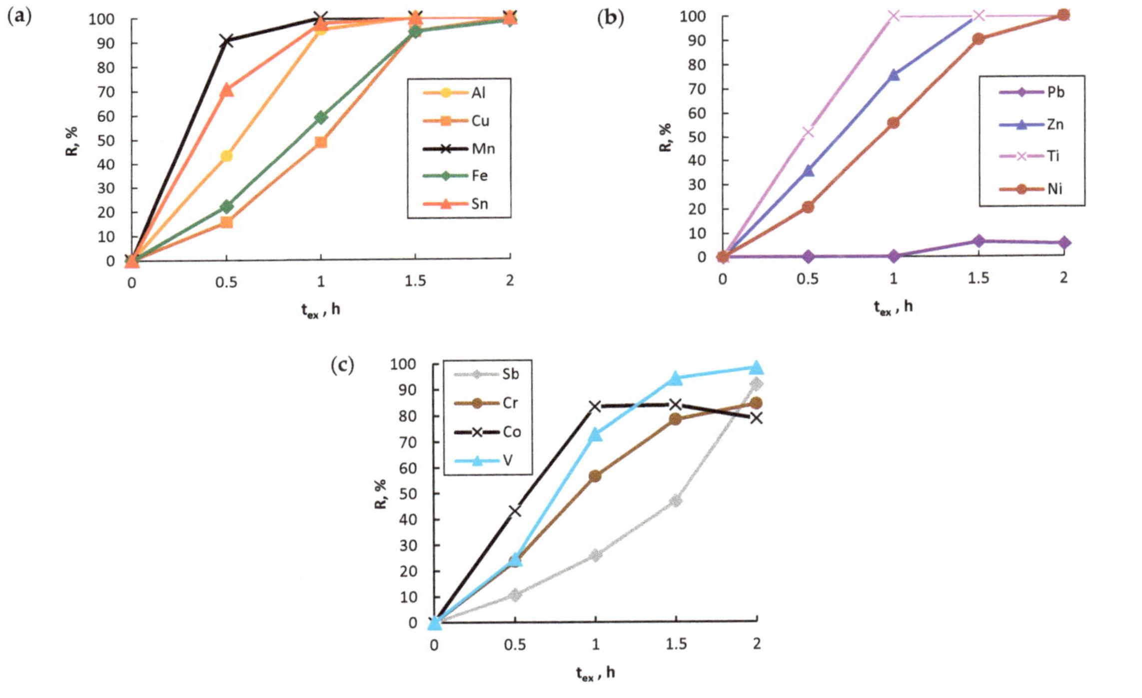

Figure 8. Impact of the experiment duration on the degree of metal leaching. Experimental conditions: raw material 1; electrochemical cell 1; loading option 2; $C_{HCl} = 6$ $mol \cdot L^{-1}$; $i = 0.80$ $A \cdot cm^{-2}$; $S/L = 8.6$ $g \cdot L^{-1}$; $t_{ex} = 2$ h.

The solid residue after completion of 0.5, 1, 1.5 and 2 h experiments is 69.8%, 39.3%, 11.7% and 5.3%, respectively. Figure 9 presents the results of the X-ray phase analysis of solid residues of PCBs after the experiments on the kinetics of the electrochemical leaching of raw material 1 (Figure 8). With the duration of the experiments increased from 0.5 to 1.5 h, the phase composition of solid residues almost does not change and is a mixture of SiO_2, Fe_3O_4 (PDF 00-019-0629), Si and α-Fe crystalline phases. After the 2 h experiment, the iron was almost completely leached into the electrolyte solution (Figure 8a), and the crystalline phases of Al_2O_3 and $BaSO_4$ were also identified in the solid residue, but the presence of magnetite was not established. The main difference in the phase composition of the sample obtained as a result of chemical treatment (Figure 3, curve 2) and the sample obtained as a result of a two-hour experiment (Figure 9, curve 4) is the presence in the latter of the crystalline phases of $BaSO_4$ (PDF 00-024-1035) and α-Fe. The crystalline phase of $BaSO_4$ was probably not found in the solid residue after chemical leaching due to the small amount of the material under study. The presence of a residual amount of iron in the solid residue indicates the incomplete leaching of metal from raw material 1. This result is also confirmed by the result of the ICP-OES analysis of the electrolyte solution: the degree of iron leaching under the experimental conditions is 98.9% (Figure 8a).

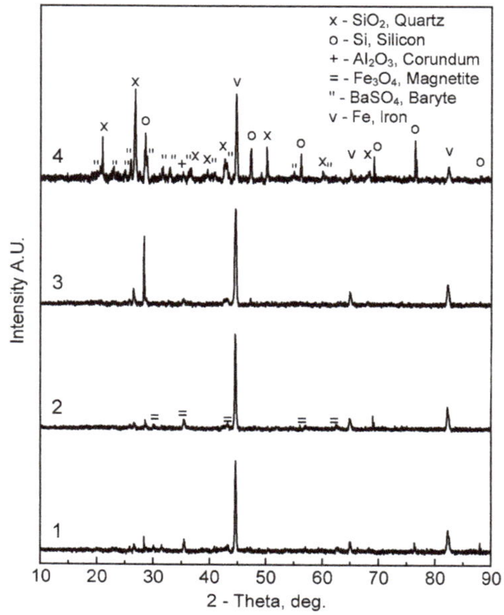

Figure 9. XRD patterns of solid residues from the experiments at different experiment duration: 1—0.5 h; 2—1 h; 3—1.5 h; and 4—2 h. Experimental conditions: electrochemical cell 1; loading option 2; C_{HCl} = 6 mol·L^{-1}; S/L = 8.6 g·L^{-1}; i = 0.80 A·cm^{-2}.

Impact of HCl Concentration in Electrolyte Solution on the Degree of Metal Electrochemical Leaching

When studying the effect of the HCl concentration in the electrolyte solution on the leaching efficiency, it was found (Figure 10) that the degree of Mn, Ti and Sn leaching reaches its maximum value at C_{HCl} = 4 mol·L^{-1}. A further increase in the concentration of hydrochloric acid in the solution to 6 mol·L^{-1} leads to an increase in the leaching efficiency of all other metals, and, for Al and Zn in particular, it leads to maximum leaching efficiency. With a further increase in the HCl concentration to 8 mol·L^{-1}, the degree of Cu, Ni, V, and Cr leaching continues to increase; however, in the case of Co, Sb, Fe and Pb, a slight decrease in the dissolution efficiency is observed. This is probably due to a reduction in the solubility of complex salts of these metals.

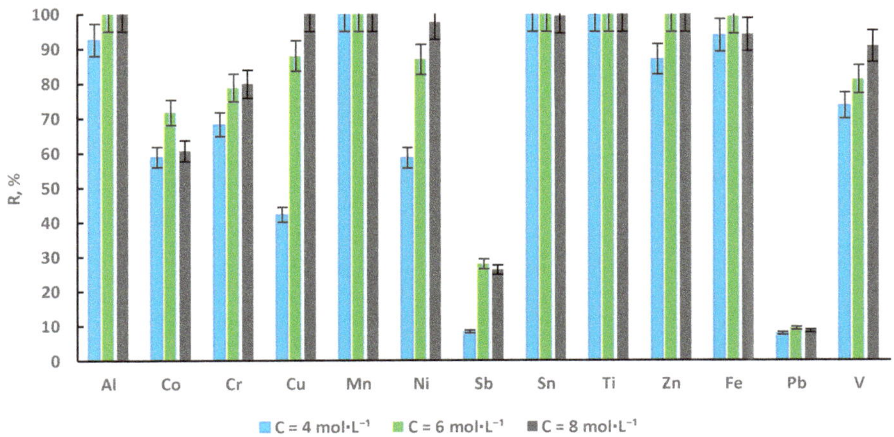

Figure 10. Impact of HCl concentration on the degree of metal leaching. Experimental conditions: electrochemical cell 2; loading option 2; raw material 1; i = 0.72 A·cm^{-2}; t_{ex} = 2 h. The electrolyte temperature during processes increased from 20 °C to 97 °C.

Impact of Electrolyte Temperature on the Degree of Metal Electrochemical Leaching

To study the effect of electrolyte temperature on the efficiency of metal leaching, experiments were carried out under the following three conditions: temperature control of the electrolyte at 70 °C, without forced temperature control (20–97 °C) and with water cooling of the electrolyte (20–44 °C). According to the results obtained (Figure 11), an increase in the electrolyte temperature contributes to an increase in the leaching efficiency of all studied metals.

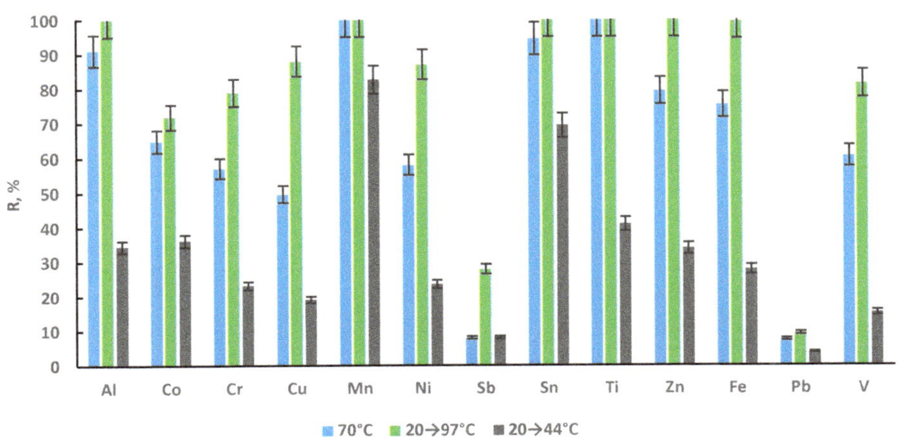

Figure 11. Impact of electrolyte temperature on the degree of metal leaching. Experimental conditions: electrochemical cell 2; loading option 2; C_{HCl} = 6 mol·L^{-1}; raw material fraction 1; i = 0.72 A·cm^{-2}; t_{ex} = 2 h.

3.2.2. Raw Material 2

Impact of Experiment Duration on Metal Chemical Leaching Degree

In the study on metal leaching from raw material 2, only loading option 1 was used (Figure 1a). Increasing the experiment duration from 2 to 5 h in the absence of an electric current led to an increase in the degree of extraction of almost all metals (Figure 12).

Moreover, after a 2 h experiment in the electrolyte solution, the presence of such noble metals as Au [36], Ag and Pd were also established. A further increase in the experiment duration has practically no effect on the degree of gold leaching, but leads to a noticeable increase in the efficiency of Ag leaching from 66.0 to 100% and Pd leaching from 29.7 to 95.6%. At the same time, an increase in the degree of Cu leaching from 78.6 to 85.6% is also observed. Under the conditions of these experiments, cupric ions can play the role of an oxidising agent for silver and palladium (Equation (2)) [15,16].

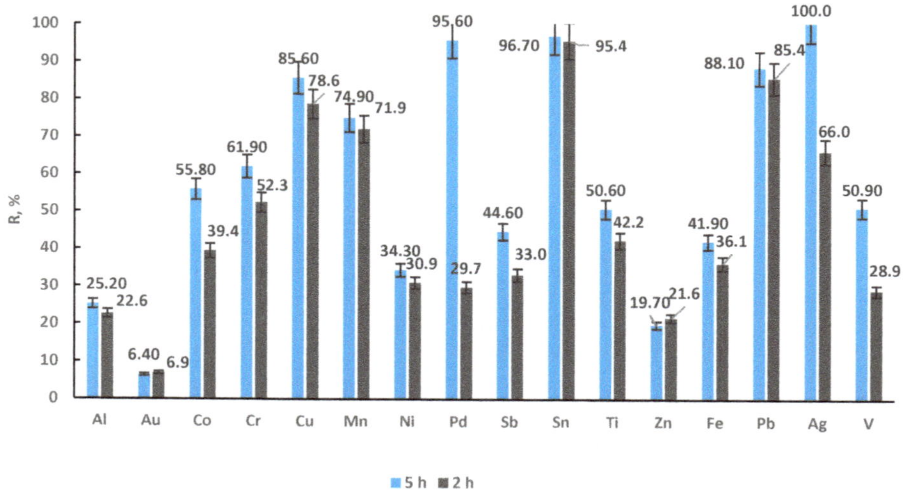

Figure 12. Impact of the experiment duration on the degree of metal leaching. Experimental conditions: raw material 2; electrochemical cell 1; loading option 1; C_{HCl} = 6 mol·L^{-1}; i = 0; S/L = 8.6 g·L^{-1}; ambient temperature.

A comparison of these results with those presented above using raw material 1 (Figure 4) under similar experimental conditions (t_{ex} = 2 h, loading option 1) demonstrates differences in metal leaching efficiency. In the case of raw material 2, the degree of Pb leaching reaches 85.4%, while metal leaching from raw material 1 is not observed at all. The lead content, however, in the latter is less than ~8 times, which is likely because, in raw material 2, lead is a component of a tin–lead alloy (solder), while in raw material 1, it is a part of the $PbO{\cdot}TiO_2$ mixed oxide. Meanwhile, for such metals as Cr, Cu and V, despite their higher content in raw material 2 compared to raw material 1 (Table 1), higher leaching degrees are established. In all other cases, with a lower content of metals in the raw material, higher values of the degree of metal leaching are also observed.

Impact of Experiment Duration on Metal Electrochemical Leaching Degree

The active electrochemical leaching of all metals is observed during the first 30 min of the experiment (Figure 13). A further increase in the experiment duration has a different effect on the efficiency of metal leaching. In the case of V and Co, an active transition of metals into the solution is observed within 1 h from the beginning of the experiment (Figure 13c), and, for Al, within 2 h (Figure 13b). Complete leaching of Ag is achieved in 1 h of the process and of Pd in 2 h (Figure 13d). Sn and Pb thoroughly leach into the solution in 3 h (Figure 13a), and Al and V in 4 h. A relatively low efficiency of leaching characterises Ni, Zn and Fe, which take 5h to leach into the electrolyte solution, with leaching degrees of R_{Ni} = 42.6%, R_{Zn} = 30.8% and R_{Fe} = 50%, respectively. During this time, the degree to which other metals leach reaches sufficiently high values (≥80%); however, their complete leaching is not achieved. As a result of a 2 h experiment, the concentrations of metals in the electrolyte solution (mg·L^{-1}) are Al—107.2; Cu—103.6; Mn—12.4; Ni—33.8; Ti—137.7; Zn—

41.7; Co—1.41; Cr—1.16; Fe—265.3; Pb—652.2; Sb—8.14; V—0.236; Au—5.99; Pd—13.2; and Sn—408.1. Following a 5 h experiment, the concentrations are Al—113.3; Cu—105.0; Mn—12.6; Ni—37.5; Ti—146.4; Zn—52.9; Co—1.54; Cr—1.24; Fe—324.4; Pb—684.1; Sb—9.28; V—0.269; Au—5.86; Pd—13.2; and Sn—408.1. A comparison of these results (Figure 13) and the results obtained under similar experimental conditions, but in the absence of AC (Figure 12), allows us to conclude that the AC superimposition on the system under study intensifies the process of leaching metals from raw material 2.

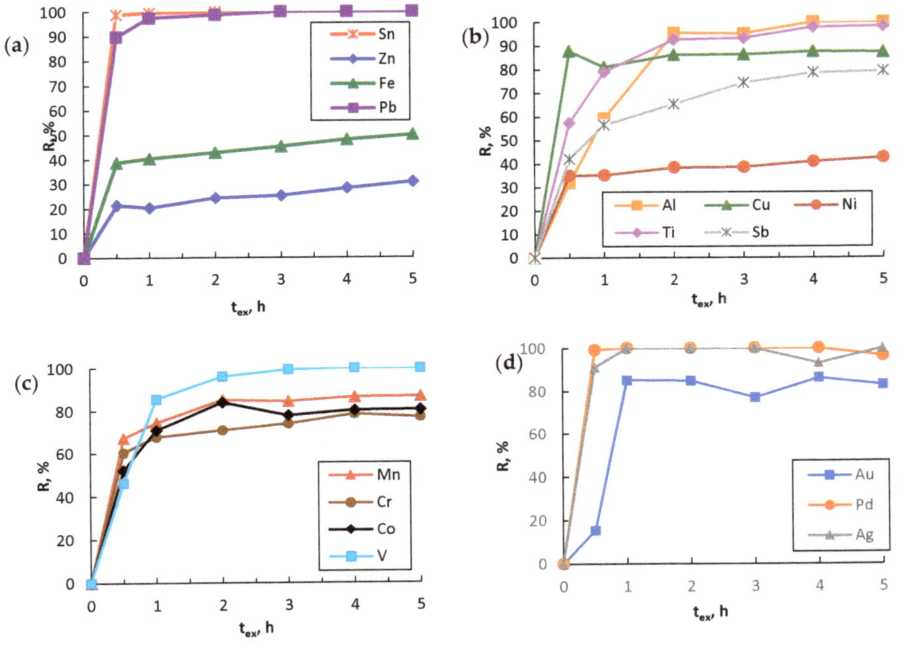

Figure 13. Impact of the experiment duration on the degree of metal leaching. Experimental conditions: raw material 2; electrochemical cell 1; loading option 1; C_{HCl} = 6 mol·L^{-1}; i = 0.88 A·cm^{-2}; S/L = 8.6 g·L^{-1}.

A comparison of the results of 2 h experiments carried out using different loading options and different raw materials (Figures 8 and 13) showed that the degree of all metal leaching from raw material 1 with loading option 2 (Figure 8) is higher than in the case of raw material 2 and loading option 1, except for Pb (Figure 13). The significant difference in the lead leaching efficiency of Pb (R_{Pb} = 98.7% for loading option 1 and R_{Pb} = 5.1% for loading option 2) is most likely due to the different chemical forms of the metal in which it is present in the raw materials.

During the research, after completing all experiments and separating the electrochemically treated powders of PCBs, the transparent electrolyte solutions were obtained. However, when spent electrolytes (after experiments with a duration from 3 to 5 h using raw material 2) were stored in a closed glass container, a flaky lemon-coloured precipitate gradually appeared in them. A month later, the mass of precipitate in electrolyte solutions after completing 4 and 5 h experiments was 0.005 g each. The amount of precipitate in the electrolyte from the 3 h experiment was significantly less and was not weighed. The result of the X-ray phase analysis of the precipitate (Figure 14) showed the presence of a crystalline phase of chloranil.

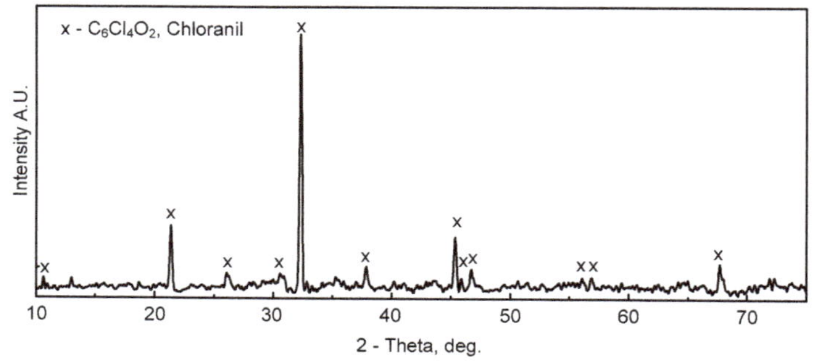

Figure 14. XRD pattern of a precipitate formed in the electrolyte solution during storage (1 month after the experiment). Experimental conditions: raw material 2; electrochemical cell 1; loading option 1; C_{HCl} = 6 mol·L^{-1}; t_{ex} = 4 h; S/L = 8.6 g·L^{-1}; i = 0.88 A·cm^{-2}.

The results of the XRF analysis of the precipitate presented in Figure 15 do not contradict the XRD analysis data. The presence of any metals in the sample was not established. Chloranil can be formed during an electrochemical process due to the interaction of phenol in the PCB's material with chlorine [40]. It should be noted that the precipitate was easily separated from the solution, and its formation in the electrolyte solution was not observed during further storage.

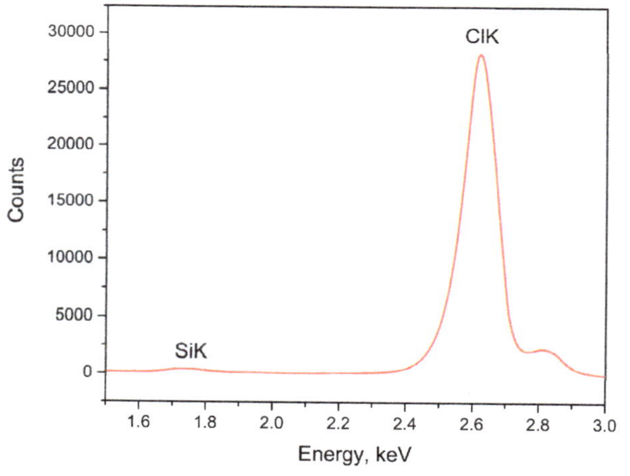

Figure 15. Fluorescence spectrum of the precipitate formed in an electrolyte solution during storage (1 month after the experiment). Experimental conditions are provided in the Figure 14 caption.

Thus, as a result of the experiments carried out, multicomponent solutions containing all the studied metals were obtained. It is shown that, under the same conditions of the electrochemical hydrochlorination process, the loading option of the raw material affects the efficiency of Fe, Sn, Mn, Al, Cu, Zn, Pb, Ni, Ti, Sb, Cr, Co and V leaching. The use of loading option 2 increases the completeness of the extraction of almost all metals compared to loading option 1. However, loading option 2 for raw material 2 is limited by a significant amount of non-conductive matrix. Therefore, to reduce its content in the raw material prior to leaching, pretreatment is required (electrostatic or air separation).

Nevertheless, regardless of the composition of the raw material and its loading option, the application of AC intensifies the metal leaching process. According to the results obtained, in the absence of current for 5 h of the experiment, 100% of Ag and 95.6% of Pd are leached into a solution of 6 mol·L^{-1} of HCl from raw material 2. Under the conditions of AC superimposition (i = 0.88 $A \cdot cm^{-2}$), complete leaching of silver and palladium is achieved in 1 h and 2 h, respectively. The transition of these metals into a chloride multicomponent solution in the absence of chlorine does not contradict the results of studies presented in the research [14–16].

4. Conclusions

The results of the research have shown that the process parameters with the most significant impact on the leaching degree of such metals as Fe, Sn, Mn, Al, Cu, Zn, Pb, Ni, Ti, Sb, Cr, Co and V from powders of the raw material with d < 90 μm, obtained from the twofold crushing of computer motherboards in a disintegrator, during the process of electrochemical hydrochlorination, were the AC density on the graphite electrodes, electrolyte temperature, and hydrochloric acid concentration. An increase in the electrolyte temperature contributes to a rise in the leaching efficiency of all the studied metals. In contrast, an increase in the AC density from 0.40 to 0.80 $A \cdot cm^{-2}$ enhances the degree of all metal leaching, except for Sb and Pb. Under the same experimental conditions ($i = 0.72\ A \cdot cm^{-2}$, $t_{ex} = 2$ h), the use of an electrolyte with the concentration of HCl 4 $mol \cdot L^{-1}$ leads to the complete leaching of Mn, Ti and Sn, while an increase in the concentration of hydrochloric acid to 6 $mol \cdot L^{-1}$ allows complete leaching of Al and Zn and increases the leaching efficiency of other metals. A further increase in hydrochloric acid content in the solution to 8 $mol \cdot L^{-1}$ not only raises the leaching efficiency of Cu, Ni, V and Cr but also leads to some decrease in the leaching efficiency of Co, Sb, Fe and Pb.

A study on the impact of the raw material loading option on the leaching efficiency of all the metals above, with powders obtained from different source materials (computer motherboards and mixed PCBs), showed that loading the samples into filtering containers is preferable.

Based on this research, the results have determined the experimental conditions needed to ensure the complete leaching of Ag and Pd ($t_{ex} = 1$ h and $t_{ex} = 2$ h, respectively) from the powders of the raw material, with d < 90 μm obtained as a result of a single crushing of mixed PCBs in a disintegrator, and loading the raw material into the electrolyte volume, $C_{HCl} = 6\ mol \cdot L^{-1}$, $i = 0.88\ A \cdot cm^{-2}$, $S/L = 8.6\ g \cdot L^{-1}$. It has been established that the complete leaching of Cu from the raw material obtained from computer motherboards is observed under the following experimental conditions: the loading of the raw material into filter containers; $C_{HCl} = 6\ mol \cdot L^{-1}$; $i = 0.80\ A \cdot cm^{-2}$; $S/L = 8.6\ g \cdot L^{-1}$; $t_{ex} = 2$ h.

Author Contributions: Conceptualization, V.S. and A.S.; methodology, V.S., A.Z., E.E. and M.M. (Mikhail Maiorov); validation E.E. and V.P.; investigation V.S., M.M. (Maksims Melnichuks), A.Z., E.E., E.B. and M.M. (Mikhail Maiorov); resources, E.B., V.P. and A.S.; data curation M.M. (Maksims Melnichuks), A.S. and V.S.; writing—original draft preparation, V.S. and A.Z.; writing—review and editing, A.S. and V.P.; visualization, A.Z. and M.M. (Maksims Melnichuks); supervision, E.B. and V.P.; project administration, E.B. and V.P.; funding acquisition, E.B. and V.P. All authors have read and agreed to the published version of the manuscript.

Funding: This research was supported by ERDF project no. 1.1.1.1/20/A/139 "Development of sustainable recycling technology of electronic scrap for precious and non-ferrous metals extraction". The project was co-financed by REACT-EU funding to mitigate the effects of the pandemic crisis. The article was published with financial support from the Riga Technical University Research Support Fund. This research was also supported by the Institute of Solid State Physics, University of Latvia, as the Center of Excellence has received funding from the European Union's Horizon 2020 Framework Program H2020-WIDESPREAD-01-2016-2017-TeamingPhase2 under grant agreement No. 739508, project CAMART2.

Data Availability Statement: Not applicable.

Conflicts of Interest: The authors declare no conflict of interest. The funders had no role in the design of the study; in the collection, analyses, or interpretation of data; in the writing of the manuscript; or in the decision to publish the results.

References

1. Akcil, A.; Erust, C.; Gahan, C.S.; Ozgun, M.; Sahin, M.; Tuncuk, A. Precious Metal Recovery from Waste Printed Circuit Boards Using Cyanide and Non-Cyanide Lixiviants—A Review. *Waste Manag.* **2015**, *45*, 258–271. [CrossRef] [PubMed]
2. Barnwal, A.; Dhawan, N. Recycling of Discarded Mobile Printed Circuit Boards for Extraction of Gold and Copper. *Sustain. Mater. Technol.* **2020**, *25*, e00164. [CrossRef]
3. Neag, E.; Kovacs, E.; Dinca, Z.; Török, A.I.; Varaticeanu, C.; Levei, E.A. Hydrometallurgical Recovery of Gold from Mining Wastes. In *Strategies of Sustainable Solid Waste Management*; IntechOpen: London, UK, 2021.
4. Kim, K.; Candeago, R.; Rim, G.; Raymond, D.; Park, A.-H.A.; Su, X. Electrochemical Approaches for Selective Recovery of Critical Elements in Hydrometallurgical Processes of Complex Feedstocks. *iScience* **2021**, *24*, 102374. [CrossRef]
5. Wu, Z.; Yuan, W.; Li, J.; Wang, X.; Liu, L.; Wang, J. A Critical Review on the Recycling of Copper and Precious Metals from Waste Printed Circuit Boards Using Hydrometallurgy. *Front. Environ. Sci. Eng.* **2017**, *11*, 8. [CrossRef]
6. Birich, A.; Stopic, S.; Friedrich, B. Kinetic Investigation and Dissolution Behavior of Cyanide Alternative Gold Leaching Reagents. *Sci. Rep.* **2019**, *9*, 7191. [CrossRef] [PubMed]
7. Konyratbekova, S.S.; Baikonurova, A.; Akcil, A. Non-Cyanide Leaching Processes in Gold Hydrometallurgy and Iodine-Iodide Applications: A Review. *Miner. Process. Extr. Metall. Rev.* **2015**, *36*, 198–212. [CrossRef]
8. Aylmore, M.G. Alternative Lixiviants to Cyanide for Leaching Gold Ores. In *Gold Ore Processing*; Elsevier: Amsterdam, The Netherlands, 2016; pp. 447–484.
9. Gökelma, M.; Birich, A.; Stopic, S.; Friedrich, B. A Review on Alternative Gold Recovery Re-Agents to Cyanide. *J. Mater. Sci. Chem. Eng.* **2016**, *4*, 8–17. [CrossRef]
10. Birloaga, I.; De Michelis, I.; Ferella, F.; Buzatu, M.; Vegliò, F. Study on the Influence of Various Factors in the Hydrometallurgical Processing of Waste Printed Circuit Boards for Copper and Gold Recovery. *Waste Manag.* **2013**, *33*, 935–941. [CrossRef]
11. Xu, B.; Kong, W.; Li, Q.; Yang, Y.; Jiang, T.; Liu, X. A Review of Thiosulfate Leaching of Gold: Focus on Thiosulfate Consumption and Gold Recovery from Pregnant Solution. *Metals* **2017**, *7*, 222. [CrossRef]
12. Sethurajan, M.; van Hullebusch, E.D.; Fontana, D.; Akcil, A.; Deveci, H.; Batinic, B.; Leal, J.P.; Gasche, T.A.; Ali Kucuker, M.; Kuchta, K.; et al. Recent Advances on Hydrometallurgical Recovery of Critical and Precious Elements from End of Life Electronic Wastes—A Review. *Crit. Rev. Environ. Sci. Technol.* **2019**, *49*, 212–275. [CrossRef]
13. Nikoloski, A.; Stockton, B. Application of Alternative Lixiviants for Secondary Heap Leaching of Gold. In Proceedings of the Seventh Mill Operators Conference, Kalgoorlie, Australia, 12–14 October 2000; Australasian Institute of Mining and Metallurgy Publication Series: Kalgoorlie, Australia, 2000; pp. 281–286.
14. Ding, Y.; Zheng, H.; Li, J.; Zhang, S.; Liu, B.; Ekberg, C. An Efficient Leaching of Palladium from Spent Catalysts through Oxidation with Fe(III). *Materials* **2019**, *12*, 1205. [CrossRef] [PubMed]
15. Nogueira, C.A.; Paiva, A.P.; Costa, M.C.; da Costa, A.M.R. Leaching Efficiency and Kinetics of the Recovery of Palladium and Rhodium from a Spent Auto-Catalyst in $HCl/CuCl_2$ Media. *Environ. Technol.* **2020**, *41*, 2293–2304. [CrossRef] [PubMed]
16. Zhang, Z.; Zhang, F.-S. Synthesis of Cuprous Chloride and Simultaneous Recovery of Ag and Pd from Waste Printed Circuit Boards. *J. Hazard. Mater.* **2013**, *261*, 398–404. [CrossRef] [PubMed]
17. Pereira, M.M.; Costa, F.O.; Gomes, R.F.; Rodrigues, M.L.M.; da Silva, G.A.; Leão, V.A. Multivariate Study of a Novel Hydrometallurgical Route Employing Chloride/Hypochlorite for Leaching Silver from Printed Circuit Boards. *Chem. Eng. Res. Des.* **2020**, *163*, 115–124. [CrossRef]
18. Herreros, O.; Quiroz, R.; Viñals, J. Dissolution Kinetics of Copper, White Metal and Natural Chalcocite in Cl2/Cl−Media. *Hydrometallurgy* **1999**, *51*, 345–357. [CrossRef]
19. Tran, T.; Lee, K.; Fernando, K. Halide as an Alternative Lixiviant for Gold Processing—An Update. In *Cyanide: Social, Industrial and Economic Aspects*; Young, C.A., Twidwell, L.G., Anderson, C.G., Eds.; The Minerals, Metals and Materials Society: Warrendale, PA, USA, 2001; pp. 501–508.
20. Filmer, A.O.; Lawrence, P.R.; Hoffman, W.A. *A Comparison of Cyanide, Thiourea, and Chlorine as Lixiviants for Gold*; Australasian Institute of Mining and Metallurgy: Melbourne, Australia, 1984; pp. 279–287.
21. Hine, F.; Inuta, S. Solubility of Chlorine in Hydrochloric Acid. *Bull. Chem. Soc. Jpn.* **1968**, *41*, 71–75. [CrossRef]
22. Kim, M.; Lee, J.; Park, S.; Jeong, J.; Kumar, V. Dissolution Behaviour of Platinum by Electro-Generated Chlorine in Hydrochloric Acid Solution. *J. Chem. Technol. Biotechnol.* **2013**, *88*, 1212–1219. [CrossRef]
23. Kim, E.; Kim, M.; Lee, J.; Yoo, K.; Jeong, J. Leaching Behavior of Copper Using Electro-Generated Chlorine in Hydrochloric Acid Solution. *Hydrometallurgy* **2010**, *100*, 95–102. [CrossRef]
24. Kim, E.; Kim, M.; Lee, J.; Jeong, J.; Pandey, B.D. Leaching Kinetics of Copper from Waste Printed Circuit Boards by Electro-Generated Chlorine in HCl Solution. *Hydrometallurgy* **2011**, *107*, 124–132. [CrossRef]
25. Kim, E.; Kim, M.; Lee, J.; Pandey, B.D. Selective Recovery of Gold from Waste Mobile Phone PCBs by Hydrometallurgical Process. *J. Hazard. Mater.* **2011**, *198*, 206–215. [CrossRef]
26. Pilone, D.; Kelsall, G.H. Metal Recovery from Electronic Scrap by Leaching and Electrowinning IV. In *Electrometallurgy and Environmental Hydrometallurgy*; John Wiley & Sons, Inc.: Hoboken, NJ, USA, 2013; pp. 1565–1575.
27. Winand, R. Chloride Hydrometallurgy. *Hydrometallurgy* **1991**, *27*, 285–316. [CrossRef]
28. Zhao, J.; Liu, Z.; He, C.; Yang, Y.; Li, J.; Fujita, T.; Wang, G.; Shen, F. Improved Leaching of Cu, Sn, Pb, Zn, and Al from Waste Printed Circuit Boards by Electro-Generated Cl2 in HCl Solution. *Waste Manag.* **2022**, *153*, 386–396. [CrossRef] [PubMed]

29. Ilyas, S.; Srivastava, R.R.; Kim, H. Gold Recovery from Secondary Waste of PCBs by Electro-Cl2 Leaching in Brine Solution and Solvo-Chemical Separation with Tri-Butyl Phosphate. *J. Clean. Prod.* **2021**, *295*, 126389. [CrossRef]
30. Lister, T.E.; Wang, P.; Anderko, A. Recovery of Critical and Value Metals from Mobile Electronics Enabled by Electrochemical Processing. *Hydrometallurgy* **2014**, *149*, 228–237. [CrossRef]
31. Bagotsky, V.S. *Fundamentals of Electrochemistry*; Khimiya Press: Moscow, Russia, 1988; p. 400.
32. Shulgin, L.P. *Electrochemical Processes on Alternating Current*; Nauka: Leningrad, Russia, 1974; p. 70.
33. Shulgin, L.P. Overvoltage of Electrode Reactions in Solutions during the Passage of a Symmetrical Alternative Current. *J. Phys. Chem.* **1979**, *3*, 2048–2051.
34. Myrzabekov, B.E.; Bayeshov, A.B.; Makhanbetov, A.B.; Mishra, B.; Baigenzhenov, O.S. Dissolution of Platinum in Hydrochloric Acid Under Industrial-Scale Alternating Current Polarization. *Metall. Mater. Trans. B* **2018**, *49*, 23–27. [CrossRef]
35. Sarbayeva, M.T.; Bayeshev, A.B.; Sarbayeva, G.T.; Sarbayeva, K.T.; Bayesheva, A.K.; Zhurinov, M.Z. Dissolution of Iron Electrodes during Polarization with Three-Phase AC in Hydrochloric and Sulfuric Acids Solutions. *Am. J. Adv. Drug Deliv.* **2014**, *2*, 454–462.
36. Serga, V.; Zarkov, A.; Blumbergs, E.; Shishkin, A.; Baronins, J.; Elsts, E.; Pankratov, V. Leaching of Gold and Copper from Printed Circuit Boards under the Alternating Current Action in Hydrochloric Acid Electrolytes. *Metals* **2022**, *12*, 1953. [CrossRef]
37. Blumbergs, E.; Serga, V.; Shishkin, A.; Goljandin, D.; Shishko, A.; Zemcenkovs, V.; Markus, K.; Baronins, J.; Pankratov, V. Selective Disintegration–Milling to Obtain Metal-Rich Particle Fractions from E-Waste. *Metals* **2022**, *12*, 1468. [CrossRef]
38. Frumina, N.S.; Lisenko, N.F.; Chernova, M.A. *Analytical Chemistry of Elements. Chlorine*; Nauka: Moscow, Russia, 1983; p. 200.
39. Lobko, S.V.; Kuzas, E.A.; Naboychenko, S.S.; Voinov, V.N. Electrochlorination of Secondary Raw Materials Containing Noble Metals Using Bulk Current Lead. *Tsvetnye Metall.* **2017**, *3*, 45–49. [CrossRef]
40. Muller, F.; Caillard, L. Chlorophenols. In *Ullmann's Encyclopedia of Industrial Chemistry*; Wiley-VCH Verlag GmbH & Co. KGaA: Weinheim, Germany, 2011.

metals MDPI

Correction

Correction: Xhaferaj, N.; Ferella, F. Extraction and Recovery of Metals from Spent HDS Catalysts: Lab- and Pilot-Scale Results of the Overall Process. *Metals* 2022, *12*, 2162

Nertil Xhaferaj [1,2] and Francesco Ferella [3,*]

1 Department of Food Technology, Faculty of Food and Biotechnology, Agricultural University of Tirana, KodërKamëz, SH1, 1000 Tirana, Albania
2 School of Pharmacy, University of Camerino, Via S. Agostino 1, 62032 Camerino, Italy
3 Department of Industrial and Computer Engineering and Economics, University of L'Aquila, P.le E. Pontieri 1, Monteluco di Roio, 67100 L'Aquila, Italy
* Correspondence: francesco.ferella@univaq.it; Tel.: +39-0862-434238

Citation: Xhaferaj, N.; Ferella, F. Correction: Xhaferaj, N.; Ferella, F. Extraction and Recovery of Metals from Spent HDS Catalysts: Lab- and Pilot-Scale Results of the Overall Process. *Metals* 2022, *12*, 2162. *Metals* **2023**, *13*, 1254. https://doi.org/10.3390/met13071254

Received: 17 May 2023
Accepted: 18 May 2023
Published: 10 July 2023

The authors would like to make the following corrections about the published paper [1]. The changes are as follows:

(1) Removal of Authors

The names of Fabio Maggiore and Nicolò M. Ippolito were removed because these authors did not want to be included in the new version due to a precedent that occurred with the Orim company. The corrected Author Contributions statement appears here.

Author Contributions: Conceptualization, N.X.; methodology, F.F.; data curation, N.X.; writing—original draft preparation, N.X.; writing—review and editing, F.F.; supervision, F.F. All authors have read and agreed to the published version of the manuscript.

(2) Affiliations Correction

In the publication [1], there were errors regarding the affiliations. In the new version, affiliation 4 has been removed and affiliation 2 has been replaced with the following: "School of Pharmacy, University of Camerino, Via S. Agostino 1, 62032 Camerino, Italy". The updated affiliation information can be seen in the affiliation part of this correction.

(3) Removal of Acknowledgments

In the original publication [1], acknowledgments to the Orim company were included. In the new version, that paragraph was removed.

(4) Text Correction

A correction was made to the Abstract section. The following lines were added to the Abstract, Lines 2–4: "The process put together stages already known in the technical literature, tested again with samples coming from the roasting stage in a pilot kiln, which is the most limiting stage of metal recovery from spent catalysts".

A correction was made to the Section 2.1, Paragraph 1. The following lines were removed: "The catalyst samples used in this work were provided by Orim SpA, a company located in Macerata, Italy".

(5) References

We removed the old references [20,21], and added a new reference [22]. The citation has now been inserted in Section 1, Paragraph 2 and should read:

Various methods for recovery of HDS catalysts have been proposed in the literature: acid leaching followed by solvent extraction [11], primary leaching of spent catalyst, and then separation of metals through selective precipitation [12] recovery from biotechnological routes [13,14], carbon adsorption [15], polyelectrolyte extraction [16] and solvent extraction [17–19]. To all methods mentioned above, the recovery of metals was at different rates; however, the extraction of metals can achieve around 90% [11–13]. Presently, vanadium recovery is always greater than 90% concerning the concentration in spent catalysts [8,20]. Considering that these catalysts contain sulfur, coke, and sometimes hydrocarbons like naphtha, a thermal pre-treatment is required to convert these metals into oxides and enhance the leaching yield. The hydrometallurgical way is still the most used over the pyrometallurgical one [21]. Spent HDS catalysts are classified as hazardous materials: their disposal causes environmental threats because of the potential release of heavy metals into the soil and the groundwater table [20–22].

An old reference was replaced:

10. Le, M.N.; Lee, M.S. A Review on Hydrometallurgical Processes for the Recovery of Valuable Metals from Spent Catalysts and Life Cycle Analysis Perspective. *Min. Proc. Ext. Met. Rev.* **2021**, *42*, 335–354.

replace to

10. Feng, J.; Yan, S.; Zhang, R.; Gu, S.; Qu, X.; Bi, J. Recycling and reuse performance of cobalt catalyst for coal hydrogasification. *Fuel* **2023**, *335*, 126939. https://doi.org/10.1016/j.fuel.2022.126939.

A new reference was added:

22. Le, M.N.; Lee, M.S. A Review on Hydrometallurgical Processes for the Recovery of Valuable Metals from Spent Catalysts and Life Cycle Analysis Perspective. *Min. Proc. Ext. Met. Rev.* **2021**, *42*, 335–354.

With these corrections, the order of some references have been adjusted accordingly. Therefore, the total number of references is now 56.

The authors state that the scientific conclusions are unaffected. This correction was approved by the Academic Editor. The original publication has also been updated.

Reference

1. Xhaferaj, N.; Ferella, F. Extraction and Recovery of Metals from Spent HDS Catalysts: Lab- and Pilot-Scale Results of the Overall Process. *Metals* **2022**, *12*, 2162. [CrossRef]

MDPI
St. Alban-Anlage 66
4052 Basel
Switzerland
www.mdpi.com

Metals Editorial Office
E-mail: metals@mdpi.com
www.mdpi.com/journal/metals

www.ingramcontent.com/pod-product-compliance
Lightning Source LLC
Chambersburg PA
CBHW061435210926
51850CB00008B/1522
* 9 7 8 3 0 3 6 5 8 6 8 8 5 *